Trigonometric Fourier Series and Their Conjugates

Mathematics and Its Applications

Volume 372

Trigonometric Fourier Series and Their Conjugates

by

Levan Zhizhiashvili

Department of Mechanics and Mathematics,
Tbilisi State University,
Tbilisi, Georgia

KLUWER ACADEMIC PUBLISHERS

DORDRECHT / BOSTON / LONDON

A C.I.P. Catalogue record for this book is available from the Library of Congress.

ISBN-13: 978-94-010-6612-9 e-ISBN-13: 978-94-009-0283-1
DOI: 10.1007/978-94-009-0283-1

Published by Kluwer Academic Publishers,
P.O. Box 17, 3300 AA Dordrecht, The Netherlands.

Kluwer Academic Publishers incorporates
the publishing programmes of
D. Reidel, Martinus Nijhoff, Dr W. Junk and MTP Press.

Sold and distributed in the U.S.A. and Canada
by Kluwer Academic Publishers,
101 Philip Drive, Norwell, MA 02061, U.S.A.

In all other countries, sold and distributed
by Kluwer Academic Publishers Group,
P.O. Box 322, 3300 AH Dordrecht, The Netherlands.

This is a revised and updated translation of the original Russian work
Some Problems of the Theory of Trigonometric Fourier Series and Their Conjugate Series,
Tbilisi State University Press, Tbilisi, Georgia © 1993.
Translated by George Kvinikadze.

In fond memory of

V. G. CHELIDZE AND D. E. MEN'SHOV

Table of Contents

Part 2
MULTIPLE TRIGONOMETRIC SERIES

Preface

Research in the theory of trigonometric series has been carried out for over two centuries. The results obtained have greatly influenced various fields of mathematics, mechanics, and physics. Nowadays, the theory of simple trigonometric series has been developed fully enough (we will only mention the monographs by Zygmund [15, 16] and Bari [2]). The achievements in the theory of multiple trigonometric series look rather modest as compared to those in the one-dimensional case though multiple trigonometric series seem to be a natural, interesting and promising object of investigation. We should say, however, that the past few decades have seen a more intensive development of the theory in this field. To form an idea about the theory of multiple trigonometric series, the reader can refer to the surveys by Shapiro [1], Zhizhiashvili [16], [46], Golubov [1], D'yachenko [3]. As to monographs on this topic, only that of Yanushauskas [1] is known to me.

This book covers several aspects of the theory of multiple trigonometric Fourier series: the existence and properties of the conjugates and Hilbert transforms of integrable functions; convergence (pointwise and in the L^p-norm, $p > 0$) of Fourier series and their conjugates, as well as their summability by the Cesàro (C,α), $\alpha > -1$, and Abel–Poisson methods; approximating properties of Cesàro means of Fourier series and their conjugates. The structure of the book is motivated by my wish to place a special emphasis on the new effects which arise while dealing with multiple series and which are not inherent in the one-dimensional case. In Part 1, the familiar results concerning functions of one variable are discussed, while in Part 2, the material is parallelled for the multidimensional case. There the results obtained by the author are presented, with many proofs being published for the first time.

At the beginning of each chapter, a brief survey of known results is given, and at the end, unsolved problems are formulated whose solution, in my opinion, will greatly contribute to the development of the theory.

I hope the book will prove useful both to research workers and graduate students whose scientific interests lie in this field. It should be noted that the book is not self-contained and assumes the reader has sufficient background knowledge.

The Russian original of this book was first published in 1993 in Tbilisi (Georgia), but it has not really become widely available to the majority of potential readers. So I am very grateful to Kluwer Academic Publishers for their interest in the English translation of the book. The English version contains some minor changes and improvements.

I wish to express my sincere appreciation to Prof. P. L. Ul'yanov, corresponding member of the Russian Academy of Sciences, Prof. B. I. Golubov and Prof. V. Kokilashvili who read the manuscript and offered valuable advice.

I am also grateful to my colleagues from Tbilisi State University A. Ambroladze, T. Akhobadze, G. Bareladze, V. Bugadze, R. Getsadze, M. Lekishvili, D. Leladze, L. Panjikidze, T. Tevzadze, G. Tkebuchava for help during the work on the manuscript and proofreading both of the original and of the translation.

I am especially thankful to my colleagues G. Kvinikadze and M. Kvinikadze, the former for translating the book and the latter for preparing the camera-ready typescript.

Tbilisi, November 1995 LEVAN ZHIZHIASHVILI

Part 1

SIMPLE TRIGONOMETRIC SERIES

CHAPTER I

The Conjugation Operator and the Hilbert Transform

1.1. Some Definitions and Well-known Results

Throughout the book we use the notation $T = [-\pi, \pi]$, $\mathbb{R} =]-\infty, +\infty[$, and $\mathbb{R}_+ = [0, +\infty[$. Given $p \in]0, +\infty[$, $L^p(T)$ will stand for the set of all 2π-periodic measurable functions $f : \mathbb{R} \to \mathbb{R}$ for which the expression

$$\|f\|_p = \left\{ \frac{1}{2\pi} \int_T |f(x)|^p dx \right\}^{1/p}$$

is finite, while for $p = +\infty$, we will assume

$$L^\infty(T) = C(T), \quad \|f\|_\infty = \sup_{x \in T} |f(x)|.$$

In the case $p \in [1, +\infty]$, $(L^p(T), \|\cdot\|_p) \equiv L^p(T)$ is a Banach space. If, however, $p \in]0, 1[$, then $L^p(T)$ is not normable (see, e.g., Rudin [2, Ch. I]). In that case, by setting

$$\rho(f_1, f_2) = \|f_1 - f_2\|_p^p$$

for any $f_i \in L^p(T)$ $(i = 1, 2)$, we define a complete invariant metric ρ on L^p.

We denote by $B(T)$ $(B(\mathbb{R}))$ the space of all measurable bounded 2π-periodic functions on T (measurable bounded functions on $\mathbb{R}$), and we set for them

$$\|f\|_B = \sup_{x \in T} |f(x)| \quad \Big(\|f\|_B = \sup_{x \in \mathbb{R}} |f(x)| \Big).$$

Let $U = \{z : |z| < 1\}$ be the unit disc of the complex plane $\mathbb{C}$. Given $p \in]0, +\infty]$, $H^p(U) \equiv H^p$ will denote the set of all functions f which are regular in U and satisfy

$$\sup_{r \in [0,1[} \int_T |f(re^{it})|^p dt < +\infty$$

if p is finite, and

$$\sup_{z \in U} |f(z)| < +\infty$$

if $p = +\infty$. By setting

$$\|f\|_p = \sup_r \left\{ \frac{1}{2\pi} \int_T |f(re^{it})|^p dt \right\}^{1/p} \tag{1.1.1}$$

if p is finite and $\|f\|_\infty = \sup_{z\in U} |f(z)|$, we turn $H^p \equiv (H^p(T), \|\cdot\|_p) \equiv H^p(U)$ into a normed space for $p \in [1, +\infty]$, while for $p \in]0,1[$, the expression (1.1.1) does not have the convexity property. This means that, in that case, the function $\|\cdot\|_p$ defined on H^p is not a norm. If, however, for any $f_i \in H^p$ $(i = 1, 2)$ with $p \in]0,1[$ we put

$$\rho(f_1, f_2) = \|f_1 - f_2\|_p^p,$$

then $H^p \equiv (H^p, \rho)$ will become a complete linear metric space. As Livingston [1] showed in 1953, the space H^p, $p \in]0,1[$, is not normable.

The spaces H^p, introduced by Hardy [2] in 1915, play a considerable role in various fields of modern mathematics (in particular, in harmonic analysis). They have been investigated by a lot of mathematicians. The important results obtained in this direction can be found, for example, in the monographs by Koosis [1], Duren [1], Garnett [1], Khavin [1], Privalov [3], Shvedenko [1].

Suppose that $f \in L^p(T)$, $p \in [1, \infty]$, $\delta \in [0, 2\pi]$, and

$$\Delta_h^k(f, x) = \sum_{j=0}^{k} (-1)^{k-j} \binom{k}{j} f(x + jh).$$

Then the expression

$$\omega^{(k)}(\delta, f)_p = \sup_{|h|\le\delta} \|\Delta_h^k(f, \cdot)\|_p$$

is called the k-th order modulus of smoothness of f (see, e.g., Dzyadyk [1, Ch. III], Nikol'skiĭ [5, Ch. IV], Timan [1, Ch. III]). We denote

$$\omega^{(1)}(\delta, f)_p \equiv \omega(\delta, f)_p, \quad \omega^{(k)}(\delta, f)_\infty \equiv \omega^{(k)}(\delta, f)_C.$$

Let ω be a modulus of continuity, i.e., $\omega : [0, 2\pi[\to \mathbb{R}_+$ is nondecreasing, continuous, $\omega(0) = 0$, and $\omega(t_1 + t_2) \le \omega(t_1) + \omega(t_2)$ for $0 < t_1 < t_1 + t_2 < 2\pi$ (see, e.g., Nikol'skiĭ [5]). For any $p \in [1, +\infty]$, we assume that

$$H_p^\omega \equiv H_p^\omega(T) = \left\{ f : \omega(\delta, f)_p \le A(f, p)\omega(\delta) \right\}, \quad H_\infty^\omega \equiv H^\omega,$$

where $A(f, p)$ is a positive constant. If $\omega(\delta) = \delta^\alpha$ with $\alpha \in]0,1]$, then $H_p^\omega \equiv Lip\,(\alpha, p)$ and $H^\omega \equiv Lip\,\alpha$.

Let $f \in L(T)$. As usually done, we will denote by $\sigma[f]$ and $\overline{\sigma}[f]$ the trigonometric Fourier series and its conjugate, respectively, i.e.,

$$\sigma[f] = \frac{a_0(f)}{2} + \sum_{k=1}^{\infty} a_k(f) \cos kx + b_k(f) \sin kx, \tag{1.1.2}$$

$$\overline{\sigma}[f] = \sum_{k=1}^{\infty} -b_k(f) \cos kx + a_k(f) \sin kx, \tag{1.1.3}$$

where

$$a_k(f) = \frac{1}{\pi} \int\limits_T f(t) \cos kt \, dt \quad \text{for } k \in \mathbb{N}_0,$$
$$b_k(f) = \frac{1}{\pi} \int\limits_T f(t) \sin kt \, dt \quad \text{for } k \in \mathbb{N}. \tag{1.1.4}$$

The series (1.1.2) and (1.1.3) can be rewritten in the complex form

$$\sigma[f] = \sum_{h \in \mathbb{Z}} c_k(f) e^{ikx},$$
$$\overline{\sigma}[f] = \sum_{k \in \mathbb{Z}} \Big(-i \operatorname{sign} k \Big) c_k(f) e^{ikx},$$

where

$$c_k(f) = \frac{1}{2\pi} \int\limits_T f(t) e^{-ikt} \, dt \quad \text{for } k \in \mathbb{Z}.$$

A lot of problems of the modern harmonic analysis are connected with the behavior (in the sense of convergence and summability in various metrics) of the series $\overline{\sigma}[f]$. The investigation of this matter requires, in its turn, the study of various properties of the function $\overline{f}$ called the conjugate to f and defined by

$$\overline{f}(x) = -\frac{1}{2\pi} \lim_{\varepsilon \to 0+} \int\limits_\varepsilon^\pi \Big[f(x+t) - f(x-t) \Big] \operatorname{ctg} \frac{t}{2} \, dt =$$
$$= -\frac{1}{2\pi} \int\limits_0^\pi \Big[f(x+t) - f(x-t) \Big] \operatorname{ctg} \frac{t}{2} \, dt. \tag{1.1.5}$$

Luzin [2, Ch. V] established in 1915 that for any $f \in L^2(T)$, the function $\overline{f}$ exists almost everywhere, and in 1919 Privalov [2] showed that Luzin's assertion is true for any $f \in L(T)$. This fundamental fact is known as Luzin–Privalov theorem (see, e.g., Bari [2, Ch. VIII]).

For the spaces $L^p(\mathbb{R})$, $p \in [1, +\infty[$, of all measurable functions $f : \mathbb{R} \to \mathbb{R}$ for which the expression

$$\|f\|_p = \left\{ \int\limits_{\mathbb{R}} |f(x)|^p dx \right\}^{1/p}$$

is finite, the role analogous to the conjugation operator $f \to \overline{f}$ is played by the mapping $f \to \widetilde{f}$ defined as

$$\widetilde{f}(x) = -\frac{1}{\pi} \lim_{\varepsilon \to 0+} \int\limits_\varepsilon^\infty t^{-1} \Big[f(x+t) - f(x-t) \Big] \, dt =$$
$$= -\frac{1}{\pi} \int\limits_0^\infty t^{-1} \Big[f(x+t) - f(x-t) \Big] \, dt. \tag{1.1.6}$$

This mapping is called the Hilbert transform, and $\tilde{f}$ is called the Hilbert transform of f.

In 1923, Plessner [1] proved that if $f \in L(\mathbb{R})$, then its Hilbert transform exists almost everywhere.

Various properties of the functions $\overline{f}$ and $\tilde{f}$ as well as of more general singular operators have been intensively studied. Important results obtained in this direction can be found in a series of monographs and surveys, for example, Bari [2], Bari and Stechkin [1], A. Calderón [2], Koosis [1], Danilyuk [1], Dunford and Schwartz [2], Duren [1], Dyn'kin [1], [2], Dyn'kin and Osilenker [1], Edwards [1], [2], Gamelin [1], Garnett [1], Kashin and Saakyan [1], Katznelson [1], Khvedelidze [1], Kislyakov [1], Kokilashvili [1], Luzin [2], Muskhelishvili [1], Neri [1], Stein [3], Stein and Weiss [1], Timan [1], Titchmarsh [3], Torchinsky [1], Ul'yanov [8], Zhizhiashvili [13], [16], [35], [46], Zygmund [15], [16].

We are not going to present here in detail all the results concerning the properties of the functions $\overline{f}$ and $\tilde{f}$, giving only those which are necessary for the material of the book. We have tried to indicate the precise dates and sources of the first publication and to keep to the chronology.

Everywhere below, A, A_1, $A_2, \ldots$ will denote positive finite constants, while $A(\alpha)$, $A(n,p), \ldots$ will denote positive finite constants depending only on the indicated parameters. These constants, though having the same notation, may have different values at different places.

In 1916, Privalov [1] proved that if $f \in Lip\,\alpha$ on T for some $\alpha \in]0,1[$, i.e.,

$$\omega(\delta, f)_C \leq A(f)\delta^\alpha,$$

then so does $\overline{f}$, the assertion becoming invalid for $\alpha = 1$.

Later, in 1924, Zygmund [1] established a stronger statement: if $f \in C(T)$ and

$$\int_0^\pi t^{-1}\omega(t, f)\,dt < +\infty, \tag{1.1.7}$$

then $\overline{f}$ exists everywhere, $\overline{f} \in C(T)$ and

$$\omega(\delta, \overline{f})_C \leq A\Big[\int_0^\delta t^{-1}\omega(t, f)_C\,dt + \delta\int_\delta^\pi t^{-2}\omega(t, f)_C\,dt\Big] \quad \text{for } \delta \in \Big]0, \frac{\pi}{2}\Big]. \tag{1.1.8}$$

In 1945, Zygmund [9] further showed that for second order moduli of smoothness, the analogue of Privalov's theorem is true for $\alpha = 1$ as well, i.e, if

$$\omega^{(2)}(\delta, f)_C \leq A(f)\delta,$$

then

$$\omega^{(2)}(\delta, \overline{f})_C \leq A(f)\delta.$$

In 1955 Bari [1], and in 1956 Stechkin [1], and Bari and Stechkin [1] established a series of new facts concerning the behavior of k-th order moduli of smoothness of

f and $\overline{f}$. For example, it follows from a general theorem of Bari and Stechkin [1] that if $f \in C(T)$ and

$$\sum_{\nu=1}^{\infty} \nu^{-1}\omega^{(k)}\left(\frac{1}{\nu}, f\right)_C < +\infty,$$

then $\overline{f} \in C(T)$ and

$$\omega^{(k)}\left(\frac{1}{n}, \overline{f}\right)_C \le A(k)\left[n^{-k}\sum_{\nu=1}^{n} \nu^{k-1}\omega^{(k)}\left(\frac{1}{\nu}, f\right)_C + \sum_{\nu=n+1}^{\infty} \nu^{-1}\omega^{(k)}\left(\frac{1}{\nu}, f\right)_C\right].$$

In the same paper it is shown that for $k = 1$, this estimate implies (1.1.8). Moreover, the following problem is posed and solved: let a continuous increasing function φ be defined on $[0, \pi]$ and $\varphi(\delta) \to 0$ as $\delta \to 0+$. What necessary and sufficient conditions should be imposed on φ for the relations

$$\omega(\delta, f)_C \le A(f)\varphi(\delta) \quad \text{and} \quad \omega(\delta, \overline{f})_C \le A(f)\varphi(\delta)$$

to be equivalent? While solving this problem, it was found that the estimate (1.1.7) of Zygmund cannot be improved in the following sense.

Let $\omega : [0, \pi] \to [0, \omega(\pi)]$ be a modulus of continuity satisfying

$$\int_0^{\pi} t^{-1}\omega(t)\, dt < +\infty.$$

Then there exists a function $f \in C(T)$ such that

$$A_1(f)\omega(\delta) \le \omega(\delta, f)_C \le A(f)\omega(\delta), \quad \overline{f} \in C(T),$$

and

$$\omega(\delta, \overline{f})_C \ge \frac{1}{\pi}\left[\int_0^{\delta} t^{-1}\omega(t)\, dt + \delta\int_{\delta}^{\pi} t^{-2}\omega(t)\, dt\right].$$

Similar results are established for the space $L(T)$. In this connection, the reader can be referred to Zygmund [9] and Stechkin [3] (see also Timan [1, Ch. V], Geĭt [1]).

In 1976, Oskolkov [1] obtained quantitative estimates of Luzin's C-property and of the rate of approximation of 1-periodic functions on the sets of full measure in terms of their integral moduli of continuity. These results with detailed proofs are presented in his own work [2] as well as in the survey of Ul'yanov [9] (the latter contains mainly those concerning Luzin's C-property). By a quantitative estimate of Luzin's C-property we mean an estimate of differences $|f(x) - f(y)|$ which is uniformly small for small $|x - y|$ when a set of arbitrarily small measure is removed from a period. To be more precise, we are concerned with the estimates of the form

$$|f(x) - f(y)| \le [C_f(x) + C_f(y)]w^*(|x - y|, f),$$

where $C_f \ge 0$ is a measurable function which is finite almost everywhere over a period and

$$w^*(\delta, f) \downarrow 0+ \quad \text{as} \quad \delta \downarrow 0+.$$

In 1983, Oskolkov [3] studied the behavior of moduli of C-continuity of conjugate functions. For $f \in H^{\omega}$, he established certain estimates of $w^*(\delta, \overline{f})$ in terms of ω and investigated the question of their exactness.

In some of the above mentioned works (Zygmund [9], Bari [1], Stechkin [3], Bari and Stechkin [1], Timan [1, Ch. V]), the results are presented dealing with the behavior of $E_n(f)_p$ and $E_n(\overline{f})_p$, the best L^p-approximations of the functions f and $\overline{f}$, respectively, by trigonometric polynomials of order $\leq n$, where $p \in [1, +\infty]$.

For the Hilbert transform, Titchmarsh [1] in 1924 proved the following analogue of the above mentioned Privalov's theorem.

Let $p \in]1, +\infty[$ and $\alpha \in]0, 1[$. If $f \in L^p(\mathbb{R})$ and

$$\sup_{x \in \mathbb{R}} |f(x+h) - f(x)| \leq A(f)|h|^{\alpha} \quad \text{for } h \in [-1, 1], \tag{1.1.9}$$

then so does $\tilde{f}$, and the mutually dual Hilbert formulas

$$\tilde{f}(x) = -\frac{1}{\pi} \int_0^{+\infty} t^{-1}[f(x+t) - f(x-t)]\, dt$$

and

$$f(x) = \frac{1}{\pi} \int_0^{+\infty} t^{-1}[\tilde{f}(x+t) - \tilde{f}(x-t)]\, dt$$

are valid for any $x \in \mathbb{R}$.

Moreover, Titchmarsh found that this assertion becomes invalid for $\alpha = 1$.

Some questions concerning certain properties of the Hilbert transform (namely, estimates for best approximations) can be found in the monograph by Timan [1].

In 1924, M. Riesz [2] established that:

a) if $f \in L^p(T)$ for some $p \in]1, +\infty[$, then $\overline{f}$ exists almost everywhere on T, $\overline{f} \in L^p(T)$, and

$$\|\overline{f}\|_p \leq A(p)\|f\|_p; \tag{1.1.10}$$

b) if $f \in L^p(\mathbb{R})$ for some $p \in]1, +\infty[$, then $\tilde{f}$ exists almost everywhere on $\mathbb{R}$, $\tilde{f} \in L^p(\mathbb{R})$, and

$$\|\tilde{f}\|_p \leq A(p)\|f\|_p. \tag{1.1.11}$$

In 1927, he [3] presented the detailed proofs of these fundamental theorems.

In [2] and [3], M. Riesz also studied the discrete Hilbert transform.

Later on, these basic results of M. Riesz were generalized in various directions. We will dwell on those which are most characteristic (for the circle and the real axis).

Let $\Phi : \mathbb{R}_+ \to \mathbb{R}_+$ be a nonincreasing function satisfying

$$t^{-1}\Phi(t) \to +\infty \quad \text{as} \quad t \to +\infty.$$

As usually done, we define

$$L_\Phi(T) = \left\{ f : \mathbf{R} \to \mathbf{R} \,\middle|\, f \text{ is } 2\pi\text{-periodic and } \int_T \Phi(|f(x)|)\,dx < +\infty \right\},$$

and

$$L_\Phi(\mathbf{R}) = \left\{ f : \mathbf{R} \to \mathbf{R} \,\middle|\, \int_{\mathbf{R}} \Phi(|f(x)|)\,dx < +\infty \right\}.$$

For the classes L_Φ, the problem arose: what necessary and sufficient conditions should be imposed on Φ for the following statements to be true:

a) if $f \in L_\Phi(T)$, then $\overline{f} \in L_\Phi(T)$ and

$$\int_T \Phi(|\overline{f}(x)|)\,dx \le A(\Phi) \int_T \Phi(|f(x)|)\,dx; \tag{1.1.12}$$

b) if $f \in L_\Phi(\mathbf{R})$, then $\tilde{f} \in L_\Phi(\mathbf{R})$ and

$$\int_{\mathbf{R}} \Phi(|\tilde{f}(x)|)\,dx \le A(\Phi) \int_{\mathbf{R}} \Phi(|f(x)|)\,dx. \tag{1.1.13}$$

The first result in this direction was appearently published in the monograph by Zygmund [6, Ch. VIII]. We set forth this result.

Let $s : \mathbf{R}_+ \to \mathbf{R}_+$ be a convex function whose derivative is continuous on $]0, +\infty[$ and tends to $+\infty$ as $x \to +\infty$. Put

$$S(t) = \int_0^t s(x)\,dx.$$

Let, moreover, $\Psi : \mathbf{R}_+ \to \mathbf{R}_+$ be a convex function with $\Psi \to +\infty$ as $x \to +\infty$ and Ψ', Ψ'' continuous on $]0, +\infty[$. Suppose that

$$S''(t) + t^{-1}S'(t) \le A(s, \Psi)\Psi''(t) \quad \text{for } t \in]0, +\infty[.$$

Then

a) if $f \in L_\Psi(T)$, then $\overline{f} \in L_\Psi(T)$;

b) if $S(t) \le \Psi(t)$ for $t \in [0, +\infty[$, then

$$\int_T S(|\overline{f}(x)|)dx \le A(s, \Psi) \int_T \Psi(|f(x)|)\,dx.$$

Later on, in Marcinkiewicz [5], Lozinskiĭ [1] and Zygmund [3], stronger assertions were obtained.

In 1959, Lamperti [1] considered the space $L_\Phi(T)$, where $\Phi : \mathbf{R}_+ \to \mathbf{R}_+$ is a convex function, $\Phi(0) = 0$, and

$$(\Delta_2) \qquad \Phi(2t) \le A(\Phi)\Phi(t) \quad \text{for } 0 < t_0 \le t < +\infty. \tag{1.1.14}$$

He introduced the following classes of functions Φ satisfying some additional conditions:

$A^* = \{\Phi:\ \Phi'(t)$ is concave and $\Phi(t^\theta)$ is convex for some $\theta \in]0,1[\}$;

$B^* = \{\Phi:\ \Phi'(t)$ is convex and $\Phi(t^\theta)$ is concave for some $\theta \in]0,1[\}$;

$C^* = \{\Phi:\ \Phi(t) = \Phi_1(t) + \Phi_2(t),\ t \in \mathbb{R}_+,\ \Phi_1 \in A^*,\ \Phi_2 \in B^*\}$;

$D^* = \{\Phi:\ 0 < a \le \Phi(t)[\Phi_1(t)]^{-1} \le b < +\infty,\ 0 < t_0 \le t < +\infty,\ \Phi_1 \in C^*\}$;

$E^* = \{\Phi:\ \Phi(t) = t^p\Psi(t),\ p \in]1,+\infty[,\ p \ne 2$, and for any $\lambda \in]0,+\infty[$,

$$\lim_{t\to+\infty} \Psi(\lambda t)[\Psi(t)]^{-1} = 1\}.$$

It is not difficult to check that $A^* \cup B^* \subset C^* \subset D^*$. Lamperti [1] also proved that

a) $E^* \subset D^*$;

b) if Φ belongs to one of the classes A^*—E^* and $f \in L_\Phi(T)$, then $\overline{f}$ exists almost everywhere and (1.1.12) is fulfilled.

If $\Phi \in A^*$ or $\Phi \in B^*$, then the conclusion of b) follows from the corresponding results of Lozinskiĭ [1], and if $\Phi \in E^*$, it is a consequence of results of Marcinkiewicz [5]. In this connection, the works by Y. Chen [1], [3] are also relevant.

In 1963, Ryan [1] gave a full solution of the problem as posed. In particular, he proved that $f \to \overline{f}$ is a bounded map from the Orlicz space $L_\Phi(T)$ into itself iff $L_\Phi(T)$ is reflexive and (1.1.12) is fulfilled.

Note that this assertion of Ryan is valid for the Hilbert transform as well, if it is considered as acting from the space $L_\Phi(\mathbb{R})$ into itself and (1.1.12) is replaced by (1.1.13).

Now let us consider another generalization of the classes L^p. If $\omega: T \to \mathbb{R}_+$ is a measurable function which is almost everywhere nonzero and $p \in [1,+\infty[$, we put

$$L^p(T,\omega) = \left\{f: T\to\mathbb{R} \,\middle|\, f \text{ is } 2\pi\text{-periodic and } \int_T |f(x)|^p\omega(x)\,dx < +\infty\right\}.$$

The class $L^p(\mathbb{R},\omega)$ with $\omega: \mathbb{R} \to \mathbb{R}_+$ is defined quite analogously (naturally, the 2π-periodicity is not required here).

The problem arose of finding necessary and sufficient conditions which should be imposed on the "weight" ω in order that

a) for any $f \in L^p(T,\omega)$ with $p \in]1,+\infty[$, we would always have $\overline{f} \in L^p(T,\omega)$ and

$$\int_T |\overline{f}(x)|^p\omega(x)\,dx \le A(p,\omega)\int_T |f(x)|^p\omega(x)\,dx; \tag{1.1.15}$$

b) for any $f \in L^p(\mathbb{R},\omega)$ with $p \in]1,+\infty[$, we would always have $\tilde{f} \in L^p(\mathbb{R},\omega)$ and

$$\int_{\mathbb{R}} |\tilde{f}(x)|^p\omega(x)\,dx \le A(p,\omega)\int_{\mathbb{R}} |f(x)|^p\omega(x)\,dx. \tag{1.1.16}$$

First some necessary conditions were obtained. In 1936, Hardy and Littlewood [8] proved that if either

a) $f \in L^p(T,\omega)$ with $p \in]1,+\infty[$,

$$\alpha \in]-1,p-1[\quad\text{and}\quad \omega(x) = |x|^\alpha \quad \text{for } x \in T, \tag{1.1.17}$$

or

b) $f \in L^p(\mathbb{R},\omega)$ with $p \in]1,+\infty[$,

$$\alpha \in]-1,p-1[\quad \text{and} \quad \omega(x) = |x|^\alpha \quad \text{for } x \in \mathbb{R}, \tag{1.1.18}$$

then (1.1.15) or (1.1.16) respectively hold.

Further investigation of the weights ω of the types (1.1.17) and (1.1.18) showed that the restrictions imposed on α are essential.

In 1948, the assertions of Hardy and Littlewood were proved anew by Babenko [1] who used another method.

In 1958, Gaposhkin [1] found sufficient conditions to be imposed on ω for the relation (1.1.15) to be valid for any $f \in L^p(T,\omega)$. Moreover, these conditions imply $f \in L(T)$. In the book [13, Ch. I] of the author, Theorem 1 of Gaposhkin is extended to the Hilbert transform.

It should be noted that if $\alpha \in]-1,1[$ and $p \in [2,+\infty[$, or if $\alpha \in]-p+1,p-1[$ and $p \in]1,2[$, the assertion of Hardy and Littlewood concerning $\overline{f}$ is a consequence of Theorem 1 of Gaposhkin [1] (this follows from Gaposhkin's reasoning), but, in general, this theorem does not cover completely the statement of Hardy and Littlewood.

In 1960, Helson and Szegö [1] solved the case a) of the problem for $p = 2$. They established that ω must be representable as

$$\omega(x) = e^{u(x)+\overline{v}(x)},$$

where u and v are real essentially bounded functions, and

$$\|v\|_B < \frac{\pi}{2}.$$

The statement of Helson and Szegö solves the posed problem in the case b) as well (see, e.g., Garnett [1, Ch. VI]).

From the further results connected with the problem in question, we should mention those of Y. Chen [1] and Widom [1].

In 1963, Forelli [1] solved the case a) of the problem for arbitrary p but, as noted in Hunt, Muckenhoupt and Wheeden [1], Forelli's conditions are difficult both to check and to use.

In the mentioned work, Hunt, Muckenhoupt and Wheeden completely solved the problem. They showed that the "weight" ω has to satisfy the so-called A_p-condition

$$(A_p) \qquad \sup_I \left[\frac{1}{|I|}\int_I \omega(x)\,dx\right]\left[\frac{1}{|I|}\int_I \omega(x)^{-\frac{1}{p-1}}\,dx\right]^{\frac{p-1}{p}} \le A(p,\omega), \tag{1.1.19}$$

where I is an interval of positive finite length.

It is easy to verify that the function

$$\omega(x) = |x|^\alpha \quad \text{with } \alpha \in]0,+\infty[$$

satisfies (1.1.19) iff $\alpha \in]-1,p-1[$. Thus the conditions (1.1.17) and (1.1.18) of Hardy and Littlewood are essential.

Hunt, Muckenhoupt and Wheeden also described discrete "weights" for discrete Hilbert transforms.

In the case where f is either even or odd, the interval $]-1,p-1[$ in which the exponent α from the Hardy–Littlewood conditions takes its value, can be extended. From the relevant results we should mention those of K. Chen [1], Y. Chen [1], [3], Flett [2], Gurielashvili [3]–[5] and Hirschman [1].

Tkebuchava [1] established some weighted integral inequalities in reflexive Orlicz spaces. In particular, his result implies the corresponding result of Ryan [1].

A peculiar generalization of M. Riesz's inequality (1.11) was obtained by Pichorides [2]:

Let $f \in L^p(\mathbb{R})$ for some $p \in]2,+\infty[$. Then for any α and β from $[0,p]$,

$$\int_{\mathbb{R}} |\tilde{f}(x)|^\alpha |f(x)|^{p-\alpha} dx \leq A(p,\alpha,\beta) \int_{\mathbb{R}} |\tilde{f}(x)|^\beta |f(x)|^{p-\beta} dx. \tag{1.1.20}$$

As Pichorides himself noted, (1.1.20) is not true for $p \in]1,2]$ (if α and β are allowed to vary arbitrarily in $[0,p]$). For $\beta = 0$ and $\alpha = p$, (1.1.20) obviously implies (1.1.11).

Now consider the cases where either $f \in L(T)$ or $f \in L(\mathbb{R})$. First introduce the well-known notation.

For $f : T \to \mathbb{R}$ and $\varphi : \mathbb{R} \to \mathbb{R}$, we put

$$E_y(f,T) = \Big\{x : |f(x)| > y > 0, \ \ x \in T\Big\},$$

and

$$E_y(\varphi,\mathbb{R}) = \Big\{x : |\varphi(x)| > y > 0, \ \ x \in \mathbb{R}\Big\}.$$

As usual (see, e.g., Saks [2, Ch. VII], Chelidze and Jvarsheishvili [1, Ch. I]), given a set $E \subset \mathbb{R}$, the symbols $VB(E)$ and $VB^*(E)$ denote the classes of all real finite functions of bounded variation on E in broad and restricted sense, respectively; the expressions $Var(f,E)$ and $Var^*(f,E)$ denote respectively the broad and the restricted variation of f on E.

In 1936, Marcinkiewicz [1] established that if $F \in L(T)$ and the finite F' exists on a set $E \subset T$ with $\mu_1(E) > 0$, then the function F_1^* defined by

$$F_1^*(x) = -\frac{1}{\pi} \lim_{\varepsilon \to 0+} \int_{\varepsilon}^{\pi} \frac{F(x+t) + F(x-t) - 2F(x)}{2\sin^2 \frac{t}{2}} dt =$$
$$= -\frac{1}{\pi} \int_0^{\pi} \frac{F(x+t) + F(x-t) - 2F(x)}{2\sin^2 \frac{t}{2}} dt$$

exists almost everywhere on E. In particular, it is the case if $f \in V(T)$ (i.e., is of bounded variation on T).

In 1946, Loomis [1] showed that if $F \in V(\mathbb{R})$, then the function F_2^* defined by

$$F_2^*(x) = V.P.\frac{1}{\pi} \int_{\mathbb{R}} \frac{dF(t)}{t-x}$$

exists almost everywhere.

As far back as in 1915, Luzin ([2, Ch. V]) noted that there exists a function $f \in L(T)$ whose conjugate is not integrable on any interval $]a, b[\subset T$ with $b - a > 0$.

In 1925, Kolmogorov [3] proved that

a) if $f \in L(T)$, then for any $p \in]0,1[$, we have $\overline{f} \in L^p(T)$ and

$$\|\overline{f}\|_p \leq A(p)\|f\|_1; \tag{1.1.21}$$

b) if $f \in L(T)$, then

$$\mu_1[E_y(\overline{f}, T)] \leq Ay^{-1}\|f\|_1. \tag{1.1.22}$$

An analogue of (1.1.22)$'$ for the Hilbert transform follows from Loomis [1]: if $f \in L(\mathbb{R})$, then

$$\mu_1[E_y(\tilde{f}, \mathbb{R})] \leq Ay^{-1}\|f\|_1. \tag{1.1.23}$$

Loomis [1] also proved inequalities of the type (1.1.22) and (1.1.23) for functions of bounded variation.

In connection with the above results, the works of Besicovitch [1], [2] and of Littlewood [1] should also be mentioned.

In 1983, T. Lukashenko [3] (see also his paper [2]) established various estimates of the type (1.1.22) for the functions $f \in VB^*(E)$. We give some of them which are characteristic:

a) if $F \in VB^*(E)$, $E \subset T$, then F_1^* exists almost everywhere on E and

$$\mu_1[E_y(F_1^*, E)] \leq Ay^{-1} Var^*(F, E);$$

b) for the functions $F \in VB(E)$, the statement is not true.

Hunt, Muckenhoupt and Wheeden [1] described the weight functions ω for which

$$\int\limits_{E_y(\overline{f},T)} \omega(x)\,dx \leq A(\omega)y^{-1} \int\limits_T |f(x)|\omega(x)\,dx$$

and

$$\int\limits_{E_y(\tilde{f},\mathbb{R})} \omega(x)\,dx \leq A(\omega)y^{-1} \int\limits_{\mathbb{R}} |f(x)|\omega(x)\,dx.$$

Concerning the generalizations of Kolmogorov's inequality (1.1.21), we should mention the following.

In 1966, Lesniewicz [1] proved that the linear map $f \to \overline{f}$ is bounded from the space $L(T)$ into the nonreflexive Orlicz space $L^*_\Phi(T)$ iff the function $\Phi : \mathbb{R}_+ \to \mathbb{R}_+$ satisfies

$$\int\limits_1^\infty t^{-2}\Phi(t)\,dt < +\infty.$$

On the other hand, in 1958 Flett [2] showed that if $\alpha \in]-1, 0[$,

$$\omega(x) = |x|^\alpha \quad \text{for} \quad x \in T,$$

and $f \in L(T,\omega)$, then for any $p \in]0,1[$, the inequality

$$\left\{\int_T |\overline{f}(x)|^p |x|^{\alpha p} dx\right\}^{1/p} \leq A(p,\alpha)\int_T |f(x)|\,|x|^\alpha dx$$

is valid.

It is not difficult to see that the conditions of Flett's theorem imply $f \in L(T)$. In the same work, Flett noted that when f is even or odd, the interval $]-1,0[$ of values of α can be extended. In this connection, the results of Y. Chen [3], [4] and the author [13] should be mentioned.

In 1937, Titchmarsh [3, Ch. V] proved that if $f \in L(\mathbb{R})$, then for any $p \in]0,1[$, we have

$$\int_{\mathbb{R}} \frac{|\tilde{f}(x)|^p}{1+x^2}\,dx < +\infty.$$

Later, in 1943, Kober [2] improved Titchmarsh's statement. Namely, he showed that if $f \in L(\mathbb{R})$, $p \in]0,1[$ and $\beta > 1/2 - p/2$, then

$$\int_{\mathbb{R}} \frac{|\tilde{f}(x)|^p}{(1+x^2)^\beta}\,dx \leq A(p,\beta)\int_{\mathbb{R}} |f(x)|\,dx, \tag{1.1.24}$$

the condition $\beta > 1/2 - p/2$ being essential. Thus the analogue of Kolmogorov's inequality (1.1.21) was established for the Hilbert transform. Kober's paper also contains some other results concerning the behavior of $\tilde{f}$. Some relevant assertions can be found in the book of the author [13, Ch. I].

As already pointed out, the condition $f \in L(T)$ does not guarantee the summability of $\overline{f}$ on T. However, as Zygmund [4] showed in 1929, if $f \in L\log^+ L(T)$, then $\overline{f} \in L(T)$ and

$$\int_T |\overline{f}(x)|\,dx \leq A\int_T |f(x)|\log^+|f(x)|\,dx + 2\pi. \tag{1.1.25}$$

On the other hand, in 1927 M. Riesz proved that if $f \geq 0$ on T, and f and $\overline{f}$ are integrable on T, then $f \in L\log^+ L(T)$. Thus, in the case $f \geq 0$, Zygmund's theorem is invertible.

Note that the straightforward analogue of Zygmund's theorem for the Hilbert transform is not true.

Here a result of Kober [1] should be mentioned which was proved in 1942: if f and $\tilde{f}$ are summable on $\mathbb{R}$, then

$$\int_{\mathbb{R}} f(x)\,dx = 0.$$

For the generalizations of (1.1.25) for the cases of different weight functions ω, see Flett [2] and Y. Chen [1], [2].

In 1971, Burkholder, Gundy and Silverstein [1] (see also Garnett [1, Ch. III], Koosis [1, Ch. VIII]) investigated from another viewpoint the problem of summability of $\overline{f}$ and $\tilde{f}$. The results they obtained are currently considered to be highly important.

In 1972, Pichorides [1] studied the question of exactness of the constants in the inequalities of M. Riesz (1.1.10), Kolmogorov (1.1.21), and Zygmund (1.1.25). He found the optimal value of $A(p)$ in M. Riesz's inequality. Later in Davis [1], [2], the exact constants were found for Kolmogorov's inequalities (1.1.24), (1.1.25) as well.

Various metric properties of conjugate functions connected with the results of M. Riesz [2], [3], Kolmogorov [3], Zygmund [4], and other authors, can be found in Tsereteli [1].

Note at last that Zygmund's theorem (see (1.1.24)) was generalized for nonreflexive Orlicz spaces by Marcinkiewicz [5] in 1939, Lozinskiĭ [1] in 1944, and Zygmund [13] in 1956.

1.2. On the Theorems of Kolmogorov, Titchmarsh and Kober

In what follows, we will need some auxiliary statements

Lemma 1.2.1. *Let two analytic functions*

$$f(z) = u(z) + iv(z) \quad \textit{and} \quad t(z) = \omega(z) + i\varphi(z)$$

be defined in the circle $|z| < 1$, *and let* $v(0) = \varphi(0) = 0$. *If* $p \in]0,1[$, $z = re^{ix}$, $r \in [0,1[$, *and*

a) $\omega(z) > 0$,

b) $\omega(z) \geq q|\varphi(z)|$ *with* $q > \operatorname{tg} p\frac{\pi}{2}$,

then

$$\left\{\int_T |v(z)|^p \omega(z)\, dx\right\}^{1/p} \leq A(p,q) \int_T |u(z)|\, dx \left\{\int_T \omega(z)\, dx\right\}^{1/p}. \tag{1.2.1}$$

Proof. First we will assume that $u(z) > 0$. Put

$$\theta = \arg f(z), \quad \psi = \arg t(z),$$

where

$$-\frac{\pi}{2} < \theta < \frac{\pi}{2}, \quad -\frac{\pi}{2} < \psi < \frac{\pi}{2}.$$

Using the conditions of the lemma, we obtain

$$\begin{aligned} p\frac{\pi}{2} + \psi &= p\frac{\pi}{2} + \operatorname{arctg}\frac{\varphi}{\omega} \leq p\frac{\pi}{2} + \operatorname{arctg}\frac{1}{q} < \\ &< p\frac{\pi}{2} + \operatorname{arctg}\left(\operatorname{ctg} p\frac{\pi}{2}\right) = \\ &= p\frac{\pi}{2} + \operatorname{arctg}\left[\operatorname{tg}\left(\frac{\pi}{2} - p\frac{\pi}{2}\right)\right] = \frac{\pi}{2}; \end{aligned}$$

$$-p\frac{\pi}{2}+\psi=-p\frac{\pi}{2}+\operatorname{arctg}\frac{\varphi}{\omega}\geq -p\frac{\pi}{2}-\operatorname{arctg}\frac{1}{q}>$$
$$>-p\frac{\pi}{2}-\operatorname{arctg}\left[\operatorname{tg}\left(\frac{\pi}{2}-p\frac{\pi}{2}\right)\right]=-\frac{\pi}{2}.$$

Therefore we can conclude that

$$\cos(p\theta+\psi)\geq\cos\left(p\frac{\pi}{2}+\operatorname{arctg}\frac{1}{q}\right)>0.$$

So putting

$$A(p,q)=\frac{1}{\cos(p\frac{\pi}{2}+\operatorname{arctg}\frac{1}{q})},$$

we can see that

$$|\sin\theta|^p\cos\psi\leq A(p,q)\cos(p\theta+\psi). \tag{1.2.2}$$

Assume that

$$f(z)=\rho_1\cos\theta+i\rho_1\sin\theta,\quad t(z)=\rho\cos\psi+i\rho\sin\psi. \tag{1.2.3}$$

The Cauchy formula (see, e.g., Titchmarsh [4, Ch. II]) implies

$$[f(0)]^p t(0)=\frac{1}{2\pi i}\int\limits_{|z|=r}\frac{[f(z)]^p t(z)}{z}\,dz=$$
$$=\frac{1}{2\pi}\int\limits_T[u(z)+iv(z)]^p[\omega(z)+i\varphi(z)]\,dx.$$

Hence, taking into account that $v(0)=\varphi(0)=0$, according to (1.2.3) we find

$$[u(0)]^p\omega(0)=\frac{1}{2\pi}\int\limits_T\rho_1^p\rho\cos(p\theta+\psi)dx. \tag{1.2.4}$$

Multiplying both sides of (1.2.2) by $\rho_1^p\rho$ and integrating by x over T, we obtain

$$\int\limits_T|\rho_1\sin\theta|^p\rho\cos\psi\,dx\leq A(p,q)\int\limits_T\rho_1^p\rho\cos(p\theta+\psi)dx.$$

Then (1.2.4) implies

$$\left\{\int\limits_T|v(z)|^p\omega(z)\,dx\right\}^{1/p}\leq A(p,q)\int\limits_T|u(z)|\,dx\left\{\omega(z)\,dx\right\}^{1/p}$$

since

$$u(0)=\frac{1}{2\pi}\int\limits_T u(z)\,dx,\quad \omega(0)=\frac{1}{2\pi}\int\limits_T\omega(z)\,dx.$$

This means that (1.2.1) is valid. It remains now to get rid of the restriction $u(z)>0$.

Put

$$g(x)=u(re^{ix}),\quad \lambda(x)=\omega(re^{ix}),$$

$$g_1(x) = \begin{cases} g(x), & g(x) \geq 0, \\ 0, & g(x) < 0, \end{cases}$$

$$g_2(x) = g(x) - g_1(x).$$

The functions λ, g_1 and g_2 are clearly nonnegative and continuous. We will assume that $g(x) \not\equiv 0$ and $g_i(x) \not\equiv 0$ $(i = 1,2)$ on T (otherwise the lemma is obvious). Denote by $g(\tau, x)$, $\lambda(\tau, x)$ and $g_i(\tau, x)$ $(i = 1,2)$, $\tau \in [0,1[$, the harmonic functions corresponding to g, λ and g_i $(i = 1,2)$, and by $\overline{g}(\tau, x)$, $\overline{\lambda}(\tau, x)$ and $\overline{g}_i(\tau, x)$ $(i = 1,2)$ their conjugates, i.e.,

$$g(\tau, x) = \frac{1}{2\pi} \int\limits_T g(s) P(\tau, s - x)\, ds,$$

$$\overline{g}(\tau, x) = -\frac{1}{\pi} \int\limits_T g(s) Q(\tau, s - x)\, ds$$

(the functions $\lambda(\tau, x)$, $\overline{\lambda}(\tau, x)$ and $g_i(\tau, x)$, $\overline{g}_i(\tau, x)$ $(i = 1,2)$ are defined analogously), where, as usual,

$$P(\tau, x) = \frac{1 - \tau^2}{1 + \tau^2 - 2\tau \cos x}, \tag{1.2.5}$$

$$Q(\tau, x) = \frac{\tau \sin x}{1 + \tau^2 - 2\tau \cos x}. \tag{1.2.6}$$

It is not difficult to see that

$$\lambda(\tau, x) \geq q |\overline{\lambda}(\tau, x)|.$$

Then, according to the already proven, we have

$$\left\{ \int\limits_T |\overline{g}_i(\tau, x)|^p \lambda(\tau, x)\, dx \right\}^{1/p} \leq$$

$$\leq A(p, q) \int\limits_T |g_i(\tau, x)|\, dx \left\{ \int\limits_T \lambda(\tau, x)\, dx \right\}^{\frac{1}{p}} \quad (i = 1,2).$$

Hence, taking into account that

$$\|g_i(\tau, \cdot) - g_i(\cdot)\|_C \to 0, \quad \|\lambda_i(\tau, \cdot) - \lambda_i(\cdot)\|_C \to 0 \text{ as } \tau \to 1-$$

(see, e.g., Tolstov [1, Ch. VI]), according to the Lebesgue theorem on the passage to the limit (see, e.g., Natanson [3, Ch. VI]) we obtain

$$\limsup_{\tau \to 1-} \left\{ \int\limits_T |\overline{g}_i(\tau, x)|^p \lambda(\tau, x)\, dx \right\}^{1/p} \leq$$

$$\leq A(p, q) \int\limits_T |g_i(x)|\, dx \left\{ \int\limits_T \lambda(x)\, dx \right\}^{\frac{1}{p}} \quad (i = 1,2). \tag{1.2.7}$$

It is clear that

$$\overline{g}(\tau,x)=\overline{g}_1(\tau,x)+\overline{g}_2(\tau,x).$$

On the other hand (see, e.g., Hardy, Littlewood, Pólya [1, Ch. II]), for any real a and b the inequalities

$$|a+b|^p\le|a|^p+|b|^p \text{ if } p\in]0,1],$$
$$|a+b|^p\le 2^p(|a|^p+|b|^p) \text{ if } p\in]1,+\infty[$$

are valid. So

$$\left\{\int_T|\overline{g}(\tau,x)|^p\lambda(\tau,x)\,dx\right\}^{1/p}\le$$
$$\le\int_T|\overline{g}_1(\tau,x)|^p\lambda(\tau,x)\,dx+\int_T|\overline{g}_2(\tau,x)|^p\lambda(\tau,x)\,dx\Bigg\}^{1/p}\le$$
$$\le 2^{1/p}\left\{\left[\int_T|\overline{g}_1(\tau,x)|^p\lambda(\tau,x)\,dx\right]^{1/p}+\right.$$
$$\left.+\left[\int_T|\overline{g}_2(\tau,x)|^p\lambda(\tau,x)\,dx\right]^{1/p}\right\}. \tag{1.2.8}$$

It is well-known (see, e.g., Bari [2, Ch. VIII]) that

$$\lim_{t\to1-}\overline{g}(\tau,x)=\overline{g}(x)$$

almost everywhere on T. Therefore, applying Fatou's theorem (see, e.g., Natanson [3, Ch. VI]) and using (1.2.7) and (1.2.8), we find

$$\left\{\int_T|\overline{g}(x)|^p\lambda(x)\,dx\right\}^{1/p}\le A(p,q)\int_T|g(x)|\,dx\left\{\int_T\lambda(x)\,dx\right\}^{2/p},$$

that is,

$$\left\{\int_T|v(z)|^p\omega(z)\,dx\right\}^{1/p}\le A(p,q)\int_T|u(z)|\,dx\left\{\int_T\omega(z)\,dx\right\}^{2/p}. \qquad \blacksquare$$

Note that in (1.2.1), the constant $A(p,q)$ does not exceed

$$\left[\frac{2}{\cos\left(p\frac{\pi}{2}+\operatorname{arctg}\frac{1}{q}\right)}\right]^{1/p}.$$

Lemma 1.2.2. *Let two analytic functions*

$$f(z)=u(z)+iv(z) \quad \textit{and} \quad t(z)=\omega(z)+i\varphi(z)$$

be defined in the circle $|z|<1$ *and let* $v(0)=\varphi(0)=0$. *If* $p\in]0,1[$, $z=re^{ix}$, $r\in[0,1[$, *and*

a) $\omega(z)>0$,

b) $\omega(z) \geq q|\varphi(z)|$, *where*

$$q > 0 \quad \textit{for } p \in \left]0, \frac{1}{2}\right] \textit{ and } q > \left|\operatorname{tg}\frac{\pi}{2p}\right| \quad \textit{for } p \in \left]\frac{1}{2}, 1\right[,$$

then

$$\left\{\int_T |v(z)\omega(z)|^p \, dx\right\}^{1/p} \leq A(p,q) \int_T |u(z)| \, dx \int_T \omega(z) \, dx. \tag{1.2.9}$$

Proof. As while proving Lemma 1.2.1, we will assume that

$$\theta = \arg f(z), \quad \psi = \arg t(z)$$

and

$$f(z) = \rho_1 \cos\theta + i\rho_1 \sin\theta, \quad t(z) = \rho\cos\psi + i\rho\sin\psi.$$

As in the previous lemma, without loss of generality we can assume that $u(z) > 0$, so that

$$-\frac{\pi}{2} < \theta < \frac{\pi}{2}, \quad -\frac{\pi}{2} < \psi < \frac{\pi}{2}.$$

By the conditions of the lemma, we find

$$p\frac{\pi}{2} + p\psi \leq p\frac{\pi}{2} + p\operatorname{arctg}\frac{1}{q} < p\pi \leq \frac{\pi}{2} \quad \text{for } p \in \left]0, \frac{1}{2}\right];$$

$$-p\frac{\pi}{2} + p\psi \geq -p\frac{\pi}{2} + -p\operatorname{arctg}\frac{1}{q} > -p\pi \geq -\frac{\pi}{2} \quad \text{for } p \in \left]0, \frac{1}{2}\right].$$

Analogously, if $p \in]\frac{1}{2}, 1[$, then

$$\begin{aligned} p\frac{\pi}{2} + p\psi &\leq p\frac{\pi}{2} + p\operatorname{arctg}\frac{1}{q} < \\ &< p\frac{\pi}{2} + p\operatorname{arctg}\left(-\operatorname{ctg}\frac{\pi}{2p}\right) = \frac{\pi}{2}; \\ -p\frac{\pi}{2} + p\psi &= -p\frac{\pi}{2} - p\frac{\pi}{2} - p\operatorname{arctg}\frac{1}{q} > \\ &> -p\frac{\pi}{2} - p\operatorname{arctg}\left(-\operatorname{ctg}\frac{\pi}{2p}\right) = -\frac{\pi}{2}. \end{aligned}$$

Therefore

$$\cos(p\theta + p\psi) \geq \cos\left(p\frac{\pi}{2} + p\operatorname{arctg}\frac{1}{q}\right) > 0.$$

Denoting

$$A(p,q) = \frac{1}{\cos\left(p\frac{\pi}{2} + p\operatorname{arctg}\frac{1}{q}\right)},$$

we have

$$|\sin\theta\cos\psi|^p \leq A(p,q)\cos(p\theta + p\psi). \tag{1.2.10}$$

On the other hand, by the Cauchy formula (see, e.g., Titchmarsh [4, Ch. II]) we find

$$[f(0)t(0)]^p = \frac{1}{2\pi} \int\limits_{|z|=r} \frac{[f(z)t(z)]^p}{z}\, dz.$$

Hence, taking into account that $v(0) = \varphi(0) = 0$, we obtain

$$[u(0)\omega(0)]^p = \frac{1}{2\pi} \int\limits_T \rho_1^p \rho^p \cos(p\theta + p\psi)\, dx. \tag{1.2.11}$$

Multiplying (1.2.10) by $\rho_1^p\rho$ and integrating by x over T, in view of (1.2.11) we conclude

$$\frac{1}{2\pi} \int\limits_T |v(z)\omega(z)|^p \le A(p,q)[u(0)\omega(0)]^p,$$

so

$$\left\{ \int\limits_T |v(z)\omega(z)|^p dx \right\}^{1/p} \le A(p,q) \int\limits_T |u(z)|\, dx \int\limits_T \omega(z)\, dx. \quad \blacksquare$$

Note that in (1.2.9), the constant $A(p,q)$ does not exceed

$$\left[\frac{4\pi}{\cos\left(p\frac{\pi}{2} + p \operatorname{arctg} \frac{1}{q}\right)} \right]^{1/p}$$

Consider now a function $\omega \in L(T)$ such that $\omega > 0$ almost everywhere on T. Put

$$\begin{aligned} \omega(r,x) &= \frac{1}{2\pi} \int\limits_T \omega(t) P(r, t-x)\, dt, \\ \overline{\omega}(r,x) &= -\frac{1}{\pi} \int\limits_T \omega(t) Q(r, t-x)\, dt, \quad r \in [0,1[, \end{aligned} \tag{1.2.12}$$

where $P(r,x)$ and $Q(r,x)$ are defined by (1.2.5) and (1.2.6), respectively.

Theorem 1.2.1. *Let $f \in L(T)$ and $p \in]0,1[$. If a function ω satifies*

$$\omega(r,x) \ge q|\overline{\omega}(r,x)| \quad \textit{with} \quad q > \operatorname{tg} p\frac{\pi}{2},$$

then

$$\left\{ \int\limits_T |\overline{f}(x)|^p \omega(x)\, dx \right\}^{1/p} \le A(p,q) \int\limits_T |f(x)|\, dx \left\{ \int\limits_T \omega(x)\, dx \right\}^{1/p}. \tag{1.2.13}$$

Proof. Consider two analytic functions

$$f(z) = f(r,x) + i\overline{f}(r,x), \quad t(z) = \omega(r,x) + i\overline{\omega}(r,x),$$
$$z = re^{ix}, \quad r \in [0,1[.$$

They obviously satisfy all the conditions of Lemma 1.1.1. Then by (1.2.1), we have

$$\left\{\int_T |\overline{f}(r,x)|^p \omega(r,x)\,dx\right\}^{1/p} \le$$

$$\le A(p,q)\int_T |f(r,x)|\,dx\left\{\int_T \omega(r,x)\,dx\right\}^{1/p}. \tag{1.2.14}$$

But it is well-known (see, e.g., Bari [2. Ch. I]) that

$$\lim_{r\to 1-} f(r,x) = f(x), \quad \lim_{r\to 1-} \omega(r,x) = \omega(x),$$

$$\lim_{r\to 1-} \overline{f}(r,x) = \overline{f}(x)$$

almost everywhere on T. Then, applying Fatou's theorem (see, e.g., Natanson [3, Ch. VI]) and using (1.2.14), we get (1.2.13). ■

Using Lemma 1.2.2, we can quite analogously prove

Theorem 1.2.2. *Let $f \in L(T)$ and $p \in]0,1[$. If a function ω satisfies*

$$\omega(r,x) \ge q|\overline{\omega}(r,x)|,$$

where

$$\begin{aligned} &q > 0 \quad \textit{for} \quad p \in \left]0, \frac{1}{2}\right] \quad \textit{and} \\ &q > \left|\operatorname{tg}\frac{\pi}{2p}\right| \quad \textit{for} \quad p \in \left]\frac{1}{2}, 1\right[, \end{aligned} \tag{1.2.15}$$

then

$$\left\{\int_T |\overline{f}(x)\omega(x)|^p\,dx\right\}^{1/p} \le A(p,q)\int_T |f(x)|\,dx \int_T \omega(x)\,dx.$$

Remark 1.2.1. The method we have made use of in the proof of Theorems 1.2.1 and 1.2.2 is due to A. Calderón [1]. He applied this method in 1950 to prove the theorems of M. Riesz (see (1.1.10)) and Zygmund (see (1.1.24)). Note also that Theorems 1.2.1 and 1.2.2 generalize Kolmogorov's inequalities (see (1.2.21)).

Now let us study the question of analogues of Theorems 1.2.1 and 1.2.2 for the Hilbert transform.

Let a function $\omega : \mathbb{R} \to \mathbb{R}_+$ satisfy the following conditions:

a) $\omega > 0$ almost everywhere on $\mathbb{R}$;

b) ω is measurable on $\mathbb{R}$;

c) $\displaystyle\int_{\mathbb{R}} \frac{\omega(x)}{1+x^2}\,dx < +\infty.$ (1.2.16)

Put

$$\omega(a,z) = \frac{y}{\pi}\int_{-a}^{a} \frac{\omega(t)}{(t-x)^2+y^2}\,dt,$$

$$\omega_1(a,z) = -\frac{1}{\pi}\int_{-a}^{a} \frac{\omega(t)(t-x)}{(t-x)^2+y^2}\,dt, \quad a>0, \tag{1.2.17}$$

$$y>0, \quad z = x+iy.$$

Theorem 1.2.3. *Let* $f \in L(T)$, $p \in]0,1[$, *and a function* ω *satisfy conditions* a), b) *and* c). *Suppose, moreover, that* ω *is even for any* $\alpha \in]0,+\infty[$ *and*

$$\omega(a,z) \geq q|\omega_1(a,z)| \quad \text{for } p \in \left]0,\frac{1}{2}\right[,$$

$$q > \left|\operatorname{tg}\frac{\pi}{2p}\right| \quad \text{for } q \in \left]\frac{1}{2},1\right[.$$

Then

$$\left\{\int_{\mathbf{R}} \frac{|\tilde{f}(x)\omega(x)|^p}{1+x^2}\,dx\right\} \leq A(p,q)\int_{\mathbf{R}} \frac{|f(x)|}{1+|x|}\,dx \int_{\mathbf{R}} \frac{\omega(x)}{1+x^2}\,dx. \tag{1.2.18}$$

If, in addition, f *is even, then*

$$\left\{\int_{\mathbf{R}} \frac{|\tilde{f}(x)\omega(x)|^p}{1+x^2}\,dx\right\}^{1/p} \leq A(p,q)\int_{\mathbf{R}} \frac{|f(x)|}{1+x^2}\,dx \int_{\mathbf{R}} \frac{\omega(x)}{1+x^2}\,dx. \tag{1.2.19}$$

Proof. Without loss of generality, we can assume that f is not equivalent to zero and $f \geq 0$ on $\mathbf{R}$. Then it is clear that for $y \in]0,+\infty[$ and $0 < a_0 \leq a < +\infty$, we have $f(a,z) > 0$ on $\mathbf{R}$. Denote

$$F(a,z) = \frac{y}{\pi i}\int_{-a}^{a} \frac{f(t)}{t-z}\,dt = f(a,z) + if_1(a,z), \tag{1.2.20}$$

$$\Phi(a,z) = \frac{1}{\pi i}\int_{-a}^{a} \frac{\omega(t)}{t-z}\,dt = \omega(a,z) + i\omega_1(a,z) \tag{1.2.21}$$

and $\theta = \arg F$, $\psi = \arg \Phi$, where

$$-\frac{\pi}{2} < \theta < \frac{\pi}{2}, \quad -\frac{\pi}{2} < \psi < \frac{\pi}{2}. \tag{1.2.22}$$

Let now $y \in]0,1]$ and $K \in]1,+\infty[$ be some fixed numbers, and let γ be the contour consisting of the linear segment connecting the points $-K+iy$ and $K+iy$, and of the semicircle of radius K over it.

Consider the integral

$$\int_{\gamma} \frac{[F(a,z)\Phi(a,z)]^p}{1+z^2}\,dz.$$

For each fixed $a \in [a_0, +\infty[$, we have

$$|F(a,z)| \le A(f,z)\,\frac{1}{|z|},$$
$$|\Phi(a,z)| \le A(\omega,a), \quad 0 < |z_0| \le |z| < +\infty. \tag{1.2.23}$$

The function f given by

$$f(z) = \frac{1}{1+z^2}$$

has a simple pole at the point $z = i$. Therefore, by the theorem on residues (see, e.g., Titchmarsh [4, Ch. III]), we obtain

$$\int\limits_{\gamma} \frac{[F(a,z)\Phi(a,z)]^p}{1+x^2}\,dz = \pi[F(a,i)\Phi(a,i)]^p.$$

Hence, using (1.2.23) and passing to a limit as $K \to +\infty$, we get

$$\int\limits_{\mathbb{R}} \frac{[F(a,z)\Phi(a,z)]^p}{1+x^2}\,dx = \pi[F(a,i)\Phi(a,i)]^p.$$

The last equality can be written as

$$\int\limits_{\mathbb{R}} \frac{\left[F(a,z)\Phi(a,z)\right]^p (x^2-y^2+1-2ixy)}{(x^2-y^2+1)^2+4x^2y^2}\,dx =$$
$$= \int\limits_{\mathbb{R}} \frac{\left[|F(a,z)|\,|\Phi(a,z)|\right]^p |(\eta(z)|e^{i(p\theta+p\psi+\varphi)}}{(x^2-y^2+1)^2+4x^2y^2}\,dx =$$
$$= \pi[F(a,i)\Phi(a,i)]^p,$$

where

$$\eta(z) = x^2 - y^2 + 1 - 2ixy, \quad \varphi = \arg\eta,$$

so that

$$\int\limits_{\mathbb{R}} \frac{\left[|F(a,z)|\,|\Phi(a,z)|\right]^p (\eta(z)|\cos(p\theta+p\psi+\varphi)}{(x^2-y^2+1)^2+4x^2y^2}\,dx =$$
$$= \operatorname{Re}\left\{\pi[F(a,i)\Phi(a,i)]^p\right\}. \tag{1.2.24}$$

Since ω is even, we have

$$\int\limits_{-a}^{a} \omega(t)\,\frac{t}{1+t^2}\,dt = 0.$$

Therefore, by (1.2.17) and (1.2.21),

$$\Phi(a,i) = \frac{1}{\pi i}\int\limits_{-a}^{a} \frac{\omega(t)}{t-i}\,dt = \frac{1}{\pi}\int\limits_{-a}^{a} \frac{\omega(t)}{1+t^2}\,dt. \tag{1.2.25}$$

Moreover,

$$F(a,z)=\frac{1}{\pi i}\int\limits_{-a}^{a}\frac{f(t)}{t-i}\,dt=\frac{1}{\pi}\int\limits_{-a}^{a}\frac{f(t)}{1+t^2}\,dt-\frac{i}{\pi}\int\limits_{-a}^{a}f(t)\frac{t}{1+t^2}\,dt.$$

Hence it follows that

$$|F(a,z)|<\frac{2}{\pi}\int\limits_{\mathbb{R}}\frac{f(t)}{1+|t|}\,dt, \tag{1.2.26}$$

$$|F(a,i)|\le\frac{1}{\pi}\int\limits_{\mathbb{R}}\frac{f(t)}{1+t^2}\,dt, \tag{1.2.27}$$

when f is even.

If $y\in]0,1[$, then

$$(x^2-y^2+1)^2+4x^2y^2<5(x^2+1)^2$$

for all $x\in\mathbb{R}$. But then we easily see that

$$\frac{1}{|\eta(z)|}>\frac{1}{3(x^2+1)}. \tag{1.2.28}$$

On the other hand, taking into account the conditions of the theorem and (1.2.22), as while establishing the relation (1.2.10) in the proof of Lemma 1.2.2, we obtain

$$\cos(p\theta+p\psi+\varphi)\ge\cos\left(p\,\frac{\pi}{2}+p\operatorname{arctg}\frac{1}{q}\right)>0. \tag{1.2.29}$$

It is clear that the absolute value of the expression

$$\frac{-2xy}{x^2-y^2+1},\qquad y\in\left]0,\frac{1}{2}\right[,$$

does not exceed $2y$ and tends to zero as $|z|\to+\infty$ or $y\to0+$. This means that $\varphi\to0+$ as $|z|\to\infty$ or $y\to0+$. Therefore, using (1.2.29), we get

$$\cos(p\theta+p\psi+\varphi)\ge$$
$$\ge\frac{1}{2}\cos\left(p\,\frac{\pi}{2}+p\operatorname{arctg}\frac{1}{q}\right),\qquad 0<y\le y_0<1. \tag{1.2.30}$$

The last relation along with (1.2.24) and (1.2.28) implies

$$\int\limits_{\mathbb{R}}\frac{\big[|F(a,z)|\,|\Phi(a,z)|\big]^p}{1+x^2}\,dx\le A(p,q)|F(a,i)\Phi(a,i)|^p.$$

Hence, using (1.2.17), (1.2.26), and (1.2.27), we find

$$\left\{\int\limits_{\mathbb{R}}\frac{|F(a,z)\Phi(a,z)|^p}{1+x^2}\,dx\right\}^{1/p}\le A(p,q)\int\limits_{\mathbb{R}}\frac{f(x)}{1+|x|}\,dx\int\limits_{\mathbb{R}}\frac{\omega(x)}{1+x^2}\,dx, \tag{1.2.31}$$

$$\left\{\int\limits_{\mathbb{R}}\frac{|F(a,z)\Phi(a,z)|^p}{1+x^2}\,dx\right\}^{1/p}\le A(p,q)\int\limits_{\mathbb{R}}\frac{f(x)}{1+x^2}\,dx\int\limits_{\mathbb{R}}\frac{\omega(x)}{1+x^2}\,dx. \tag{1.2.32}$$

But it is well-known that (see, e.g., Titchmarsh [3, Ch. V])

$$\lim_{y\to 0+}\lim_{a\to+\infty}\omega(a,z)=\omega(x),$$
$$\lim_{y\to 0+}\lim_{a\to+\infty}f_1(a,z)=\tilde{f}(x)$$

almost everywhere on $\mathbb{R}$. Therefore by (1.2.31), (1.2.32), and Fatou's theorem, the inequalities (1.2.18) and (1.2.19) are valid. ∎

Analyzing the proof of the above theorem and that of relation (1.2.2), we can conclude that it holds

Theorem 1.2.4. *Let $f\in L(T)$, $p\in]0,1[$, and ω satisfy the conditions* a), b) *and* c). *Let, moreover, ω be even and for any $a\in]0,+\infty[$ satisfy*

$$\omega(a,z)\geq q|\omega_1(a,z)| \text{ with } q>\operatorname{tg} p\frac{\pi}{2}.$$

Then

$$\left\{\int_{\mathbb{R}}\frac{|\tilde{f}(x)|^p\omega(x)}{1+x^2}\,dx\right\}^{1/p}\leq A(p,q)\int_{\mathbb{R}}\frac{|f(x)|}{1+|x|}\,dx\left\{\int_{\mathbb{R}}\frac{\omega(x)}{1+x^2}\,dx\right\}^{1/p}. \tag{1.2.33}$$

If, in addition, f is even, then

$$\left\{\int_{\mathbb{R}}\frac{|\tilde{f}(x)|^p\omega(x)}{1+x^2}\,dx\right\}^{1/p}\leq A(p,q)\int_{\mathbb{R}}\frac{|f(x)|}{1+x^2}\,dx\left\{\int_{\mathbb{R}}\frac{\omega(x)}{1+x^2}\,dx\right\}^{1/p}. \tag{1.2.34}$$

The scheme of the proof of Theorem 1.2.3 imply the following statements.

1) If in the conditions of the theorem f and ω are not necessarily even, then

$$\left\{\int_{\mathbb{R}}\frac{|\tilde{f}(x)\omega(x)|^p}{1+x^2}\,dx\right\}^{1/p}\leq A(p,q)\int_{\mathbb{R}}\frac{|f(x)|}{1+|x|}\,dx\int_{\mathbb{R}}\frac{\omega(x)}{1+|x|}\,dx;$$

2) If f is even and ω is not necessarily even, then

$$\left\{\int_{\mathbb{R}}\frac{|\tilde{f}(x)\omega(x)|^p}{1+x^2}\,dx\right\}^{1/p}\leq A(p,q)\int_{\mathbb{R}}\frac{|f(x)|}{1+x^2}\,dx\int_{\mathbb{R}}\frac{\omega(x)}{1+x^2}\,dx.$$

Similar remarks are valid for Theorem 1.2.4 as well (see (1.2.33) and (1.2.34)).

Finally note that Theorems 1.2.3 and 1.2.4 generalize the theorems of Titchmarsh and Kober (see p. 14) and improve the relation (1.2.24) for $\beta\in[1,+\infty[$.

Remark 1.2.2. A large class of the weight functions ω satisfy the conditions given in the theorems of this section. This fact is confirmed by corresponding statements of Gaposhkin [1] and properties of sums of trigonometric series with monotone coefficients(see, e.g., Bari [2, Ch. X], Zygmund [15, Ch. V], Pak [1]).

If in the conditions of Theorems 1.2.3 and 1.2.4 we suppose that

$$\omega(x) \equiv 1, \quad x \in \mathbb{R},$$

then

$$\omega(a,z) > 0, \quad \omega_1(a,z) = 0.$$

Therefore the conditions of these theorems (those imposed on ω) are fulfilled for any $q \in]0,+\infty[$, if only $f \in L(T)$. Then

$$\left\{ \int_{\mathbb{R}} \frac{|\tilde{f}(x)|^p}{1+x^2}\,dx \right\}^{1/p} \le A(p) \int_{\mathbb{R}} \frac{|f(x)|}{1+|x|}\,dx.$$

If, in addition, f is even, then

$$\left\{ \int_{\mathbb{R}} \frac{|\tilde{f}(x)|^p}{1+x^2}\,dx \right\}^{1/p} \le A(p) \int_{\mathbb{R}} \frac{|f(x)|}{1+x^2}\,dx,$$

and in these inequalities, in general, we cannot take $p = 1$.

It should be noted once more that the question of summability on $\mathbb{R}$ of the Hilbert transform is essentialy different from the question of summability on T of the conjugate function $\overline{f}$. For example, for any $p \in]0,1]$, there exists a function $f \in \bigcap_{q \ge p} L^q(\mathbb{R})$ (depending on p) such that $\tilde{f} \notin L(\mathbb{R})$.

Indeed, let $f : \mathbb{R} \to \mathbb{R}_+$ be defined by

$$f(x) = \begin{cases} \dfrac{1}{(1+x)^{2/p}}, & x \in [0,+\infty[, \\ 0, & x \in]-\infty, 0[. \end{cases}$$

It is clear that $f \in L^q(\mathbb{R})$ for any $q \in [p,+\infty[$. Further, if $x \in]-\infty,0[$, then

$$|\tilde{f}(x)| = \frac{1}{\pi}\left| \int_{\mathbb{R}} \frac{f(t)}{t-x}\,dt \right| > \frac{1}{\pi} \int_0^{|x|} \frac{f(t)}{t-x}\,dt > \frac{A(p)}{|x|}, \quad |x| \in [1,+\infty[.$$

Therefore

$$\int_{\mathbb{R}} |\tilde{f}(x)|\,dx = +\infty.$$

If $f : \mathbb{R} \to \mathbb{R}_+$ and both f and $\tilde{f}$ are summable on $\mathbb{R}$, then in view of an assertion of Kober [1, Theorem 1] (see also p. 14), we have

$$\int_{\mathbb{R}} f(x)\,dx = 0. \tag{1.2.35}$$

Therefore, for the Hilbert transform, an analogue of the well-known theorem of M. Riesz [2] is true only in the trivial case.

Note that, as it was shown by Kober [1], the condition (1.2.35) is not sufficient for the summability of $\tilde{f}$ on $\mathbb{R}$.

1.3. On a Theorem of Hardy and Littlewood

In 1928, Hardy and Littlewood [6] proved that if either $p \in]1,+\infty[$, $\alpha \in]0,1]$ or $p = 1$, $\alpha \in]0,1[$, and $f \in Lip(\alpha,p)$, then $\overline{f} \in Lip(\alpha,p)$.

It should be noted that if $p \in]1,+\infty[$, then the result of Hardy and Littlewood follows from (1.1.10), i.e., from M. Riesz's theorem [2].

Below we present statements improving the theorem of Hardy and Littlewood.

First of all, we give some definitions.

Definition 1.3.1. Let $p \in [1,+\infty[$, $\omega : T \to \mathbb{R}_+$, and $\omega \in L(T)$. Suppose that a function φ is defined on T, is even, nondecreasing on $[0,\pi]$ and satisfies $\lim_{t\to 0}\varphi(t) = 0$. We say that a measurable function $f : T \to \mathbb{R}$ belongs to the class $Lip(\varphi,p,\omega)$ if there exists a positive $\lambda_0 \equiv \lambda_0(f,\omega,p)$ such that

$$\left\{\int_T |f(x+h+\eta) - f(x+\eta)|^p \omega(x)\,dx\right\}^{1/p} \le A(f,\omega,p,\varphi)\varphi(h) \qquad (1.3.1)$$

for any $h \in [-\lambda_0,\lambda_0]$, $\eta \in [-\lambda_0,\lambda_0]$.

Definition 1.3.2. Let a function φ satisfy the conditions of Definition 1.3.1. Suppose that $a \in]0,+\infty[$ and $b \in]0,+\infty[$. We say that $\varphi \in N(a,b)$ if there exists a positive $\lambda_1 \equiv \lambda_1(a,b,\varphi)$ such that

$$\begin{aligned} &\int_0^u \frac{\varphi(t)}{t^{1+a}}\,dt \le A(a,\varphi)\,\frac{\varphi(u)}{u^a}, \\ &\int_u^\delta \frac{\varphi(t)}{t^{1+b}}\,dt \le A(b,\varphi)\,\frac{\varphi(u)}{u^b} \quad \text{for} \quad u \in]0,\lambda_1], \end{aligned} \qquad (1.3.2)$$

where δ is a certain number from $]\lambda_1,\pi]$.

Theorem 1.3.1. *Let $f \in L(T) \cap Lip(\varphi,p,\omega)$ and*

$$\int_T |f(x+\tau)|^p \omega(x)\,dx \le A(f,\omega,p) \qquad (1.3.3)$$

for any $\tau \in [-\lambda_2,\lambda_2]$, where $\lambda_2 \equiv \lambda_2(f,\omega,p)$ is some positive constant. Let, moreover, $\varphi \in N(0,1)$. Then $\overline{f} \in Lip(\varphi,p,\omega)$.

Proof. Choose a positive number δ_0 such that for $h \in [-\delta_0,\delta_0]$, $\eta \in [-\delta_0,\delta_0]$, $\tau \in [-\delta_0,\delta_0]$, and $u \in]0,\delta_0]$, the conditions (1.3.1)–(1.3.3) be fulfilled. Since for almost all x and $x+h$ the relations

$$\overline{f}(x) = -\frac{1}{2\pi}\int_T [f(x+t) - f(x)]\operatorname{ctg}\frac{t}{2}\,dt$$

and

$$\overline{f}(x+h) = -\frac{1}{2\pi}\int_T [f(x+t) - f(x+h)]\operatorname{ctg}\frac{t-h}{2}\,dt$$

are valid, for $h\in]0,\frac{\delta_0}{4}]$, $\eta\in[-\frac{\delta_0}{4},\frac{\delta_0}{4}]$, $u\in]0,\frac{\delta_0}{4}]$ we have

$$\overline{f}(x+h)-\overline{f}(x)=\frac{1}{2\pi}\int_{-2h}^{2h}[f(x+t)-f(x)]\operatorname{ctg}\frac{t}{2}\,dt-$$
$$-\frac{1}{2\pi}\int_{-2h}^{2h}[f(x+t)-f(x+h)]\operatorname{ctg}\frac{t-h}{2}\,dt+$$
$$+\frac{1}{2\pi}[f(x+h)-f(x)]\left[\Big(\int_{-\pi}^{-2h}+\int_{2h}^{\pi}\Big)\operatorname{ctg}\frac{t-h}{2}\,dt\right]+$$
$$+\frac{1}{2\pi}\left[\Big(\int_{-\pi}^{-2h}+\int_{2h}^{\pi}\Big)\big[f(x+t)-f(x)\big]\Big(\operatorname{ctg}\frac{t}{2}-\operatorname{ctg}\frac{t-h}{2}\Big)dt\right]\equiv$$
$$\equiv\sum_{k=1}^{4}I_k(x,f,h). \tag{1.3.4}$$

Applying Minkowski's inequality (see, e.g., Hardy, Littlewood, Pólya [1, Ch. VI]) for $p>1$ and Fubini's theorem (see, e.g., Natanson [3, Ch. XII]) for $p=1$, according to (1.3.1) and (1.3.2) we obtain

$$\Big\{\int_T|I_1(x,f,h)|^p\omega(x+\eta)\,dx\Big\}^{1/p}=\Big\{\int_T\Big[|I_1(x,f,h)|^p\omega^{1/p}(x+\eta)\Big]^p dx\Big\}^{1/p}\le$$
$$\le\Big\{\int_T\Big[\int_{-2h}^{2h}|f(x+t)-f(x)|\omega^{1/p}(x+\eta)\operatorname{ctg}\Big|\frac{t}{2}\Big|\,dt\Big]^p dx\Big\}^{1/p}\le$$
$$\le\int_{-2h}^{2h}\Big|\operatorname{ctg}\frac{t}{2}\Big|\Big\{\int_T|f(x+t)-f(x)|^p\omega(x+\eta)\,dx\Big\}^{1/p}dt\le$$
$$\le A\int_0^{2h}t^{-1}\Big\{\int_T|f(x+t-\eta)-f(x-\eta)|^p\omega(x)\,dx\Big\}^{1/p}dt\le$$
$$\le A(f,\omega,p,\varphi)\int_0^{2h}t^{-1}\varphi(t)\,dt\le A_1(f,\omega,p,\varphi)\varphi(h), \tag{1.3.5}$$

since

$$\sin x\ge\frac{2}{\pi}x,\quad x\in\Big[0,\frac{\pi}{2}\Big]$$

and the conditions of the theorem imply that

$$\varphi(2t)\le A(\varphi)\varphi(t),\quad t\in[0,\pi]$$

(see, e.g., Bari and Stechkin [1]).

Analogously, in view of (1.3.4), we get

$$\left\{\int_T |I_2(x,f,h)|^p \omega(x+\eta)\,dx\right\}^{1/p} \le$$

$$\le \int_{-2h}^{2h} \left|\operatorname{ctg}\frac{t-h}{2}\right| \left\{\int_T |f(x+t-\eta)-f(x+h-\eta)|^p \omega(x)\,dx\right\}^{1/p} dt \le$$

$$\le A(f,\omega,p,\varphi)\int_{-2h}^{2h} \frac{\varphi(t-h)}{|t-h|}\,dt \le A_1(f,\omega,p,\varphi)\varphi(h). \tag{1.3.6}$$

Since the inequality $2|t-h| \ge |t|$ is valid for $|t| \ge 2h$, we have

$$\left|\operatorname{ctg}\frac{t-h}{2} - \operatorname{ctg}\frac{t+h}{2}\right| \le At^{-2}h.$$

Then, by (1.3.4) and (1.3.1), we obtain

$$\left\{\int_T |I_3(x,f,h)|^p \omega(x+\eta)\,dx\right\}^{1/p} \le$$

$$\le A\left\{\int_T |f(x+h)-f(x)|^p \omega(x+\eta)\,dx \times\right.$$

$$\left.\times\left(\int_{2h}^{\pi} \left|\operatorname{ctg}\frac{t-h}{2} - \operatorname{ctg}\frac{t+h}{2}\right| dt\right)^p\right\}^{1/p} dt \le$$

$$\le A\left\{\int_T |f(x+t-\eta)-f(x-\eta)|^p \omega(x)\,dx\right\}^{1/p} \le A(f,\omega,p,\varphi)\varphi(h). \tag{1.3.7}$$

Now consider the expression $I_4(x,f,h)$ in (1.3.4). It suffices to establish the required estimates for the expressions

$$I_4^{(1)}(x,f,h) \equiv \int_{2h}^{\pi} [f(x+t)-f(x)]\left(\operatorname{ctg}\frac{t}{2} - \operatorname{ctg}\frac{t-h}{2}\right) dt$$

and

$$I_4^{(2)}(x,f,h) \equiv \int_{2h}^{\pi} [f(x-t)-f(x)]\left(\operatorname{ctg}\frac{t}{2} - \operatorname{ctg}\frac{t+h}{2}\right) dt.$$

We confine ourselves to $I_4^{(1)}(x,f,h)$. Taking into account (1.3.1)–(1.3.3), we find

$$\left\{\int_T |I_4^{(1)}(x,f,h)|^p \omega(x+\eta)\,dx\right\}^{1/p} \le$$

$$\leq A\Big\{\int\limits_T\Big[\Big(\int\limits_{2h}^{\frac{3}{4}\delta_0}+\int\limits_{\frac{3}{4}\delta_0}^{\pi}\Big)|f(x+h)-f(x)|\omega^{1/p}(x+\eta)t^{-2}h\,dt\Big]^p dx\Big\}^{1/p}\leq$$

$$\leq A\Big\{\Big[\int\limits_T\Big[\int\limits_{2h}^{\frac{3}{4}\delta_0}|f(x+h)-f(x)|\omega^{1/p}(x+\eta)t^{-2}h\,dt\Big]^p dx\Big\}^{1/p}+$$

$$+A\Big\{\Big[\int\limits_T\Big[\int\limits_{\frac{3}{4}\delta_0}^{\pi}|f(x+h)-f(x)|\omega^{1/p}(x+\eta)t^{-2}h\,dt\Big]^p dx\Big\}^{1/p}\leq$$

$$\leq Ah\int\limits_{2h}^{\pi}t^{-2}\Big\{\int\limits_T|f(x+t-\eta)-f(x-\eta)|^p\omega(x)\,dx\Big\}^{1/p}dt+$$

$$\leq A(\delta_0)\Big\{\int\limits_T\Big[h\int\limits_{\frac{3}{4}\delta_0}^{\pi}|f(x+h)-f(x)|\omega^{1/p}(x+\eta)\,dt\Big]^p dx\Big\}^{1/p}\leq$$

$$\leq A(f,\omega,p,\varphi)h\int\limits_h^{\pi}t^{-2}\varphi(t)\,dt+$$

$$+A(\delta_0)\Big\{\int\limits_T\Big[\int\limits_T\big[|f(x+h)|-|f(x)|\big]\omega^{1/p}(x+\eta)\,dt\Big]^p dx\Big\}^{1/p}\leq$$

$$\leq A(f,\omega,p,\varphi)\varphi(h)+A(\delta_0)h\Big\{2\pi\Big[\int\limits_T[f(x-\eta)]^p\omega(x)\,dx\Big]^{1/p}+$$

$$+\int\limits_T|f(x)|\,dx\Big[\int\limits_T\omega(x)\,dt\Big]^{1/p}dx\Big\}\leq A_1(f,\omega,p,\varphi)\varphi(h) \tag{1.3.8}$$

since the conditions of the theorem imply that $\varphi(h)\geq Ah$.

Thus, according to (1.3.4)–(1.3.8), we obtain

$$\Big\{\int\limits_T|\overline{f}(x+h-\eta)-\overline{f}(x-\eta)|^p\omega(x)\,dx\Big\}^{1/p}\leq A(f,\omega,p,\varphi)\varphi(h),$$

where $\eta\in[-\frac{\delta_0}{4},\frac{\delta_0}{4}]$ and $h\in]0,\frac{\delta_0}{4}]$. Therefore the theorem is proved for $h\in[0,\frac{\delta_0}{4}]$. The scheme of the proof shows that the theorem is valid in the case $h\in[-\frac{\delta_0}{4},0]$ as well. ■

If $\varphi(t)=|t|^\alpha$, $\alpha\in]0,1[$, $t\in T$, and $\omega(x)\equiv1$, then the theorem just proved implies the corresponding assertion of Hardy and Littlewood.

If, however, $\varphi(t)=|t|$ for $t\in T$, then $\varphi\notin N(0,1)$. Therefore, in this case, Theorem 1.3.1 cannot be applied which is natural since for $p=1$ and $\omega(x)\equiv1$, the condition (1.3.1) turns into

$$\int_T |f(x+h) - f(x)|\, dx \le A(f)|h|. \tag{1.3.9}$$

But it is well-known (see, e.g., Titchmarsh [4, Ch. XI]) that $\overline{f}$, in general, does not satisfy (1.3.9). However, it holds

Theorem 1.3.2. *Let f and ω satisfy the conditions of Theorem* 1.3.1 *with* $\varphi(t) = |t|$, $t \in T$. *Then*

$$\left\{ \int_T |\overline{f}(x+h+\eta) + \overline{f}(x-h+\eta) - 2\overline{f}(x+\eta)|^p \omega(x)\, dx \right\}^{1/p} \le A(f,\omega,p,\varphi)\varphi(h).$$

Theorem 1.3.2 can be proved, in the main, by the same method as Theorem 1.3.1.

It should be noted that one can consider the differences of higher orders (appropriately changing the class $N(a,b)$) and prove more general statements.

The theorem of M. Riesz (see (1.1.10)) has no local analogue, i.e., if $f \in L(T)$ and for some $p \in]1,+\infty[$ and an interval $[a,b] \subset T$ with $b-a > 0$ we have $f \in L^p([a,b])$, then, genarally speaking, the inequality

$$\int_{a_1}^{b_1} |\overline{f}(x)|^p dx \le A(p,a,b,a_1,b_1) \int_a^b |f(x)|^p dx$$

is not valid for any interval $[a_1,b_1]$ satisfying $a < a_1 < b_1 < b$. However, it follows from Theorem 1.3.1 that a local analogue of Hardy and Littlewood's theorem is true.

Indeed, let $p \in]1,+\infty[$, $[a,b] \subset T$, $b-a > 0$, and $f \in L(T) \cap L^p([a,b])$. Moreover, let

$$\left\{ \int_a^b |f(x+h) - f(x)|^p\, dx \right\}^{1/p} \le A(f,p,\varphi,a,b)\varphi(h)$$

for $h \in [-\lambda_0, \lambda_0]$, where $\varphi \in N(0,1)$. If $a < a_1 < b_1 < b$, then put

$$\lambda = \min\left(\frac{b-b_1}{2}, \frac{a_1-a}{2}\right), \quad a_2 = a+\lambda, \quad b_2 = b-\lambda.$$

Define the function ω on T as follows:

$$\omega(x) = \begin{cases} 1, & x \in [a_2,b_2], \\ 0, & x \in T \setminus [a_2,b_2], \\ \omega(x+2\pi) & \text{for all } x. \end{cases}$$

We have

$$\int_T |f(x+\eta)|^p \omega(x)\, dx = \int_{a_2}^{b_2} |f(x+\eta)|^p dx \le \int_a^b |f(x)|^p dx \quad \text{for } \eta \in \left[-\frac{\lambda}{2}, \frac{\lambda}{2}\right],$$

so that (1.3.3) is fulfilled. Further,

$$\left\{\int_T |f(x+h+\eta)-f(x+\eta)|^p\omega(x)\,dx\right\}^{1/p} =$$

$$= \left\{\int_{a_2}^{b_2} |f(x+h+\eta)-f(x+\eta)|^p dx\right\}^{1/p} \le$$

$$\le \left\{\int_a^b |f(x+h)-f(x)|^p dx\right\}^{1/p} \le A(f,p,\varphi,a,b)\varphi(h)$$

for $h \in [-\lambda_0, \lambda_0]$ and $\eta \in [-\frac{\lambda}{2}, \frac{\lambda}{2}]$. Therefore (1.3.1) holds as well. Then, according to Theorem 1.3.1, we get

$$\left\{\int_{a_1}^{b_1} |\overline{f}(x+h)-\overline{f}(x)|^p dx\right\}^{1/p} \le A(f,p,\varphi,a,b,a_1,b_1)\varphi(h)$$

for $h \in [-h_1, h_1]$. Hence, putting $\varphi(t) = |t|^\alpha$ for $t \in T$ with $\alpha \in]0,1[$, we obtain a local analogue of the corresponding result of Hardy and Littlewood. To be more precise, the following assertion is true.

Suppose that $f \in L(T)$ and there exist numbers $p \in [1,+\infty[$, $\alpha \in]0,1[$ and a segment $[a,b] \subset T$ such that

$$\left\{\int_a^b |f(x+h)-f(x)|^p dx\right\}^{1/p} \le A(f,p,a,b)|h|^\alpha, \quad h \in [-h_0, h_0]$$

(if $p \in]1,+\infty[$, for the sake of simplicity it is assumed that $f \in L^p([a,b])$). Then, for any segment $[a_1,b_1]$ with $a < a_1 < b_1 < b$, we have

$$\left\{\int_{a_1}^{b_1} |\overline{f}(x+h)-\overline{f}(x)|^p dx\right\}^{1/p} \le A(f,p,\varphi,a,b,a_1,b_1)|h|^\alpha, \quad h \in [-h_1, h_1].$$

Remark 1.3.1. If we suppose that (1.3.1) holds for any $h \in]-\pi,\pi[$ and (1.3.2) holds for any $\lambda_1 \in]0,\pi[$ and $\delta = \pi$, then condition (1.3.3) will become unnecessary, and in this case, $\overline{f} \in Lip(\varphi,p,\omega)$ for all $h \in]-\pi,\pi[$.

Remark 1.3.2. Theorems 1.3.1 and 1.3.2 cannot be proved using those of the results of Section 1.1 which are connected with boundedness in the space $L^p(T,\omega)$, $p \in [1,+\infty[$, of the conjugation operator. This is confirmed by the fact that in Theorems 1.3.1 and 1.3.2, the function ω can be zero on a set of positive measure.

1.4. Unsolved Problems

1.4.1. Is it possible to "repair" an arbitrary function $f \in H_1^\omega$ on a set of arbitrarily small measure so that the conjugate to the "repaired" function would also belong to H_1^ω?

It should be noted that according to M. Riesz's theorem (see (1.1.10)), for a function from H_p^ω, $p \in]1, +\infty[$, the similar problem has a trivial solution. However, the problem makes sense for nonreflexive Orlicz spaces.

1.4.2. Give a full description of weight functions ω for which Theorems 1.2.1–1.2.4 are valid.

1.4.3. Give a full description of weight functions ω for which the inequality

$$\left\{ \int_T |\overline{f}(x)\omega(x)|^p dx \right\}^{1/p} \leq A(p,\omega) \int_T |f(x)|\omega(x)\,dx$$

holds for any $p \in]0, 1[$.

1.4.4. Give a full description of weight functions ω for which the inequality

$$\left\{ \int_{\mathbb{R}} \frac{|\tilde{f}(x)\omega(x)|^p}{1+x^2}\,dx \right\}^{1/p} \leq A(p,\omega) \int_{\mathbb{R}} |f(x)|\omega(x)\,dx$$

holds for any $p \in]0, 1[$.

CHAPTER II

Pointwise Convergence and Summability of Trigonometric Series

2.1. Some Definitions and Well-known Results

Let

$$\frac{a_0}{2} + \sum_{k=1}^{\infty} a_k \cos kx + b_k \sin kx \tag{2.1.1}$$

be a trigonometric series and

$$\sum_{k=1}^{\infty} b_k \cos kx - a_k \sin kx \tag{2.1.2}$$

be its conjugate. We will denote by $S_n(x)$ and $\overline{S}_n(x)$ the partial sums of (2.1.1) and (2.1.2), respectively.

If $c_k = a_k - ib_k$ and $z = e^{ix}$ for $x \in T$, then

$$\frac{a_0}{2} + \sum_{k=1}^{\infty} c_k z^k \tag{2.1.3}$$

is a series whose real and imaginary parts are (formally) (2.1.1) and (2.1.2), respectively. Everywhere below it is assumed that $\Gamma = \{z \in \mathbb{C} : |z| = 1\}$.

If $f \in L(T)$, then, as usual, $S_n(x, f)$ and $\overline{S}_n(x, f)$, $n \in \mathbb{N}_0$, denote the partial sums of the series $\sigma[f]$ and $\overline{\sigma}[f]$, respectively.

Everywhere below for $p \in [1, +\infty[$ the symbols $L^p(T_+)$ and $L^p(T_-)$ denote, respectively, the classes of all even and all odd functions from $L^p(T)$. Analogously, $B(T_+)$ and $B(T_-)$ will denote, respectively, the classes of all bounded even and all bounded odd functions $f : T \to \mathbb{R}$.

If $f \in L(T_+)$, $g \in L(T_-)$, and

$$\sigma[f] = \sum_{k=1}^{\infty} a_k(f) \cos kx,$$

$$\sigma[g] = \sum_{k=1}^{\infty} b_k(g) \sin kx,$$

then we put

$$U_k = k^{-1}\sum_{i=1}^{k} a_i(f), \quad M_k = k^{-1}\sum_{i=1}^{k} b_k(g), \tag{2.1.4}$$

$$U_k^* = \sum_{i=k}^{\infty} i^{-1}a_i(f), \quad M_k^* = \sum_{i=k}^{\infty} i^{-1}b_k(g), \quad k \in \mathbb{N}. \tag{2.1.5}$$

Remark 2.1.1. It is clear that the expressions (2.1.5) not always have meaning. U_k, M_k, U_k^* and M_k^* $(k = 1, 2, \dots)$, in general, need not be Fourier coefficients of some functions, and if they are, the functions may be different from f and g.

Consider the following trigonometric series

$$\sum_{k=1}^{\infty} U_k \cos kx, \quad \sum_{k=1}^{\infty} U_k \sin kx, \tag{2.1.6}$$

$$\sum_{k=1}^{\infty} M_k \cos kx, \quad \sum_{k=1}^{\infty} M_k \sin kx, \tag{2.1.7}$$

$$\sum_{k=1}^{\infty} U_k^* \cos kx, \quad \sum_{k=1}^{\infty} U_k^* \sin kx, \tag{2.1.8}$$

$$\sum_{k=1}^{\infty} M_k^* \cos kx, \quad \sum_{k=1}^{\infty} M_k^* \sin kx. \tag{2.1.9}$$

Note that without additional conditions imposed on f and g, series (2.1.6)–(2.1.9) need not be Fourier series.

Further, if $f \in L(T_+)$ and $\overline{f} \in L(T_-)$, then for $x \in]0, \pi[$, put

$$\begin{aligned} F(x) &\equiv F(x, f) = \operatorname{ctg}\frac{x}{2}\int_0^x \overline{f}(t)\,dt, \\ G(x) &\equiv G(x, f) = \int_x^\pi \overline{f}(t)\operatorname{ctg}\frac{t}{2}\,dt, \\ F(0) &= G(0) = 0, \quad F(\pi) = G(\pi) = 0, \end{aligned} \tag{2.1.10}$$

and then continue F and G on $\mathbb{R}$ as odd 2π-periodic functions. It is clear how this definition should be modified in the case where $f \in L(T_-)$ and $\overline{f} \in L(T_+)$.

The investigation of the functions F and G was initiated in the paper of Hardy [4] who discovered that various properties of the sums of the series (2.1.6)–(2.1.9) are closely connected with corresponding properties of the functions F and G.

As far back as before the beginning of the 20-th century, it was well-known that the series

$$\sum_{k=1}^{\infty} \cos kx \tag{2.1.11}$$

is everywhere divergent though for $x \neq 0 \pmod{2\pi}$, its partial sums satisfy

$$\limsup_{n\to\infty} |S_n(x)| < +\infty.$$

As to the series

$$\sum_{k=1}^{\infty} \sin kx, \tag{2.1.12}$$

it is divergent for any $x \neq 0 \pmod{2\pi}$, yet

$$\limsup_{n\to\infty} |S_n(x)| < +\infty$$

for any $x \in \mathbb{R}$.

Note that the coefficients of trigonometric series (2.1.11) and (2.1.12) does not tend to zero.

After the famous doctoral degree of Lebesgue [1] was published, Fatou [1] in 1906 posed the problem whether a trigonometric series (2.1.1) exists with

$$\lim_{k\to\infty} a_k = 0, \quad \lim_{k\to\infty} b_k = 0 \tag{2.1.13}$$

which is divergent on a set $E \subset T$ of positive measure.

In 1911, Luzin [1] solved this problem in its strongest form: there exists a series (2.1.3) with coefficients tending to zero which is divergent at every point of Γ. He also proved that in his example, both (2.1.1) and (2.1.2) are divergent almost everywhere. However, as Stechkin [4] showed in 1957, these series are in fact everywhere divergent.

Examples of everywhere divergent trigonometric series with coefficients satisfying (2.1.13) were constructed by Steinhaus [1], [2]. The second one is rather simple:

$$\sum_{k=2}^{\infty} (\ln k)^{-1} \cos k(x - \ln\ln k). \tag{2.1.14}$$

This series (see, e.g., Zygmund [15, Ch. IX]), is not a Fourier series. The argument of Steinhaus does not make clear whether the "cosine-part" of (2.1.14) diverges everywhere. However, in 1959 Siobhan [1] showed that in fact this is the case. In the same paper, the question of everywhere divergence of the series

$$\sum_{k=2}^{\infty} a_k \cos kq(k) \cos kx \tag{2.1.15}$$

is investigated under certain conditions imposed on a_k and $q(k)$, $k \geq 2$.

It remains unknown to us whether a_k and $q(k)$ $(k = 2, 3, \dots)$ can be chosen such that (2.1.15) would be a Fourier series and would diverge everywhere or at least on a set $E \subset T$ with $\mu_1(E) > 0$.

It should be noted that the results of Siobhan [1] imply the existence of an everywhere divergent trigonometric series with positive coefficients and satisfying (2.1.13). This fact first was found out by Herzog [1].

According to (2.1.14), the coefficients of the Steinhaus series fulfill

$$|a_k| \ln k \leq A, \quad |b_k| \ln k \leq A, \quad k \geq 2.$$

As to the behavior of the coefficients of Luzin's power series, Kuz'min [1] showed in 1936 that they satisfy

$$|c_k|k^{1/6} \leq A, \quad k \geq 2.$$

Earlier some results were obtained concerning the rate of convergence to zero of the coefficients of an everywhere divergent trigonometric series.

In 1916, Hardy and Littlewood [2] found an everywhere divergent on Γ series (2.1.3) with

$$|c_k|k^{1/2} \leq A, \quad k \in \mathbb{N}.$$

In 1921, Neder [1] established that for any sequence $(\alpha_k)_{k\geq 1}$ satisfying $\alpha_k \downarrow 0$ as $k \uparrow \infty$ and

$$\sum_{k=1}^{\infty} \alpha_k^2 = +\infty, \tag{2.1.16}$$

there exists an everywhere divergent on Γ series (2.1.3) such that

$$|c_k| \leq \alpha_k, \quad k \in \mathbb{N}. \tag{2.1.17}$$

In 1957, Stechkin [4] proved that this series can be chosen such that (2.1.1) and (2.1.2) are everywhere divergent.

In the above mentioned work, Herzog [1] constructed the example of an everywhere divergent on Γ series (2.1.3) whose coefficients are positive and tend to zero.

In 1955, Dvoretzky and Erdös established a stronger result: for an arbitrary sequence $(c_k)_{k\geq 1}$ of complex numbers satisfying

$$|c_k| \downarrow 0 \quad \text{as } k \to \infty \quad \text{and} \quad \sum_{k=1}^{\infty} |c_k|^2 = +\infty,$$

there exists a sequence $(\varepsilon_k)_{k\geq 1}$ with $\varepsilon_k = 0$ or $\varepsilon_k = 1$ such that the series

$$\sum_{k=1}^{\infty} \varepsilon_k c_k z^k$$

diverges everywhere on Γ.

In 1964, Ul'yanov [4] noted that Herzog's theorem can be improved, and he posed the question: is the assertion of Dvoretzky and Erdös true for trigonometric series? For instance, if a sequence $(\alpha_k)_{k\geq 1}$ monotonically tends to zero and satisfies (2.1.16), then is it possible to remove some terms from the series

$$\sum_{k=1}^{\infty} a_k \cos kx$$

so that the remaining series would be everywhere divergent?

In 1985, Galstyan [2] gave a positive answer to Ul'yanov's question.

In 1965, Tandori showed that there exists a trigonometric series (2.1.1) satisfying (2.1.13) such that its partial sums $S_n(x)$ everywhere fulfill

$$-\infty \leq \liminf_{n\to\infty} S_n(x) < \limsup_{n\to\infty} S_n(x) \leq +\infty. \tag{2.1.18}$$

Thus Tandori gave a positive answer to a problem of Stechkin (see, e.g., Ul'yanov [4, p. 29]).

The works of Belov [1]–[3] should also be mentioned, where the following problem of Ul'yanov is partially solved: for which "scarce" sequences $(n_k)_{k\geq 1}$ does there exist an everywhere divergent series of the form

$$\sum_{k=1}^{\infty} a_{n_k} \cos(n_k x - b_{n_k}) \quad \text{with} \quad a_{n_k} \to 0 \quad \text{as} \quad k \to \infty?$$

Belov also studied the question of the rate of convergence to zero of the coefficients of an everywhere divergent trigonometric series. These works of Belov were published from 1971 to 1977.

Parallel to the construction of divergent general trigonometric series, the sets were studied on which a trigonometric series converges (diverges). We shall discuss the matter a little later.

All the above mentioned investigations deal with general trigonometric series.

In 1923, Kolmogorov [1] constructed an example of a function $f \in L(T)$ whose Fourier series diverges almost everywhere. Three years later, he in [3] constructed an example of a function whose Fourier series everywhere diverges to infinity.

In 1936, Marcinkiewicz [1] constructed an example of a trigonometric Fourier series which boundedly diverges almost everywhere.

However, in the examples of Kolmogorov and Marcinkiewicz, $\overline{\sigma}[f]$ was not a Fourier series (see, e.g., Ul'yanov [2]). Naturally, the problem arose: if $\sigma[f]$ diverges almost everywhere (everywhere), then may its conjugate $\overline{\sigma}[f]$ be a Fourier series?

In 1953, Sunouchi [2] showed that there exists an almost everywhere divergent series $\sigma[f]$ for which $\overline{\sigma}[f]$ also is a Fourier series (it should be said that the book by Hardy and Rogozinsky [1, Ch. VI] contained this assertion in an implicite form).

In 1968, Prokhorenko [1] established that there is a function $f \in L(T)$ with

$$\omega(\delta, f)_1 \log\log\frac{2\pi}{\delta} \leq A(f), \quad \delta \in]0, \pi], \tag{2.1.19}$$

such that the series $\sigma[f]$ diverges almost everywhere.

Applying Theorem 2 of Ul'yanov [6] and using the results of Prokhorenko, we can conclude that there exists a function f such that $f \in L(\log^+\log^+ L)^{1-\varepsilon}(T)$ for any $\varepsilon \in]0,1[$, yet $\sigma[f]$ is almost everywhere divergent.

Note that for the series $\sigma[f]$ constructed by Prokhorenko, the function $\overline{f}$ is not integrable on T.

In 1969, Tandori [2] proved that if $\varepsilon \in]0,1[$ and

$$\Phi_\varepsilon(t) = t \log\log(t + e^e)^\varepsilon, \quad t \in \mathbb{R}_+,$$

then there is a function $f \in L(T)$ such that

$$\int_T \Phi_\varepsilon(|f(x)|)\, dx < +\infty, \quad \int_T \Phi_\varepsilon(|\overline{f}(x)|)\, dx < +\infty,$$

yet the series $\sigma[f]$ and $\overline{\sigma}[f]$ are everywhere divergent.

The above mentioned statement of Tandori improves in a certain sense Prokhorenko's result, since $\overline{f} \in L(T)$ and $\sigma[f]$ and $\overline{\sigma}[f]$ are everywhere divergent. However, the rate of tending to zero of the integral modulus of continuity of the function constructed by Tandori is unknown. It is why Tandori's results cannot be considered as the generalizations of Prokhorenko's [1] ones.

In 1971, the author [15] noted that there is $f \in L(T)$ such that $\overline{f} \in L(T)$, the integral moduli of continuity of f and $\overline{f}$ satisfy (2.1.19), but $\sigma[f]$ and $\overline{\sigma}[f]$ diverge almost everywhere.

Zygmund [16, Ch. XIII] posed the following problem: does the condition

$$\omega(\delta, f)_1 \log \frac{2\pi}{\delta} \leq A(f), \quad \delta \in]0, \pi],$$

guarantee the almost everywhere convergence of $\sigma[f]$ with $f \in L(T)$?

This problem is still unsolved.

As Ul'yanov noted in his works [6], [7], if Zygmund's problem has a negative solution, then a function f can be constructed which belongs to $L(\log^+ L)^\varepsilon(T)$ for any $\varepsilon \in]0,1[$ and whose Fourier series $\sigma[f]$ diverges on a set of positive measure (almost everywhere, everywhere). Therefore, we can conclude that Zygmund's problem is wider than that of determining whether for a function f belonging to $L(\log^+ L)^\varepsilon(T)$ for some $\varepsilon \in]0,1[$, the series $\sigma[f]$ diverges on a set of positive measure (almost everywhere, everywhere).

As T. Lukashenko [1, Ch. VII] noted in 1978, if $\Phi : \mathbb{R}_+ \to \mathbb{R}_+$ and

$$\Phi(t)(t \log\log t)^{-1} \to 0 \quad \text{as} \quad t \to +\infty,$$

then, following Kolmogorov's scheme, one can construct a function $f \in L(T)$ for which

$$\int_T \Phi(|f(x)|)\, dx < +\infty,$$

yet $\sigma[f]$ is almost everywhere divergent. The papers of Körner [2] and Kheladze [1] should also be mentioned here.

In terms of the Lebesgue classes, the above results currently are the best ones.

As to the best results in terms of integral moduli of continuity, we have already discussed them above.

In 1983, Galstyan [1] solved the problem concerning the rate of convergence to zero of the coefficients of an everywhere divergent series $\sigma[f]$. In particular, he proved that if $\alpha_k \downarrow 0$ as $k \uparrow \infty$ and (2.1.16) is fulfilled, then there exists a series $\sigma[f]$ of the form (1.1.2) for which $\overline{\sigma}[f]$ is also a Fourier series and

$$|a_k(f)| \leq A\alpha_k, \quad |b_k(f)| \leq A\alpha_k \quad \text{for } k \in \mathbb{N},$$

yet $\sigma[f]$ and $\overline{\sigma}[f]$ both diverge everywhere.

In the same paper by Galstyan, the result of Tandori [1] (see (2.1.18)) is extended to trigonometric Fourier series.

In 1959, Zygmund [15, Ch. VIII] noted the following: there is a function $f \in L(T)$ such that for almost all $x \in T$ an increasing sequence $(n_k)_{k\geq 1}$ of natural numbers (depending, in general, on x) can be found such that

$$\sum_{k=1}^{\infty} \frac{1}{n_k} = +\infty$$

and

$$\limsup_{k\to\infty} |S_{n_k}(x)| = +\infty.$$

In 1958, Gosselin [1] proved that for an arbitrary sequence of natural numbers $(n_k)_{k\geq 1}$ with $n_k \uparrow \infty$ as $k \uparrow \infty$, there is a function $f \in L(T)$ such that $S_{n_k}(x, f)$ diverges almost everywhere.

In 1982, Totik [3] showed that in Gosselin's statement, the everywhere divergence of $S_{n_k}(x, f)$ can be achieved.

In 1906, the book of Lebesgue [3] appeared dealing with trigonometric series and putting together his earlier results. In this book, on the basis of a new theory of measure and integral worked out by Lebesgue himself, a series of fundamental results was set forth. In particular, Lebesgue established that if $f \in L(T)$, then almost everywhere

$$S_n(x, f)(\log n)^{-1} \to 0 \quad \text{as } n \to \infty. \tag{2.1.20}$$

In 1959, Zygmund [15, Ch. VIII] posed the problem:

Let $(\lambda_n)_{n\geq 1}$ be an arbitrary sequence of positive numbers with $\lambda_n(\log n)^{-1} \to 0$ as $n \to \infty$. Does there exist a function $f \in L(T)$ such that almost everywhere

$$\limsup_{n\to\infty} |S_n(x, f)| \lambda_n^{-1} > 0?$$

As far as we know, this problem still remains unsolved. In this direction, the following steps were made.

In 1962, Y. Chen [2] proved that if $(\lambda_n)_{n\geq 1}$ is an arbitrary sequence of positive numbers with $\lambda_n(\log\log n)^{-1} \to 0$ as $n \to \infty$, then there is a function $f \in L(T)$ such that $\overline{f} \in L(T)$ and everywhere

$$\limsup_{n\to\infty} |S_n(x, f)| \lambda_n^{-1} > 0.$$

Here the works by Stein [1], Tandori [2], and Körner [2] should also be mentioned.

In 1983, Tandori [3] considered the problem on the relation between the growth rate of the partial sums of $\sigma[f]$ and the integrability of f. In particular, he proved the following assertion.

Let Φ be the class of all $\varphi : \mathbb{R}_+ \to \mathbb{R}_+$ which are twice differentiable, increasing, concave, and satisfy $\varphi(0) = 1$, $\varphi(t) \geq A \log\log t$, $t \in]e, +\infty[$, $\varphi(t)(\log t)^{-1} \to 0$ as $t \to +\infty$. If $\varphi \in \Phi$ and there is a function $f \in L(T)$ such that

$$\limsup_{n\to\infty} S_n(x, f)[\varphi(n)]^{-1} > 0$$

on a set $E \subset T$ with $\mu_1(E) > 0$, then for any $\varepsilon \in]0, 1[$, there is a measurable function f_0 satisfying

$$\int_T |f_0(x)| \, [\varphi(|f_0(x)|)]^{1-\varepsilon} dx < +\infty$$

and such that the series $\sigma[f_0]$ diverges almost everywhere.

In 1983, Totik [4] presented some results related to the work of Tandori [3].

In 1985, Oskolkov [4] established a series of properties of subsequences of Fourier sums of functions from $L(T)$. In particular, he proved that if $f \in L(T)$ and $(n_k)_{k\geq 1}$ is an arbitrary increasing sequence of natural numbers, then almost everywhere

$$\lim_{k\to\infty} S_{n_k}(x,f)(\log k)^{-1} = 0. \tag{2.1.21}$$

It should be noted that (2.1.21) generalizes (2.1.20) for subsequences and cannot be improved.

In the monograph of Bari [2, Ch. V], some results are presented concerning the convergence on a given set and the divergence outside of it of a general trigonometric series or a Fourier series. Below we give some results in this direction.

In 1961, Taĭkov [1] proved that for any set $E \subset T$ of type F_σ, there is a pair of mutually conjugate Fourier series (2.1.1) and (2.1.2) such that they both converge on E and diverge to infinity on $T \setminus E$.

In 1971, Körner [1] found out that there is a set E of type F_σ which is not the set of divergence for no $\sigma[f]$.

In 1980, S. Lukashenko [2] established that Körner's statement is true for general trigonometric series. In [1] she proved that for any $\theta \in [0, 2\pi[$, there is $f \in L(T)$ such that

$$\limsup_{n\to\infty} |S_n(x,f)| < +\infty$$

everywhere on T, yet for the set E of divergence of $\sigma[f]$ we have $\mu_1(E) > \theta$.

In 1985, Bloshanskiĭ [2] (see also his work [3]) proved that if $E \subset T$ is an arbitrary measurable set with $\mu_1(E) > 0$, then there is $f \in L(T)$ such that $f(x) = 0$ for $x \in T \setminus E$,

$$\limsup_{n\to\infty} |S_n(x,f)| = +\infty$$

almost everywhere on E, and

$$\lim_{n\to\infty} S_n(x,f) = 0$$

almost everywhere on $T \setminus E$.

In 1966, Carleson [1] showed that if $f \in L^2(T)$, then $\sigma[f]$ is almost everywhere convergent. Thus it was proved that the Luzin hypothesis [2, Ch. V] is true.

In 1967, Hunt [1] announced a result which, in particular, implies that if $f \in L^p(T)$ for some $p \in]1, +\infty[$, then $\sigma[f]$ converges almost everywhere. A detailed proof of this result is presented in Hunt [2].

In 1969, Sjolin [1] proved that the condition $f \in L\log^+ L \log^+ \log^+ L(T)$ ensures the convergence of $\sigma[f]$ almost everywhere.

Carleson's method [1] is set forth in various papers and books: Chernykh [1], Garcia [1], Mozzochi [1], and others.

It should be noted that in 1973, Fefferman [4] proved Carleson's theorem by another method. This method is set forth in the book of T. Lukashenko [1]. Moreover, Fefferman is of opinion that by his method the results of Hunt [2] can also be proved.

Thus, in the terms of Lebesgue classes, the best result concerning the convergence almost everywhere of series $\sigma[f]$ is due to Sjolin [1]. There is neither positive nor negative result for the functions between the classes $L\log^+\log^+ L(T)$ and $L\log^+ L\log^+\log^+ L(T)$.

Various sufficient conditions are well-known (see, e.g., Bari [2, Ch. III], Zygmund [15, Ch. II]) for the convergence of $\sigma[f]$ at a given point. They have been keeping their importance up today. The same is true for a nonimprovable criterion of Marcinkiewicz [1] concerning the convergence of $\sigma[f]$ on a set $E \subset T$ with $\mu_1(E) > 0$.

In 1928, Hardy [4] proved that if $f \in L^p(T_+)$ for some $p \in [1, +\infty[$, then

$$\sum_{k=1}^{\infty} U_k \cos kx$$

(see (2.1.4)) is the Fourier series of a function $F \in L(T_+)$.

In 1944, Bellman showed that if $f \in L^p(T_+)$ for some $p \in]1, +\infty[$, then

$$\sum_{k=1}^{\infty} U_k^* \cos kx$$

also is the Fourier series of a function $F_0 \in L^p(T_+)$. Note that for $p = 1$, this statement is not true.

Using these results of Hardy and Bellman, on the basis of M. Riesz's theorem [2] (see 1.1.10)) we can conclude the following: if $f \in L^p(T_+)$ and $g \in L^p(T_-)$ for some $p \in]1, +\infty[$, then (2.1.6)–(2.1.9) are the Fourier series of certain functions from the classes $L^p(T_+)$ or $L(T_-)$ (depending on whether they are even or odd).

If either $p = 1$ or $p = +\infty$, analogous assertions are not valid. The same is true in the case where

$$f \in B(T_+) \quad \text{and} \quad g \in B(T_-).$$

As already pointed out, the studying of the questions whether the series (2.1.6)–(2.1.9) belong to various classes (in particular, to classes $B(T_+)$, $B(T_-)$, $C(T_+)$, $C(T_-)$) is strictly connected with establishing the corresponding properties of the functions F and G given by (2.1.10) (as well as of some similar functions defined for $f \in L(T_-)$ and $\overline{f} \in L(T_+)$). A lot of mathematicians devoted themselves to the matter: Hardy [4], Bellman [1], Kawata [1], Sunouchi [1], Loo [1], F. Young [1], Konyushkov [1], Kinukava and Igari [1], Igari [1], M. and S. Izumi [1], Wang [1], [2], and others. We do not dwell on this here since we are going to set forth some results in this direction in the consequent chapters.

2.2. Pointwise Convergence and Summability of Fourier Series and Conjugate Trigonometric Series

First of all, introduce some notation.

The symbols

$$\sigma_n^\alpha(x, f) \quad \text{and} \quad t_n^\alpha(x, f), \quad \alpha \in]-1, +\infty[,$$

will denote Cesàro means ((C,α)-means) of the series $\sigma[f]$ and $\overline{\sigma}[f]$, respectively, i.e.,

$$\sigma_n^\alpha(x,f) = \frac{1}{A_n^\alpha}\sum_{k=0}^{n} A_{n-k}^{\alpha-1} S_k(x,f), \tag{2.2.1}$$

$$t_n^\alpha(x,f) = \frac{1}{A_n^\alpha}\sum_{k=0}^{n} A_{n-k}^{\alpha-1} \overline{S}_k(x,f), \tag{2.2.2}$$

where

$$A_0^\alpha = 1, \quad A_k^\alpha = \frac{(\alpha+1)(\alpha+2)\cdots(\alpha+k)}{k!}, \quad k \in \mathbb{N}. \tag{2.2.3}$$

It is well-known (see, e.g., Zygmund [15, Ch. III]) that

$$A_k^\alpha = \sum_{i=0}^{k} A_{k-i}^{\alpha-1}, \quad A_k^\alpha - A_{k-1}^\alpha = A_k^{\alpha-1} \tag{2.2.4}$$

and

$$A_k^\alpha \sim k^\alpha. \tag{2.2.5}$$

For $\alpha = 0$, it is assumed that

$$\sigma_n^0(x,f) \equiv S_n(x,f), \quad t_n^0(x,f) \equiv \overline{S}_n(x,f).$$

As usual, for a fixed function $f \in L(T)$ and a fixed point $x \in T$, we put

$$\varphi(x,t) \equiv \varphi(f,x,t) = f(x+t) + f(x-t) - 2f(x),$$
$$\psi(x,t) \equiv \psi(f,x,t) = f(x+t) - f(x-t),$$
$$\Phi(x,t) \equiv \Phi(f,x,t) = \int_0^t \varphi(x,u)\,du,$$
$$\Psi(x,t) \equiv \Psi(f,x,t) = \int_0^t \psi(x,u)\,du$$

and

$$\overline{f}_n(x) = -\frac{1}{2\pi}\int_{\pi/n}^{\pi} \psi(x,f)\operatorname{ctg}\frac{t}{2}\,dt, \quad n > 1, \quad n \in \mathbb{N}, \tag{2.2.6}$$

the functions φ, ψ, Φ, Ψ and $\overline{f}_n$ being defined for those x and t for which all the applied operations are valid.

Let $t \in T$ and $|\alpha| \in]0,1[$. Put

$$K_n^\alpha(t) = \frac{1}{A_n^\alpha}\sum_{k=0}^{n} A_{n-k}^{\alpha-1} D_k(t), \tag{2.2.7}$$

$$\tau_n^\alpha(t) = \frac{1}{A_n^\alpha}\sum_{k=0}^{n} A_{n-k}^{\alpha-1} \overline{D}_k(t), \quad n \in \mathbb{N}, \tag{2.2.8}$$

where

$$D_k(t) = \frac{1}{2} + \sum_{i=1}^{k} \cos it = \frac{\sin(k+\frac{1}{2})t}{2\sin\frac{t}{2}},$$

$$\overline{D}_k(t) = \sum_{i=1}^{k} \sin it = \frac{\cos\frac{t}{2} - \cos(k+\frac{1}{2})t}{2\sin\frac{t}{2}}$$

for $|t| \in]0,\pi]$, and

$$D_k(0) = k + \frac{1}{2}, \quad \overline{D}_k(0) = 0, \quad k \in \mathbb{N},$$

i.e., D_k is the Dirichlet kernel and $\overline{D}_k$ is its conjugate.

It is well-known (see, e.g., Zygmund [15, Ch. III]) that for $|t| \in]0,\pi]$,

$$K_n^\alpha(t) = \varphi_n^\alpha(t) + r_n^\alpha(t), \tag{2.2.9}$$

$$\tau_n^\alpha(t) = \psi_n^\alpha(t) + \gamma_n^\alpha(t) + \frac{1}{2}\operatorname{ctg}\frac{t}{2}, \tag{2.2.10}$$

where

$$\varphi_n^\alpha(t) = \frac{\sin[(n+\frac{1}{2}+\frac{\alpha}{2})t - \frac{\alpha\pi}{2}]}{A_n^\alpha(2\sin\frac{t}{2})^{1+\alpha}}, \tag{2.2.11}$$

$$\psi_n^\alpha(t) = -\frac{\cos[(n+\frac{1}{2}+\frac{\alpha}{2})t - \frac{\alpha\pi}{2}]}{A_n^\alpha(2\sin\frac{t}{2})^{1+\alpha}}. \tag{2.2.12}$$

Moreover,

$$\|K_n^\alpha\|_C \le A(\alpha)n, \tag{2.2.13}$$

$$\|\tau_n^\alpha\|_C \le A(\alpha)n, \tag{2.2.14}$$

$$|r_n^\alpha(t)| \le \frac{A(\alpha)}{nt^2}, \quad t \in \left[\frac{\pi}{n},\pi\right], \tag{2.2.15}$$

$$|\gamma_n^\alpha(t)| \le \frac{A(\alpha)}{n^2t^3}, \quad t \in \left[\frac{\pi}{n},\pi\right], \quad n \ge 2. \tag{2.2.16}$$

The following inequalities are also true (see, e.g., Zygmund [15, Ch. III], [16, Ch. XI])

$$\|(K_n^\alpha)'\|_C \le A(\alpha)n^2, \tag{2.2.17}$$

$$\|(\tau_n^\alpha)'\|_C \le A(\alpha)n^2, \tag{2.2.18}$$

$$|(r_n^\alpha(t))'| \le \frac{A(\alpha)}{nt^3}, \quad t \in \left[\frac{\pi}{n},\pi\right], \tag{2.2.19}$$

$$|(\gamma_n^\alpha(t))'| \le \frac{A(\alpha)}{nt^3}, \quad t \in \left[\frac{\pi}{n},\pi\right], \quad n \ge 2. \tag{2.2.20}$$

Remark 2.2.1. Relations of the types (2.2.9)–(2.2.12) and the estimates (2.2.13)–(2.2.20) appearently for the first time showed up in Kogbetliantz [1] in 1923, and then in Zygmund [2]. The paper of Szegö [2] should also be mentioned.

If

$$\omega(\alpha,t)=\frac{\cos\frac{1+\alpha}{2}t}{(\sin\frac{t}{2})^{1+\alpha}}, \tag{2.2.21}$$

then it easily can be verified that

$$\left|\omega(\alpha,t)-\omega\left(\alpha,t+\frac{\pi}{n}\right)\right|\le\frac{A(\alpha)}{nt^{2+\alpha}},\quad t\in\left[\frac{\pi}{n},\pi\right], \tag{2.2.22}$$

$$\left|\left[\omega(\alpha,t)-\omega\left(\alpha,t+\frac{\pi}{n}\right)\right]'\right|\le\frac{A(\alpha)}{nt^{3+\alpha}},\quad t\in\left[\frac{\pi}{n},\pi\right], \tag{2.2.23}$$

$$\left|\omega(\alpha,t)-2\omega\left(\alpha,t+\frac{\pi}{n}\right)+\omega\left(\alpha,t+\frac{2\pi}{n}\right)\right|\le$$

$$\le\frac{A(\alpha)}{n^2t^{3+\alpha}},\quad t\in\left[\frac{\pi}{n},\pi\right], \tag{2.2.24}$$

$$\left|\left[\omega(\alpha,t)-2\omega\left(\alpha,t+\frac{\pi}{n}\right)+\omega\left(\alpha,t+\frac{2\pi}{n}\right)\right]'\right|\le$$

$$\le\frac{A(\alpha)}{n^2t^{4+\alpha}},\quad t\in\left[\frac{\pi}{n},\pi\right],\ n\ge2. \tag{2.2.25}$$

As already pointed out, there are various criteria of convergence of the series $\sigma[f]$ and $\overline{\sigma}[f]$ at a fixed point. Hardy [3] in 1918 and Morse and Transue [1] in 1951 investigated the problem on comparison of different criteria of the convergence of $\sigma[f]$ at a fixed point. Here the work of Ramsamujh [1] should also be mentioned.

One of the most general criteria of convergence of $\sigma[f]$ at a point is the Lebesgue–Gergen test.

In 1905, Lebesgue [2] proved that if $f\in L(T)$ and for a fixed point $x\in T$,

$$\text{a)}\quad h^{-1}\int_0^h|\varphi(x,t)|\,dt\to0\ \text{ as }\ h\to0+; \tag{2.2.26}$$

$$\text{b)}\quad \int_h^\pi t^{-1}|\varphi(x,t)-\varphi(x,t+h)|\,dt\to0\ \text{ as }\ h\to0+,$$

then $\sigma[f]$ converges at x to f.

In 1930, Gergen [1] showed that in Lebesgue's theorem the condition a) can be replaced by a weaker one:

$$h^{-1}\Phi(x,h)\to0\quad\text{as}\quad h\to0+\,.$$

In 1876, du Bois–Reymond [1] constructed the series $\sigma[f]$ of a continuous function which is divergent in some points. This fact caused the scepticism towards trigonometric Fourier series as a device of representation of functions (in the sense of pointwise convergence).

In 1904, Fejér [1] investigated the problem of representation of functions by trigonometric series from the viewpoint of summability by the method of means. This approach turned out to be fruitful. In particular, Fejér proved that if $f \in C(T)$, then

$$\|\sigma_n^1(f) - f\|_C \to 0 \quad \text{as} \quad n \to \infty.$$

In 1905, Lebesgue [2] proved the following.

1) If $f \in L(T)$, then (2.2.26) is fulfilled almost everywhere;

2) If $f \in L(T)$, then at every point where (2.2.6) is satisfied (in particular, almost everywhere), the relation

$$\left[\sigma_n^1(x,f) - f(x)\right] \to 0 \quad \text{as} \quad n \to \infty$$

also holds.

In 1923, M. Riesz [1] proved that if $f \in C(T)$, then

$$\|\sigma_n^\alpha(f) - f\|_C \to 0, \quad n \to \infty,$$

and if $f \in L(T)$, then almost everywhere

$$\left[\sigma_n^\alpha(x,f) - f(x)\right] \to 0 \quad \text{as} \quad n \to \infty, \quad \alpha \in]0,+\infty[. \tag{2.2.27}$$

In 1919, Privalov [2] showed that if $f \in L(T)$, then almost everywhere

$$\left[t_n^1(x,f) - \overline{f}_n(x)\right] \to 0 \quad \text{as} \quad n \to \infty. \tag{2.2.28}$$

Here the paper of Plessner [2] should also be mentioned.

In 1924, Zygmund [2] showed that if $f \in C(T)$, then

$$\|t_n^\alpha(f) - \overline{f}_n\|_C \to 0 \quad \text{as} \quad n \to \infty,$$

and if $f \in L(T)$, then almost everywhere

$$\left[t_n^\alpha(x,f) - \overline{f}_n(x)\right] \to 0 \quad \text{as} \quad n \to \infty, \quad \alpha \in]0,+\infty[. \tag{2.2.29}$$

We ought to say that in fact M. Riesz, Privalov, and Plessner and Zygmund established that (2.2.27)–(2.2.29) hold at the points satisfying (2.2.26).

These results especially gained in importance after Kolmogorov [1], [4] constructed an everywhere divergent series $\sigma[f]$.

On the other hand, it can be proved that there is a function $f \in C(T)$ such that for any $\alpha \in]-1,0[$, $\sigma[f]$ is summable at no point by (C,α)-method. It is why the problem arose of finding general criteria for the summability of $\sigma[f]$ and $\overline{\sigma}[f]$ at a single point by (C,α)-method with $\alpha \in]-1,0[$.

The author [3], [13, Ch. II] established some theorems in this direction. They are presented below, in some cases in a stronger form.

Theorem 2.2.1. *Let $f \in L(T)$ and $\alpha \in]-1,1[$. Suppose that at a fixed point $x \in T$ the following relations are fulfilled:*

a) $h^{-1}\Phi(x,h) \to 0$ *as* $h \to 0+$;

b) $h^{-\alpha}\int\limits_h^{\pi} t^{\alpha-1}|\varphi(x,t)-\varphi(x,t+h)|\,dt \to 0$ *as* $h \to 0+$.

Then $\sigma[f]$ *is* $(C,-\alpha)$*-summable at* x *to* f.

Proof. For $\alpha = 0$, the theorem is the above mentioned result of Lebesgue–Gergen. Therefore we will assume that $|\alpha| \in]0,1[$. It should be noted that the scheme of the proof of the theorem works in the case $\alpha = 0$ too, in some places even in a simpler form.

As usual, we put

$$A_k(x) \equiv A_k(f,x) = a_k(f)\cos kx + b_k(f)\sin kx.$$

Since φ is even with respect to t, we have

$$\frac{1}{\pi}\int\limits_0^{\pi}\varphi(x,t)\cos kt\,dt =$$

$$= \frac{1}{\pi}\int\limits_0^{\pi}[f(x+t)+f(x-t)-2f(x)]\cos kt\,dt = A_k(x), \quad k \in \mathbb{N}. \tag{2.2.30}$$

Moreover,

$$\int\limits_{\pi-h}^{\pi}\varphi(x,t)\,dt = \int\limits_{2h}^{\pi}\varphi(x,t)\,dt - \int\limits_{2h}^{\pi-h}\varphi(x,t)\,dt =$$

$$= \int\limits_h^{\pi-h}[\varphi(x,t+h)-\varphi(x,t)]\,dt + \int\limits_h^{2h}\varphi(x,t)\,dt, \quad h \in \left]0,\frac{\pi}{3}\right[.$$

Hence, by the conditions a) and b) of the theorem, we find

$$h^{-\alpha}\int\limits_{\pi-h}^{\pi}\varphi(x,t)\,dt \to 0 \quad \text{as } h \to 0. \tag{2.2.31}$$

According to (2.2.30),

$$2A_k(x) = \frac{1}{\pi}\int\limits_0^{\pi/k}\varphi(x,t)\cos kt\,dt +$$

$$+\frac{1}{\pi}\int\limits_0^{\pi/k}\left[\varphi(x,t)-\varphi\left(x,t+\frac{\pi}{k}\right)\right]\cos kt\,dt +$$

$$+\frac{1}{\pi}\int\limits_{\pi/k}^{\pi-\pi/k}\left[\varphi(x,t)-\varphi\left(x,t+\frac{\pi}{k}\right)\right]\cos kt\,dt +$$

$$+\frac{1}{\pi}\int\limits_{\pi-\pi/k}^{\pi}\varphi(x,t)\cos kt\,dt\equiv\sum_{i=1}^{4}\ell_i(x,f,k). \tag{2.2.32}$$

Integrating by parts, we obtain

$$\ell_1(x,f,k)=\frac{1}{\pi}\Big[\Phi(x,t)\cos kt\Big]_0^{\pi/k}+\frac{k}{\pi}\int\limits_0^{\pi/k}\Phi(x,t)\sin kt\,dt.$$

Hence, by virtue of the condition a), we get

$$k^\alpha\ell_1(x,f,k)\to 0\ \text{ as }\ k\to\infty.$$

Therefore

$$k^\alpha\ell_2(x,f,k)\to 0\ \text{ as }\ k\to\infty.$$

Further,

$$|\ell_3(x,f,k)|\le A(\alpha)\int\limits_{\pi/k}^{\pi}t^{\alpha-1}\Big|\varphi(x,t)-\varphi\Big(x,t+\frac{\pi}{k}\Big)\Big|\,dt.$$

Hence the condition b) of the theorem implies

$$k^\alpha\ell_3(x,f,k)\to 0\ \text{ as }\ k\to\infty.$$

Integrating by parts and using (2.2.31), we have

$$k^\alpha\ell_4(x,f,k)\to 0\ \text{ as }\ k\to\infty.$$

Therefore, adding together the obtained estimates for $\ell_i(x,f,k)$, $1\le i\le 4$, we get by (2.2.32)

$$k^\alpha A_k(x)\to 0\ \text{ as }\ k\to\infty,\quad \alpha\in]0,1[. \tag{2.2.33}$$

By virtue of (2.2.1) and (2.2.7), we obtain

$$\sigma_n^{-\alpha}(x,f)-f(x)=\frac{1}{\pi}\int\limits_0^{\pi}\varphi(x,t)K_n^{-\alpha}(t)\,dt=$$

$$=\frac{1}{\pi}\int\limits_0^{\pi/n}\varphi(x,t)K_n^{-\alpha}(t)\,dt+\frac{1}{\pi}\int\limits_{\pi/n}^{\pi}\varphi(x,t)K_n^{-\alpha}(t)\,dt\equiv$$

$$\equiv I_1(x,f,\alpha,n)+I_2(x,f,\alpha,n),\quad n>4. \tag{2.2.34}$$

Integrating by parts and using (2.2.13), (2.2.17), and the condition b) of the theorem, we find

$$|I_1(x,f,\alpha,n)|=\left|\frac{1}{\pi}\Big[\Phi(x,t)K_n^{-\alpha}(t)\Big]_0^{\pi/n}-\frac{1}{\pi}\int\limits_0^{\pi/n}\Phi(x,t)[K_n^{-\alpha}]'dt\right|\le$$

$$\le A(\alpha)\Big[\Phi\Big(x,\frac{\pi}{n}\Big)n+n^2\int\limits_0^{\pi/n}|\Phi(x,t)|\,dt\Big],$$

i.e.,

$$I_1(x,f,\alpha,n)\to 0 \quad \text{as } n\to\infty. \tag{2.2.35}$$

According to (2.2.9), we have

$$I_2(x,f,\alpha,n)=\frac{1}{\pi}\int\limits_{\pi/n}^{\pi}\varphi(x,t)\varphi_n^{-\alpha}(t)\,dt+\frac{1}{\pi}\int\limits_{\pi/n}^{\pi}\varphi(x,t)r_n^{-\alpha}(t)\,dt\equiv$$

$$\equiv P_1(x,f,\alpha,n)+P_2(x,f,\alpha,n). \tag{2.2.36}$$

As above, integrating by parts and using (2.2.15), (2.2.19), and the condition a) of the theorem, we conclude

$$|P_2(x,f,\alpha,n)|=\left|\frac{1}{\pi}\Big[\Phi(x,t)r_n^{-\alpha}(t)\Big]_{\pi/n}^{\pi}-\frac{1}{\pi}\int\limits_{\pi/n}^{\pi}\Phi(x,t)[r_n^{-\alpha}]'dt\right|\le$$

$$\le A(\alpha)\Big[n^{-1}\Phi(x,\pi)+\Phi\Big(x,\frac{\pi}{n}\Big)\Big|r_n^{-\alpha}\Big(\frac{\pi}{n}\Big)\Big|+$$

$$+n^{-1}\int\limits_{\pi/n}^{\pi}t^{-3}|\Phi(x,t)|\,dt\Big],$$

i.e.,

$$P_2(x,f,\alpha,n)\to 0 \quad \text{as } n\to\infty. \tag{2.2.37}$$

Taking into account (2.2.5), (2.2.11) and (2.2.21), we see that in order to evaluate $P_1(x,f,\alpha,n)$, we have to estimate

$$n^{\alpha}\int\limits_{\pi/n}^{\pi}\varphi(x,t)\omega(-\alpha,t)\cos nt\,dt\equiv H(x,f,\alpha,n).$$

Changing in $H(x,f,\alpha,n)$ the variable t to $t+\frac{\pi}{n}$ and summing the two expressions for $H(x,f,\alpha,n)$, we find

$$2H(x,f,\alpha,n)=n^{\alpha}\Bigg\{\int\limits_{\pi/n}^{\pi-\pi/n}\Big[\varphi(x,t)-\varphi\Big(x,t+\frac{\pi}{n}\Big)\Big]\omega(-\alpha,t)\cos nt\,dt+$$

$$+\int\limits_{\pi/n}^{\pi-\pi/n}\varphi\Big(x,t+\frac{\pi}{n}\Big)\Big[\omega(-\alpha,t)-\omega\Big(-\alpha,t+\frac{\pi}{n}\Big)\Big]\cos nt\,dt-$$

$$-\int\limits_{0}^{\pi/n}\varphi\Big(x,t+\frac{\pi}{n}\Big)\omega\Big(-\alpha,t+\frac{\pi}{n}\Big)\cos nt\,dt+$$

$$+\int\limits_{\pi-\pi/n}^{\pi}\varphi(x,t)\omega(-\alpha,t)\cos nt\,dt\Bigg\}\equiv\sum_{i=1}^{4}B_i(x,f,\alpha,n). \tag{2.2.38}$$

By (2.2.21) and the condition b) of the theorem,

$$B_1(x,f,\alpha,n)\to 0 \text{ as } n\to\infty. \tag{2.2.39}$$

Using (2.2.22), we get

$$\left|B_2(x,f,\alpha,n)-n^\alpha\int_{\pi/n}^{\pi-\pi/n}\varphi(x,t)\Big[\omega(-\alpha,t)-\omega\Big(-\alpha,t+\frac{\pi}{n}\Big)\Big]\cos nt\,dt\right|\le$$

$$\le A(\alpha)n^{\alpha-1}\int_{\pi/n}^{\pi-\pi/n}t^{\alpha-2}\left|\varphi(x,t)-\varphi\Big(x,t+\frac{\pi}{n}\Big)\right|dt\le$$

$$\le A(\alpha)n^{\alpha}\int_{\pi/n}^{\pi-\pi/n}t^{\alpha-1}\left|\varphi(x,t)-\varphi\Big(x,t+\frac{\pi}{n}\Big)\right|dt.$$

Hence, according to the condition b) of the theorem, we conclude

$$\left[B_2(x,f,\alpha,n)-n^\alpha\int_{\pi/n}^{\pi-\pi/n}\varphi(x,t)\Big[\omega(-\alpha,t)-\omega\Big(-\alpha,t+\frac{\pi}{n}\Big)\Big]\cos nt\,dt\right]\to 0 \tag{2.2.40}$$

$$\text{as } n\to\infty.$$

It is not hard to verify that

$$2n^\alpha\int_{\pi/n}^{\pi-\pi/n}\varphi(x,t)\Big[\omega(-\alpha,t)-\omega\Big(-\alpha,t+\frac{\pi}{n}\Big)\Big]\cos nt\,dt=$$

$$=2n^\alpha\int_{\pi/n}^{2\pi/n}\varphi(x,t)\Big[\omega(-\alpha,t)-\omega\Big(-\alpha,t+\frac{\pi}{n}\Big)\Big]\cos nt\,dt+$$

$$+n^\alpha\int_{2\pi/n}^{\pi-2\pi/n}\Big[\varphi(x,t)-\varphi\Big(x,t+\frac{\pi}{n}\Big)\Big]\Big[\omega(-\alpha,t)-\omega\Big(-\alpha,t+\frac{\pi}{n}\Big)\Big]\cos nt\,dt+$$

$$+n^\alpha\int_{2\pi/n}^{\pi-2\pi/n}\varphi\Big(x,t+\frac{\pi}{n}\Big)\Big[\omega(-\alpha,t)-2\omega\Big(-\alpha,t+\frac{\pi}{n}\Big)+\omega\Big(-\alpha,t+\frac{2\pi}{n}\Big)\Big]\cos nt\,dt-$$

$$-n^\alpha\int_{\pi/n}^{2\pi/n}\varphi\Big(x,t+\frac{\pi}{n}\Big)\Big[\omega\Big(-\alpha,t+\frac{\pi}{n}\Big)-\omega\Big(-\alpha,t+\frac{2\pi}{n}\Big)\Big]\cos nt\,dt+$$

$$+n^\alpha\int_{\pi-2\pi/n}^{\pi-\pi/n}\varphi(x,t)\Big[\omega(-\alpha,t)-\omega\Big(-\alpha,t+\frac{\pi}{n}\Big)\Big]\cos nt\,dt\equiv$$

$$\equiv \sum_{i=5}^{9} B_i(x,f,\alpha,n). \tag{2.2.41}$$

Integrating by parts, on the basis of (2.2.22) and (2.2.23) we find

$$|B_5(x,f,\alpha,n)| \leq A(\alpha)\cdot n\Big[\Big|\Phi\Big(x,\frac{\pi}{n}\Big)\Big| + \Big|\Phi\Big(x,\frac{2\pi}{n}\Big)\Big|\Big] +$$
$$+n^\alpha\Big[n^{-1}\int\limits_{\pi/n}^{2\pi/n} t^{\alpha-3}|\Phi(x,t)|\,dt + \int\limits_{\pi/n}^{2\pi/n} t^{\alpha-2}|\Phi(x,t)|\,dt\Big].$$

Hence the condition a) of the theorem imply

$$B_5(x,f,\alpha,n)\to 0 \ \text{ as } \ n\to\infty. \tag{2.2.42}$$

Quite analogously we can prove that

$$B_8(x,f,\alpha,n)\to 0 \ \text{ as } \ n\to\infty. \tag{2.2.43}$$

By virtue of (2.2.22) and the condition b) of the theorem, we easily conclude that

$$B_6(x,f,\alpha,n)\to 0 \ \text{ as } \ n\to\infty. \tag{2.2.44}$$

Using (2.2.24) and (2.2.25), and integrating by parts, we get

$$\Big|n^\alpha \int\limits_{\pi/n}^{\pi-2\pi/n} \varphi(x,t)\Big[\omega(-\alpha,t)-2\omega\Big(-\alpha,t+\frac{\pi}{n}\Big)+\omega\Big(-\alpha,t+\frac{2\pi}{n}\Big)\Big]\cos nt\,dt\Big| \leq$$
$$\leq A(\alpha)\Big[n^{-1}|f(x)| + n\Big|\Phi\Big(x,\frac{2\pi}{n}\Big)\Big| + n^{\alpha-1}\int\limits_{\pi/n}^{\pi} t^{\alpha-3}|\Phi(x,t)|\,dt\Big].$$

Hence, by the condition a) of the theorem, from (2.2.41) we obtain

$$B_7(x,f,\alpha,n)\to 0 \ \text{ as } \ n\to\infty. \tag{2.2.45}$$

Using (2.2.22), we have

$$B_9(x,f,\alpha,n)\to 0 \ \text{ as } \ n\to\infty. \tag{2.2.46}$$

Therefore from (2.2.40)–(2.2.46) we find

$$B_2(x,f,\alpha,n)\to 0 \ \text{ as } \ n\to\infty. \tag{2.2.47}$$

Quite similarly to (2.2.42), we obtain

$$B_3(x,f,\alpha,n)\to 0 \ \text{ as } \ n\to\infty. \tag{2.2.48}$$

Consider now the expression $B_4(x,f,\alpha,n)$ from (2.2.38). Multiplying the Fourier series of φ (see (2.2.30)) by a function ω of the bounded variation on $[\pi-\frac{\pi}{n},\pi]$ and using a result of W. Young [2] (see also Zygmund [15, Ch. IV]), we get

$$B_4(x,f,\alpha,n) = n^\alpha\Big[a_0(f)\int\limits_{\pi-\pi/n}^{\pi} \omega(-\alpha,t)\cos nt\,dt +$$

$$+\sum_{k=1}^{2n} 2A_k(x) \int_{\pi-\pi/n}^{\pi} \omega(-\alpha, t) \cos kt \cos nt\, dt +$$

$$+ \sum_{k=2n+1}^{\infty} 2A_k(x) \int_{\pi-\pi/n}^{\pi} \omega(-\alpha, t) \cos kt \cos nt\, dt \Bigg], \quad \alpha \in]0,1[. \tag{2.2.49}$$

On the other hand, for $k \neq n$

$$\int_{\pi-\pi/n}^{\pi} \omega(-\alpha, t) \cos(n \pm k)t\, dt =$$

$$= \frac{1}{n \pm k} \left\{ \left[\omega(-\alpha, t) \sin(n \pm k)t \Big|_{\pi-\pi/n}^{\pi} - \int_{\pi-\pi/n}^{\pi} [\omega(-\alpha, t)]' \sin(n \pm k)t\, dt \right] \right\},$$

so

$$\left| \int_{\pi-\pi/n}^{\pi} \omega(-\alpha, t) \cos(n \pm k)t\, dt \right| \leq \frac{A(\alpha)}{|n-k|}, \quad k \neq n.$$

Using the last relation along with (2.2.49), we have

$$|B_4(x, f, \alpha, n)| \leq A(\alpha) n^{\alpha} \Bigg[n^{-1} |a_0(f)| + n^{-1} \sum_{k=1}^{2n} |A_k(x)| +$$

$$+ \sum_{k=2n+1}^{\infty} (k-n)^{-1} |A_k(x)| \Bigg].$$

Hence, it follows by (2.2.33)

$$B_4(x, f, \alpha, n) \to 0 \ \text{ as } \ n \to \infty. \tag{2.2.50}$$

If $\alpha \in]-1, 0]$, then (2.2.50) is obvious.

Thus on the basis of (2.2.38), (2.2.39), (2.2.47), (2.2.48), and (2.2.50), we conclude

$$P_1(x, f, \alpha, n) \to 0 \ \text{ as } \ n \to \infty,$$

which along with (2.2.34)–(2.2.37) implies

$$\left[\sigma_n^{-\alpha}(x, f) - f(x) \right] \to 0 \ \text{ as } \ n \to \infty. \quad \blacksquare$$

Remark 2.2.2. In Theorem 2.2.1, it was assumed that $\alpha \in]-1,1[$. It should be noted that the theorem is true for $\alpha \in]-\infty, -1]$ as well. Consider the case $\alpha = -1$. Then by definition

$$\sigma_n^{-\alpha}(x, f) - f(x) \equiv \sigma_n^1(x, f) - f(x) = \frac{1}{\pi} \int_0^{\pi} \varphi(x, t) K_n^1(t)\, dt,$$

where K_n^1 is the Fejér kernel, i.e.,

$$K_n^1(t) \equiv K_n(t) = \frac{1 - \cos(n+1)t}{4(n+1) \sin^2 \frac{t}{2}} \quad \text{for } \ |t| \in]0, \pi], \tag{2.2.51}$$

while

$$K_n(0) = \frac{n+1}{2}.$$

It is well-known that (see, e.g., Bari [2, Ch. I])

$$\|K_n^1\|_C \le An, \quad \|(K_n^1)'\|_C \le An^2. \tag{2.2.52}$$

By virtue of Fejér's formula (see (2.2.51)), we have

$$\sigma_n^1(x,f) - f(x) = \frac{1}{\pi}\int_0^{\pi/n} \varphi(x,t)K_n^1(t)\,dt +$$

$$+\frac{1}{4\pi(n+1)}\left[\int_{\pi/n}^{\pi} \frac{\varphi(x,t)\,dt}{\sin^2\frac{t}{2}} - \int_{\pi/n}^{\pi} \frac{\varphi(x,t)\cos t\cos nt}{\sin^2\frac{t}{2}} +\right.$$

$$\left.+\int_{\pi/n}^{\pi} \frac{\varphi(x,t)\sin t\sin nt}{\sin^2\frac{t}{2}}\right] \equiv \sum_{i=10}^{13} B_i(x,f,n). \tag{2.2.53}$$

Integrating by parts, in view of (2.2.52) and the condition a) of the theorem we find

$$B_{10}(x,f,n) \to 0 \quad \text{as } n \to \infty. \tag{2.2.54}$$

Analogously, integrating by parts and using the condition a) of the theorem, we get

$$B_{11}(x,f,n) \to 0 \quad \text{as } n \to \infty. \tag{2.2.55}$$

By (2.2.21), the expression $B_{12}(x,f,n)$ can be represented as

$$2B_{12}(x,f,n) = \frac{1}{4\pi(n+1)}\left\{\int_{\pi/n}^{\pi-\pi/n} \left[\varphi(x,t) - \varphi\left(x,t+\frac{\pi}{n}\right)\right]\omega(1,t)\cos nt\,dt +\right.$$

$$+\int_{\pi/n}^{\pi-\pi/n} \varphi\left(x,t+\frac{\pi}{n}\right)\left[\omega(1,t) - \omega\left(1,t+\frac{\pi}{n}\right)\right]\cos nt\,dt -$$

$$-\int_0^{\pi/n} \varphi\left(x,t+\frac{\pi}{n}\right)\omega\left(1,t+\frac{\pi}{n}\right)\cos nt\,dt +$$

$$\left.+\int_{\pi-\pi/n}^{\pi} \varphi(x,t)\omega(1,t)\cos nt\,dt\right\} \equiv \sum_{i=14}^{17} B_i(x,f,n). \tag{2.2.56}$$

According to the condition b) of the theorem,

$$B_{14}(x,f,n) \to 0 \quad \text{as } n \to \infty.$$

Integration by parts of (2.2.22) and the condition a) of the theorem imply

$$B_{15}(x,f,n) \to 0 \quad \text{as } n \to \infty.$$

It is not difficult to see that

$$B_i(x, f, n) \to 0 \quad \text{as } n \to \infty \ (i = 16, 17).$$

Thus from (2.2.56) we conclude that

$$B_{12}(x, f, n) \to 0 \quad \text{as } n \to \infty. \tag{2.2.57}$$

Quite analogously we can get

$$B_{13}(x, f, n) \to 0 \quad \text{as } n \to \infty. \tag{2.2.58}$$

Therefore by (2.2.53), (2.2.54), (2.2.55), (2.2.57), and (2.2.58), the relation

$$\left[\sigma_n^1(x, f) - f(x)\right] \to 0 \quad \text{as } n \to \infty,$$

is true.

Suppose now that $\alpha \in]-\infty, -1[$. Then for $\alpha \in \mathbf{Z}$, the same method can be applied. If $\alpha \in]-\infty, -1[$ and differs from negative integers, then a well-known method (see, e.g., Zygmund [15, Ch. III]) of the "approximation" of $K_n^{-\alpha}$ by K_n^{β} with $\beta \in]0, 1[$ can be used. A similar argument enable us to conclude that Theorem 2.2.1 is true for $\alpha \in]1, +\infty[\setminus \mathbf{N}$ as well.

Remark 2.2.3. It is well-known that for $\alpha < \alpha'$, (C, α)-summability implies (C, α')-summability (this is a result of Chapman and Knopp, see, e.g., Hardy [6, Ch. V]).

Remark 2.2.4. In 1926, Hardy and Littlewood [4] showed that if $f \in L(T)$ and the conditions

$$\begin{aligned} h^{-1} \int_0^h |\varphi(x, t)|\, dt \le A(f, x), \quad h \in]0, \pi], \\ h^{-1}\Phi(x, h) \to 0 \quad \text{as } h \to 0+ \end{aligned} \tag{2.2.59}$$

are fulfilled at a point $x \in T$, then $\sigma[f]$ is $(C, 1)$-summable at x. Moreover, no one of the conditions from (2.2.59), taken alone, guarantees the $(C, 1)$-summability of $\sigma[f]$ at x. From the method of the proof of Theorem 2.2.1 it follows that the statement of Hardy and Littlewood is true for (C, α)-means as well, i.e., if $f \in L(T)$ and (2.2.59) is fulfilled at a point x, then for any $\alpha \in]0, +\infty[$, $\sigma[f]$ is (C, α)-summable to f at x.

Remark 2.2.5. The following theorem of Cantor and Lebesgue is well-known (see, e.g., Bari [2, Ch. I], Zygmund [15, Ch. IX]).

If a trigonometric series (2.1.1) converges on a set $E \subset T$ with $\mu_1(E) > 0$, then $a_k \to 0$, $b_k \to 0$ as $k \to \infty$.

This assertion was proved by Cantor [1] in 1872 for $E = [a, b]$, $b - a > 0$, and in 1906 Lebesgue [3] pointed out the validity of Cantor's theorem for an arbitrary $E \subset T$ with $\mu_1(E) > 0$.

On the other hand, it is well-known that (see, e.g., Zygmund [15, Ch. III]) if $\sigma[f]$ is $(C, -\alpha)$-summable, $\alpha \in]0, 1[$, to f on a set $E \subset T$ with $\mu_1(E) > 0$, then

$$k^{\alpha} A_k(x) \to 0 \quad \text{as } k \to \infty$$

(to be more precise, this result is true for arbitrary numerical series). Therefore, by Cantor–Lebesgue's theorem, in this case we have

$$k^{\alpha} a_k(x) \to 0, \quad k^{\alpha} b_k(f) \to 0 \quad \text{as } k \to \infty. \tag{2.2.60}$$

Later, in 1953, Izumi, Matsuyama and Tsutikura [1] proved that the condition a) of Theorem 2.2.1 is not necessary for the convergence of $\sigma[f]$ at a point. However, in 1960, Hsiang [1] showed that this condition is necessary for the $(C, -\alpha)$-summability of $\sigma[f]$ at a point. Further, if in Theorem 2.2.1, we replace the condition b) by

$$h^{-\alpha} \int_h^{\pi} t^{\alpha-1} |\varphi(x,t) - \varphi(x,t+h)| \, dt \leq A(f,x,\alpha), \quad h \in]0,\pi], \tag{2.2.61}$$

then $\sigma[f]$ need not be $(C,-\alpha)$-summable to f at x.

Indeed, let $\alpha \in]0,1[$ and $f : T \to \mathbb{R}$ be defined by

$$f(x) \equiv f(x,\alpha) = \sum_{k=1}^{\infty} 2^{\alpha k} \sin(2^k \pi x),$$

i.e., f is the well-known Weierstrass function. As it is well-known (see, e.g., Zygmund [15, Ch. II]), $f \in Lip\,\alpha$. Then (2.2.61) is valid for any $x \in T$; to be more precise,

$$h^{-\alpha} \int_h^{\pi} t^{\alpha-1} |\varphi(x,t) - \varphi(x,t+h)| \, dt \leq A(f), \quad h \in]0,\pi]$$

(f itself depends on α). But $\sigma[f]$ is not $(C,-\alpha)$-summable to f on a set $E \subset T$ with $\mu_1(E) > 0$ since for this function (2.2.60) is violated.

Analyzing the proof of Theorem 2.2.1, we can conclude that it holds (for $\alpha = 0$ this is the Lebesgue theorem)

Theorem 2.2.2. *Let $f \in L(T)$ and $\alpha \in [0,1[$. If f is continuous on an interval $]a,b[\subset T$, $b-a>0$, and*

$$\left[h^{-\alpha} \int_h^{\pi} t^{\alpha-1} |\varphi(x,t) - \varphi(x,t+h)| \, dt \right] \to 0 \quad as \quad h \to 0+$$

uniformly with respect to $x \in]a,b[$, then $\sigma[f]$ is $(C,-\alpha)$-summable to f uniformly within $]a,b[$.

As to the series $\overline{\sigma}[f]$, it holds

Theorem 2.2.3. *Let $f \in L(T)$ and $\alpha \in]-\infty,+\infty[\setminus \mathbb{N}$. Suppose also that at a fixed point $x \in T$, the conditions*

a) $h^{-1}\psi(x,h) \to 0$ *as* $h \to 0+$;

b) $h^{-\alpha} \int_h^{\pi} t^{\alpha-1} |\psi(x,t) - \psi(x,t+h)| \, dt \to 0$ *as* $h \to 0+$

are fulfilled. Then

$$\left[t_n^{-\alpha}(x,f)-\overline{f}_n(x)\right]\to 0 \quad as \quad n\to\infty.$$

Proof. Let $|\alpha|\in]0,1[$. According to (2.2.2), (2.2.6), (2.2.8), and (2.2.10), we have

$$t_n^{-\alpha}(x,f)-\overline{f}_n(x)=-\frac{1}{\pi}\int\limits_0^{\pi/n}\psi(x,t)\tau_n^{-\alpha}(t)\,dt-$$

$$-\frac{1}{\pi}\int\limits_{\pi/n}^{\pi}\psi(x,t)\psi_n^{-\alpha}(t)\,dt-\frac{1}{\pi}\int\limits_{\pi/n}^{\pi}\psi(x,t)\gamma_n^{-\alpha}(t)\,dt\equiv\sum_{i=1}^{3}q_i(x,f,\alpha,n).$$

The relations (2.2.12), (2.2.14), (2.2.16), (2.2.18), and (2.2.20) show that using the conditions a) and b) of the theorem, the expressions $q_i(x,f,\alpha,n)$ $(i=\overline{1,3})$ can be estimated quite similarly to the expressions $I_1(x,f,\alpha,n)$, $P_1(x,f,\alpha,n)$, $P_2(x,f,\alpha,n)$ from (2.2.34) and (2.2.36), respectively. It should be noted that since ψ is an odd function with respect to t, in this case the relation

$$\frac{1}{\pi}\int\limits_0^{\pi}\psi(x,t)\sin kt\,dt=-\left[b_k(f)\cos kx-a_k(f)\sin kx\right]\equiv-B_k(f,x),\quad k\in\mathbb{N},$$

holds instead of (2.2.30), and (2.2.33) must be replaced by

$$k^{\alpha}B_k(f,x)\to 0 \quad \text{as} \quad k\to\infty.$$

Thus, for $|\alpha|\in]0,1[$, we can consider Theorem 2.2.3 as proved.

Further, as it is known (see, e.g., Bari [2, Ch. VIII]), for $\alpha=-1$ and $|t|\in]0,\pi]$ we have

$$\tau_n^{-\alpha}(t)\equiv\overline{K}_n(t)=\frac{1}{n+1}\sum_{k=0}^{n}D_k(f)=\frac{1}{2}\operatorname{ctg}\frac{t}{2}-\frac{\sin(n+1)t}{4(n+1)\sin^2\frac{t}{2}}$$

and

$$\|\overline{K}_n\|_C\le An,\qquad \|(\overline{K}_n)'\|_C\le An^2.$$

Using the argument presented in Remark 2.2.2, we can conclude that

$$\left[t_n^{-\alpha}(x,f)-\overline{f}_n(x)\right]\to 0 \quad \text{as} \quad n\to\infty$$

for $\alpha\in]-\infty,-1]$. If $\alpha=0$, then

$$t_n^{0}(x,f)-\overline{f}_n(x)=\overline{S}_n(x,f)-\overline{f}_n(x)=$$

$$=-\frac{1}{\pi}\int\limits_0^{\pi/n}\psi(x,t)\overline{D}_n(t)\,dt+\frac{1}{\pi}\int\limits_{\pi/n}^{\pi}\psi(x,t)\,\frac{\cos(n+\frac{1}{2})t}{2\sin^2\frac{t}{2}}\,dt.$$

Using either the scheme of the proof of Theorem 2.2.1 or the one of Lebesgue–Gergen's theorem (see, e.g., Bari [2, Ch. III]), we can verify that the theorem is true for $\alpha=0$ as well. ■

Remark 2.2.6. Theorem 2.2.3 remains valid if $\alpha = 0$ and the condition a) is replaced by

$$h^{-1}\int_0^h |\psi(x,t)|\,dt \to 0 \text{ as } h \to 0+.$$

This is pointed out in the monograph by Zygmund [15, Ch. II].

Note at last that for the series $\overline{\sigma}[f]$, a theorem analogous to Theorem 2.2.2 is also valid but we do not formulate it since by Theorem 2.2.3 it is almost obvious.

From the above theorems several corollaries can be deduced concerning the problems of the convergence and summability of $\sigma[f]$ and $\overline{\sigma}[f]$ either at separate points or uniformly on an interval $]a,b[\subset T$, but we do not dwell on this. In consequent chapters we will partly return to the matter.

Now let $f \in L(T)$ and $\alpha \in]0,1[$. Consider the trigonometric series

$$\frac{a_0(f)}{2} + \sum_{k=1}^{\infty} \frac{A_k(f,x)}{k^\alpha}. \tag{2.2.62}$$

Theorem 2.2.4. *If $f \in L^p(T)$ for some $p \in]1,+\infty[$, then* (2.2.62) *is almost everywhere $(C,-\alpha)$-summable.*

Proof. Applying the theorems of Marcinkiewicz [2] on multipliers for $\sigma[f]$, we conclude that (2.2.62) is the Fourier series of a function $f_0 \in L^p(T)$, and $f_0 \equiv f_0(f,\alpha)$. Denote by $U_n^\alpha(x,f_0)$ the $(C,-\alpha)$-means of $\sigma[f_0]$ and put

$$U^*(x,f_0) = \sup_n |U_n^{-\alpha}(x,f_0)|,$$
$$S^*(x,f) = \sup_n |S_n(x,f)|.$$

By virtue of Hunt's theorem [2], we have

$$\|S^*\|_p \le A(p)\|f\|_p.$$

Using this and taking into account that $\sigma[f_0]$ and $\sigma[f]$ are connected according to (2.2.62), by Abel's transform we find

$$\|U^*\|_p \le A(p,\alpha)\|f\|_p.$$

The last inequality enables us to conclude that the theorem is true. ■

Corollary 2.2.1. *Let $f \in L^2(T)$ and $\alpha \in]0,1[$. If*

$$\sum_{k=1}^{\infty} \left[a_k^2(f) + b_k^2(f)\right] k^{2\alpha} < +\infty, \tag{2.2.63}$$

then $\sigma[f]$ is $(C,-\alpha)$-summable almost everywhere.

Remark 2.2.7. In 1950, Sunouchi and Yano [1] proved that if $(\varphi_k)_{k\ge1}$ is an arbitrary orthonormal system on $[0,1]$, $\alpha \in]0,1[$, and

$$\sum_{k=1}^{\infty} a_k^2 k^{2\alpha} < +\infty,$$

then the series

$$\sum_{k=1}^{\infty} a_k \varphi_k(x)$$

is $(C, -\alpha)$-summable almost everywhere. Bregvadze [1] showed that this assertion can not be improved.

It is clear that Corollary 2.2.1 is a consequence of the result of Sunouchi and Yano. We have just shown that it can be easily obtained using Hunt's theorem.

Remark 2.2.8. In the proof of Theorem 2.2.4, we have used the theorem on multipliers for Fourier series which was established by Marcinkiewicz in 1939 [2]. For the first time the problem of multipliers for $\sigma[f]$ was raised by Fekete [1].

Using the results of Ul'yanov [1] on the equivalence of some integral and coefficient–depending convergence conditions for $\sigma[f]$, in view of Corollary 2.2.1 we can conclude that it holds

Theorem 2.2.5. *Let $f \in L^2(T)$ and $\alpha \in]0,1[$. If*

$$\int_T\int_T \frac{\left[\sum_{i=1}^{m}(-1)^i \binom{m}{i} f[x+(m-2i)]\right]^2 dx\, dt}{|t|^{1+2\alpha}} < +\infty \quad \text{for } m \in \mathbb{N}, \tag{2.2.64}$$

then $\sigma[f]$ is almost everywhere $(C, -\alpha)$-summable.

We should once more emphasize that conditions (2.2.63) and (2.2.64) are equivalent.

Now consider the problem on a precise Weyl's multiplier for $(C, -\alpha)$-summability of $\sigma[f]$ with $\alpha \in]0,1[$.

Theorem 2.2.6. *Let $\alpha \in]0,1[$ and $(\omega_k)_{k\geq 1}$ be an arbitrary sequence of positive numbers satisfying*

$$k^{-\alpha}\omega(k) \to 0 \quad \text{as} \quad k \to \infty.$$

Then there exists a function $f \in Lip\,\alpha$ such that

$$\sum_{k=1}^{\infty} \left[a_k^2(f) + b_k^2(f)\right]\omega^2(k) < +\infty,$$

but $\sigma[f]$ is not $(C, -\alpha)$-summable on any set $E \subset T$ with $\mu_1(E) > 0$.

Proof. Put

$$\varepsilon_k \equiv k^{-\alpha}\omega(k) \quad \text{as} \quad k \in \mathbb{N}.$$

In view of the condition of the theorem, there is an increasing sequence of natural numbers $(k_i)_{i\geq 1}$ such that

$$\sum_{i=1}^{\infty} \varepsilon_{k_i} < +\infty.$$

Consider the function $f : T \to \mathbb{R}$ defined by

$$f(x) \equiv f(x,\alpha) = \sum_{i=1}^{\infty} 2^{-\alpha k_i} \sin(2^{k_i}\pi x), \tag{2.2.65}$$

i.e., f is the well-known Weierstrass function. It is clear that $f \in C(T)$. Using a well-known method (see, e.g., Zygmund [15, Ch. II]), we can verify that $f \in Lip\,\alpha$. Moreover,

$$\sum_{k=1}^{\infty} \left[a_k^2(f) + b_k^2(f)\right]\omega^2(k) = \sum_{k=1}^{\infty} b_k^2(f)\omega^2(k) = \sum_{i=1}^{\infty} \varepsilon_{2^{k_i}} < +\infty.$$

But since

$$\limsup_{k\to\infty} k^{\alpha} b_k(f) = 1,$$

the series $\sigma[f]$ is not $(C,-\alpha)$-summable on any set $E \subset T$ with $\mu_1(E) > 0$.

Note that instead of (2.2.65), we could consider an analogous series by cosines.

As already pointed out, if $f \in L^p(T_+)$ and $g \in L^p(T_-)$ for some $p \in]1,+\infty[$, then all the series (2.1.6)–(2.1.9) are the Fourier series of certain functions from the classes $L^p(T_+)$ or $L^p(T_-)$ (depending on whether they are even or odd). Therefore, according to theorems of Carleson [1] and Hunt [2], in this case all the series (2.1.6)–(2.1.9) converge almost everywhere. ■

Remark 2.2.9. As it has been noted, for $p = 1$ the situation is more complicated. In this case, coefficients of some of the series (2.1.6)–(2.1.9) may even be meaningless. The series of types (2.1.6)–(2.1.9) can be considered in the case of general trigonometric series as well, i.e. when

$$\sum_{k=1}^{\infty} a_k \cos kx \quad \text{and} \quad \sum_{k=1}^{\infty} b_k \sin kx \tag{2.2.66}$$

are arbitrary trigonometric series by sines and cosines, respectively. Simple examples show that in some cases the coefficients U_k, M_k, U_k^* and M_k^*, $k \in \mathbb{N}$, have better properties than a_k or b_k, $k \in \mathbb{N}$, in (2.2.66), and in some cases the opposite is true.

As to the $(C,-\alpha)$-summability of (2.1.6)–(2.1.9) with $\alpha \in]0,1[$, we should note the following.

1) If $f \in L^2(T_+)$ and

$$\sum_{k=1}^{\infty} a_k^2(f) k^{2\alpha} < +\infty, \tag{2.2.67}$$

then

$$\sum_{k=1}^{\infty} [U_k^*]^2 k^{2\alpha} < +\infty; \tag{2.2.68}$$

2) If $g \in L^2(T_-)$ and

$$\sum_{k=1}^{\infty} b_k^2(f) k^{2\alpha} < +\infty, \tag{2.2.69}$$

then

$$\sum_{k=1}^{\infty} [M_k^*]^2 k^{2\alpha} < +\infty. \tag{2.2.70}$$

The relations (2.2.68) and (2.2.70) are easy to verify.

Thus, by Corollary 2.2.1, (2.2.67) implies the almost everywhere $(C, -\alpha)$-summability of the series (2.1.8), and (2.2.69) guarantees that of the series (2.1.9).

Condition (2.2.67), in general, does not guarantee the convergence of the series

$$\sum_{k=1}^{\infty} U_k^2 k^{2\alpha}$$

The same conclusion can be made on the series

$$\sum_{k=1}^{\infty} M_k^2 k^{2\alpha}$$

in the case where (2.2.69) holds.

Of course, the above said does not mean that (2.1.6) and (2.1.7) may not be $(C, -\alpha)$-summable almost everywhere. Trivial examples show that this may occur when

$$\sum_{k=1}^{\infty} U_k^2 k^{2\alpha} = +\infty \quad \text{and} \quad \sum_{k=1}^{\infty} M_k^2 k^{2\alpha} = +\infty.$$

Note at last that the works of Akhobadze [2], Grigor'ev [1] and Hristov [1] are related to the questions considered in this section.

2.3. On the Divergence Almost Everywhere of Trigonometric Fourier Series

First of all, let us prove some auxiliary statements.

Lemma 2.3.1. *Let n be an arbitrary natural number and k be a nonnegative integer. If*

$$V(n,k,x) = \frac{1}{n+1}\Big[\cos kx + 2\cos(k+1)x + \cdots + (n+1)\cos(k+n)x + $$
$$+ n\cos(k+n+1) + \cdots + \cos(k+2n)x\Big], \tag{2.3.1}$$

$$\overline{V}(n,k,x) = \frac{1}{n+1}\Big[\sin kx + 2\sin(k+1)x + \cdots + (n+1)\sin(k+n)x + $$
$$+ n\sin(k+n+1) + \cdots + \sin(k+2n)x\Big], \tag{2.3.2}$$

then

$$\frac{1}{2\pi}\int_T |V(n,k,x)|\,dx \le 1, \quad \frac{1}{2\pi}\int_T |\overline{V}(n,k,x)|\,dx \le 1. \tag{2.3.3}$$

Proof. Let $t = e^{ix}$. It is not difficult to see (see, e.g., Bari [2, Ch. VIII], Zygmund [5]) that

$$\frac{1}{n+1}\,|(1 + 2t + \cdots + (n+1)t^n + nt^{n+1} + \cdots + t^{2n})| =$$

$$= \frac{1}{n+1}\,|(1+t+\cdots+t^n)^2| = \frac{1}{n+1}\left(\frac{\sin\frac{n+1}{2}x}{\sin\frac{x}{2}}\right)^2.$$

Therefore, for any nonnegative integer k, the equality

$$\frac{1}{n+1}\,|t^k(1+2t+\cdots+(n+1)t^n+nt^{n+1}+\cdots+t^{2n})| =$$
$$= \frac{1}{n+1}\left(\frac{\sin\frac{n+1}{2}x}{\sin\frac{x}{2}}\right)^2 \tag{2.3.4}$$

holds. But

$$V(n,k,x) = \operatorname{Re}\left\{\frac{1}{n+1}[t^k(1+2t+\cdots+(n+1)t^n+nt^{n+1}+\cdots+t^{2n})]\right\}$$

and

$$\overline{V}(n,k,x) = \operatorname{Im}\left\{\frac{1}{n+1}[t^k(1+2t+\cdots+(n+1)t^n+nt^{n+1}+\cdots+t^{2n})]\right\}.$$

Hence, by (2.3.4), it follows (2.3.3). ■

Lemma 2.3.2. *If*

$$\lambda_n(x) = \frac{1}{2} + \sum_{i=1}^{n}\cos ix + \sum_{i=n+1}^{2n}\left(1-\frac{i-n}{n+1}\right)\cos ix, \tag{2.3.5}$$

then

$$\|\lambda_n\|_1 \le 4. \tag{2.3.6}$$

Proof. Putting $k=0$ in (2.3.1), we have

$$V(n,0,x) = D_n(x) - K_n(x) + \frac{1}{n+1}\sum_{i=1}^{n}\cos ix +$$
$$+ \sum_{i=n+1}^{2n}\left(1-\frac{i-n}{n+1}\right)\cos ix = \lambda_n(x) - K_n(x) + \frac{1}{n+1}\sum_{i=1}^{n}\cos ix, \tag{2.3.7}$$

where D_n and K_n are Dirichlet and Fejér kernels, respectively.

Hence, by (2.3.3), it follows (2.3.6). ■

It should be noted that an assertion similar to Lemma 2.3.2 can be found in Tandori [2].

Note also that by (2.3.4) and (2.3.7), we have

$$|\lambda_n(x)| \le \frac{A}{nx^2}, \quad |x| \in]0,\pi]. \tag{2.3.8}$$

Lemma 2.3.3. *Let*

$$n \equiv n(j) = \exp_2\exp_2 j, \quad j \in \mathbb{N}, \tag{2.3.9}$$
$$m(i) \equiv m(i,n) = (4n+1)^4 4^i + n, \quad i \in \mathbb{N}. \tag{2.3.10}$$

Then for any j, there can be found a trigonometric polynomial U_j of degree $2m(2n) \equiv 2m(2n(j), n(j))$ such that

a) $\|U_j\|_1 \leq 4;$ (2.3.11)

b) *for any $\varepsilon \in]0,1]$, there is a natural $j_0(\varepsilon)$ such that for any $j \geq j_0$ a set $E_j \subset [0, 2\pi]$ with*

$$\mu_1(E_j) > 2\pi - \varepsilon \tag{2.3.12}$$

and a positive number $A(\varepsilon)$ can be found such that for any $x \in E_j$, there is an index $q \equiv q(x, j)$ satisfying

$$|S_q(x, U_j)| > A(\varepsilon)2^j. \tag{2.3.13}$$

Proof. For the proof, we will apply the scheme of Kolmogorov [1] (see, e.g., Bari [2, Ch. V]). Put

$$\alpha_\nu \equiv \alpha(\nu, n) = \frac{4\pi\nu}{4n+1}, \quad 0 \leq \nu \leq 2n, \tag{2.3.14}$$

$$\beta_\nu \equiv \beta(\nu, n) =]a_\nu + n^{-2}, a_{\nu+1} - n^{-2}[, \quad 0 \leq \nu \leq 2n-1, \tag{2.3.15}$$

$$p \equiv p(i, n) = 2m(i) = 2m(i, n),$$

$$\overline{p} \equiv \overline{p}(j) \equiv p(2n, n) = 2m(2n),$$

and

$$U_j \equiv U_j(x, \overline{p}) \equiv U_j(x) = \frac{1}{2n+1} \sum_{\nu=0}^{2n} \lambda_{m(\nu)}(x - a_\nu), \tag{2.3.16}$$

where λ_n is defined by (2.3.5).

In view of (2.3.6), we have

$$\|U_j\|_1 \leq 4,$$

i.e., (2.3.11) holds. Taking into account (2.3.5), (2.3.9), and (2.3.10), we obtain

$$S_p(x, U_j) = \frac{1}{2n+1} \sum_{\nu=0}^{i} \lambda_{m(i)}(x - a_\nu) +$$

$$+\frac{1}{2n+1} \sum_{\nu=i+1}^{2n} D_p(x - a_\nu) \equiv S_1(x, n, i) + S_2(x, n, i). \tag{2.3.17}$$

Using (2.3.14), (2.3.15), (2.3.8), and (2.3.10), we find

$$|\lambda_{m(\nu)}(x - a_\nu)| \leq A, \quad x \in \beta_\nu, \ 0 \leq \nu \leq i.$$

Then by (2.3.17), we get

$$|S_p(x, U_j)| > |S_2(x, n, i)| - A, \quad x \in \beta_\nu, \ 0 \leq \nu \leq i. \tag{2.3.18}$$

On the basis of (2.3.14) and (2.3.15), we can conclude that

$$0 < \alpha_\nu - x < \alpha_\nu - \alpha_i = \frac{4\pi(\nu - i)}{4n+1} < 2\pi$$

$$\text{for}\quad \nu = i+1, i+2, \dots, 2n, \quad x \in \beta_i. \tag{2.3.19}$$

From (2.3.10) it follows that $4m(i)+1$ is divisible by $4n+1$. Therefore, using (2.3.17) and (2.3.19), we obtain

$$\begin{aligned}
|S_{2n}(x,n,i)| &= \frac{1}{2n+1}\left|\sum_{\nu=i+1}^{2n} \frac{\sin[4m(i)+1]\frac{x-\alpha_\nu}{2}}{2\sin\frac{x-\alpha_\nu}{2}}\right| = \\
&= \frac{1}{2n+1}\left|\sin[4m(i)+1]\frac{x}{2}\right| \sum_{\nu=i+1}^{2n} \frac{1}{2\sin\frac{\alpha_\nu - x}{2}} > \\
&> \frac{1}{2n+1}\left|\sin[4m(i)+1]\frac{x}{2}\right| \sum_{\nu=i+1}^{2n} \frac{1}{\alpha_\nu - \alpha_i} = \\
&= \frac{4n+1}{4\pi(2n+1)}\left|\sin[4m(i)+1]\frac{x}{2}\right| \sum_{\nu=i+1}^{2n} \frac{1}{\nu-1} \ge \\
&\ge A\left|\sin[4m(i)+1]\frac{x}{2}\right| \log(2n-i), \quad x \in \beta_i.
\end{aligned} \tag{2.3.20}$$

Now consider all the intervals β_i with $i = 1, 2, \dots, 2n - [\sqrt{2n}]$, where $[a]$ denotes the integer part of a. It is clear that the sum of their lengths is

$$\ell_n = (2n - [\sqrt{2n}] - 1)\Big(\frac{4\pi}{4n+1} - \frac{2}{n^2}\Big). \tag{2.3.21}$$

Therefore

$$\ell_n \to 2\pi \quad \text{as} \quad n \to \infty.$$

Put

$$E_j = \bigcup_{i=1}^{2n-[\sqrt{2n}]} \Big\{x : \Big|\sin[4m(i)+1]\frac{x}{2}\Big| \ge \eta > 0, \quad x \in \beta_i\Big\},$$

$$F_j = \bigcup_{i=1}^{2n-[\sqrt{2n}]} \beta_i \setminus E_j, \quad \eta \in]0,1[.$$

As it is well-known (see, e.g., Bari [2, Ch. V]), the inequality

$$\mu_1(F_j) < 4\arcsin\eta$$

holds. Hence, by (2.3.21), we find

$$\mu_1(E_j) > (2n - [\sqrt{2n}] - 1)\Big(\frac{4\pi}{4n+1} - \frac{2}{n^2}\Big).$$

Since $\eta \in]0,1[$ is arbitrary, (2.3.12) can be considered as proved.

Using (2.3.9), (2.3.17), and (2.3.19), we get

$$|S_2(x, U_j)| > A(\varepsilon)2^j, \quad x \in E_j.$$

Therefore (2.3.13) is also true. ∎

Lemma 2.3.4. *Let* $n \equiv n(j)$ *and* $m(i) \equiv m(i,n)$ *be defined by* (2.3.9) *and* (2.3.10), *respectively. Put*

$$\nu(n) \equiv \nu[n(j)] = 2(4\exp_2\exp_2 j + 1)(\exp_4\exp_4\exp_2 j + 3\exp_2\exp_2 j) \tag{2.3.22}$$

and

$$V_1(n,x) \equiv V_1[n(j),x] = \\ = \frac{1}{2n+1}\sum_{i=1}^{2n}\lambda_{m(i)}(x-a_i)[2\cos\nu(n)x + 2\cos 2\nu(n)x]. \tag{2.3.23}$$

If

$$f(x) = \sum_{j=1}^{\infty} 2^{-j} V_1[n(j),x], \tag{2.3.24}$$

$$\varphi(x) = \sum_{j=1}^{\infty} 2^{-j} \overline{V}_1[n(j),x], \tag{2.3.25}$$

then $f \in L(T)$, $\varphi \in L(T)$ *and* $\varphi = \overline{f}$ *almost everywhere.*

Proof. It is not difficult to see that

$$\overline{V}_1(n,x) \equiv \overline{V}_1[n(j),x] = \\ = \frac{1}{2n+1}\sum_{i=1}^{2n}\lambda_{m(i)}(x-a_i)[2\sin\nu(n)x + 2\sin 2\nu(n)x]. \tag{2.3.26}$$

Therefore, by (2.3.6), we have

$$\begin{aligned} &\frac{1}{2\pi}\int_T |V_1(n,x)|\,dx \le 4, \\ &\frac{1}{2\pi}\int_T |\overline{V}_1(n,x)|\,dx \le 4. \end{aligned} \tag{2.3.27}$$

Therefore we can conclude that the functions f and φ defined by (2.3.24) and (2.3.25), respectively, are summable on T.

In view of (2.3.22), different $V(n,x)$ do not contain similar terms (the same is true for $\overline{V}(n,x)$ as well). Therefore the series (2.3.24) and (2.3.25) can be represented as

$$t_1 = \sum_{k=1}^{\infty} a_k\cos kx + b_k\sin kx, \tag{2.3.28}$$

$$t_2 = \sum_{k=1}^{\infty} -b_k\cos kx + a_k\sin kx. \tag{2.3.29}$$

Since partial sums of (2.3.28) and (2.3.29) with numbers $2\nu(n) + 2m(2n)$ converge in the norm of $L(T)$ to f and φ, respectively, we have $t_1 = \sigma[f]$, $t_2 = \overline{\sigma}[f]$, so that $\varphi = \overline{f}$ almost everywhere. ■

Theorem 2.3.1. *There is a function $f \in L(T)$ such that*

a) $\overline{f} \in L(T)$;

b) $\omega(\delta, f)_1 \log\log \frac{1}{\delta} \leq A(f), \quad \omega(\delta, \overline{f})_1 \log\log \frac{1}{\delta} \leq A(f)$ *for* $\delta \in]0, \pi]$;

c) $\sigma[f]$ *diverges almost everywhere.*

Proof. Let the function f be defined by (2.3.24). According to Lemma 2.3.4, the property a) of the theorem is obvious.

Suppose that $h \in]0, 10^{-2}[$ and $j_0 \equiv j_0(h)$ is the greatest natural number for which

$$\left(\exp_4 \exp_2 \exp_2 j\right)^{10} \leq h^{-1}, \quad 1 \leq j \leq j_0. \tag{2.3.30}$$

Therefore

$$\left(\exp_4 \exp_2 \exp_2 (j_0 + 1)\right)^{10} > h^{-1}.$$

Hence we can conclude that

$$2^{-j_0-1} \log\log \frac{1}{h} \leq A. \tag{2.3.31}$$

Using (2.3.6), (2.3.9), (2.3.10), (2.3.22), and (2.3.23), we get

$$\|V_1(n, x)\|_C < \left(\exp_4 \exp_2 \exp_2 j\right)^6. \tag{2.3.32}$$

In view of (2.3.24), we find

$$\int_T |f(x+h) - f(x)|\, dx \leq \sum_{j=1}^{j_0} 2^{-j} \int_T |V_1(n, x+h) - V_1(n, x)|\, dx +$$

$$+ \sum_{j=j_0+1}^{\infty} 2^{-j} \int_T |V_1(n, x+h) - V_1(n, x)|\, dx \equiv B_1(h, f) + B_2(h, f). \tag{2.3.33}$$

Using the Lagrange formula and an inequality of Bernstein [2] (on the estimate of the $C(T)$-norm of the derivative of a trigonometric polynomial by the $C(T)$-norm of the polynomial) as well as (2.3.23) and (2.3.33), we obtain

$$B_1(h, f) \leq Ah \sum_{j=1}^{j_0} \left(\exp_4 \exp_2 \exp_2 j\right)^6 2^{-j} <$$

$$< Ah\left(\exp_4 \exp_2 \exp_2 j_0\right)^{10} 2^{-j_0-1},$$

whence by (2.3.30) and (2.3.31), we get

$$B_1(h, f) \log\log \frac{1}{h} \leq A. \tag{2.3.34}$$

Using (2.3.33) and the first relation in (2.3.27), we have

$$B_2(h, f) \leq A 2^{-j_0-1}.$$

Then (2.3.31), (2.3.33), and (2.3.34) imply

$$\int_T |f(x+h) - f(x)|\,dx \le A(f)\Big(\log\log\frac{1}{h}\Big)^{-1}.$$

Therefore

$$\omega(\delta, f)_1 \log\log\frac{1}{\delta} \le A(f) \quad \text{for} \quad \delta \in]0,\pi].$$

Quite analogously it can be proved that

$$\omega(\delta, \overline{f})_i \log\log\frac{1}{\delta} \le A(f) \quad \text{for} \quad \delta \in]0,\pi],$$

since the expressions $\overline{V}(n,x)$ (see (2.3.25)) have all the properties of $V(n,x)$ we have made use of.

Thus the property b) is proved.

It is well-known that the almost everywhere divergence of $\sigma[f]$ can be easily verified using Lemma 2.3.3. Indeed, let $\varepsilon \in]0,1[$ be arbitrary, and consider the sets E_j from Lemma 2.3.3. Put

$$E = \limsup_{j\to\infty} E_j.$$

According to (2.3.12), we have $\mu_1(E) > 2\pi - \varepsilon$. Since ε is arbitrary, it suffices to check the divergence of $\sigma[f]$ on E. But this follows from the assertion c) of Lemma 2.3.3 (see, e.g., (2.3.13)) since for $\sigma[f]$, the Cauchy criterion is not fulfilled on E. ■

Remark 2.3.1. It immediately follows from a theorem of Plessner [4] that the series $\sigma[f]$ in Theorem 2.3.1 diverges almost everywhere. This can likewise easily be established using the same method as while proving the almost everywhere divergence of $\sigma[f]$. We will need this remark in the multi-dimensional case.

Remark 2.3.2. An expression of the type $2\cos nx + 2\cos 2nx$ whose properties we have used in the proof of Theorem 2.3.1, appearently was first introduced by Taĭkov [1].

Remark 2.3.3. We could prove Theorem 2.3.1 using the Zygmund means [7], i.e., the trigonometric polynomials

$$\lambda_n^z(x) = \frac{1}{2} + \sum_{j=1}^{n}\Big[1 - \frac{j^r}{(n+1)^r}\Big]\cos jx \quad (r = 2,3,\dots)$$

instead of λ_n (see (2.3.5)), but in this case the proof would be too long.

Finally, note that Theorem 2.3.1 improves in a certain sense a result of Prokhorenko [1].

2.4. Unsolved Problems

2.4.1. Let $\alpha \in]0,1]$. Is there a function $f \in L(T)$ such that

$$\omega(\delta, f)_1 \left(\log \frac{2\pi}{\delta}\right)^\alpha \leq A(f) \quad \text{for } \delta \in]0, \pi],$$

yet $\sigma[f]$ diverges on a set $E \subset T$ with $\mu_1(E) > 0$ (almost everywhere, everywhere)?

As already pointed out, for $\alpha = 1$ this is Zygmund's problem.

It should be noted that there is still no answer to the following weaker version of the problem.

2.4.2. Let $\alpha \in]1, +\infty[$. Is there a function $f \in L(T)$ such that

$$\omega(\delta, f)_1 \left(\log\log \frac{4\pi}{\delta}\right)^\alpha \leq A(f) \quad \text{for } \delta \in]0, \pi],$$

yet $\sigma[f]$ diverges on a set $E \subset T$ with $\mu_1(E) > 0$ (almost everywhere, everywhere)?

2.4.3. Let $\alpha \in]0,1[$. Is there a function $f \in L(\log^+ L)^\alpha(T)$ such that $\sigma[f]$ diverges on a set $E \subset T$ with $\mu_1(E) > 0$ (almost everywhere, everywhere)?

Here too we should note that still there is no progress (if it is possible at all) even for the following weaker formulation.

2.4.4. Let $\alpha \in]1, +\infty[$. Is there a function $f \in L(\log^+ \log^+ L)^\alpha(T)$ such that $\sigma[f]$ diverges on a set $E \subset T$ with $\mu_1(E) > 0$ (almost everywhere, everywhere)?

As already noted, the positive answer to the problems 1 and 2 would imply the positive answer to the problems 3 and 4.

2.4.5. Let $\alpha \in]0,1[$. Is there a function $f \in L(T)$ such that

$$\limsup_{n\to\infty} |S_n(x, f)|(\log n)^{-\alpha} > 0$$

on a set $E \subset T$ with $\mu_1(E) > 0$ (almost everywhere, everywhere)?

This problem was posed by Zygmund.

As above, we should say that still there is no answer even to the following formulation.

2.4.6. Let $\alpha \in]1, +\infty[$. Is there a function $f \in L(T)$ such that

$$\limsup_{n\to\infty} |S_n(x, f)|(\log\log n)^{-\alpha} > 0$$

on a set $E \subset T$ with $\mu_1(E) > 0$ (almost everywhere, everywhere)?

Note that if any of the above problems gets a positive answer, then the problem will arise: does the conjugate of the constructed function have the analogous properties?

Related to the classic investigations of Men'shov [1]–[4], the problem arises.

2.4.7. Let a function $f : T \to [-\infty, +\infty]$ be measurable and almost everywhere finite, and let $\alpha \in]0,1[$ and $\varepsilon \in]0,1[$ be some numbers. Is there a function $g \in C(T)$ such that

$$\mu_1\Big[\{x : f(x) \neq g(x), \ \ x \in T\}\Big] < \varepsilon$$

and $\sigma[g]$ is $(C, -\alpha)$-summable in the norm of $C(T)$? This problem was posed and partly solved by Men'shov in [5]. See also Kashin and Kosheleva [1].

Note that in this problem, the restriction $\alpha \in]0,1[$, as it follows from the corresponding results of Olevskiĭ [1], [2], is essential.

In connection with a well-known result of Bohr [1], the following problem arises.

2.4.8. Let $f \in C(T)$ and $\alpha \in]0,1[$. Does there exist a homeomorphism φ of the set T on itself such that the Fourier series of $f \circ \varphi$ is $(C,-\alpha)$-summable in the norm of $C(T)$? This problem is due to Gogoladze.

2.4.9. Let $\alpha \in]0,1[$. Denote by $U(T,\alpha)$ the class of all the functions $f \in C(T)$ for which $\sigma[f]$ is $(C,-\alpha)$-summable in the norm of $C(T)$. Describe all the Borel functions $\varphi : \mathbb{R} \to \mathbb{R}$ for which $\varphi \circ f \in U(T,\alpha)$ for any $f \in U(T,\alpha)$.

Various items concerning the "modification" and "correction" of functions are set forth in the works by Olevskiĭ [2] and Bruckner [1].

In 1981, C. Calderon [1] showed that the well-known Dini test (see, e.g., Bari [2, Ch. I], Zygmund [15, Ch. II]) is unimprovable in the following sense.

Let a continuous function $\omega : T \to \mathbb{R}_+$ satisfy $\omega(0) = 0$, $\omega \downarrow 0$ as $t \downarrow 0$, and

$$\int_0^\pi t^{-1}\omega(t)\,dt = +\infty.$$

Then there are a function $f \in L(T)$ and a set $E \subset T > 0$ with $\mu_1(E)$ such that

$$\int_0^\pi \frac{|f(x+t) - f(x)|\omega(t)\,dt}{t} < +\infty \quad \text{for } x \in E,$$

yet $\sigma[f]$ diverges almost everywhere on E.

Using this result, one can prove that if a function ω satisfies the conditions of C. Calderón's theorem, then there are a function $f \in L(T)$ and a set $E \subset T$ with $\mu_1(E) > 0$ such that

$$\int_h^\pi t^{-1}|\varphi(x,t) - \varphi(x,t+h)|\omega(t)\,dt \to 0 \quad \text{as } h \to +0, \quad x \in E,$$

yet $\sigma[f]$ diverges almost everywhere on E.

On the other hand, in 1936 Marcinkiewicz [1] proved that if $f \in L(T)$ and

$$h^{-1}\log\frac{1}{h}\int_0^h |f(x+h) - f(x)|\,dx \le A(f) \quad \text{for } h \in]0,\pi]$$

on a set $E \subset T$ with $\mu_1(E) > 0$, then $\sigma[f]$ converges almost everywhere on E. In the same paper [1], Marcinkiewicz established that his result is unimprovable in the following sense.

Let $\omega : T \to]0,+\infty[$ be even and nondecreasing on $]0,\frac{1}{3}]$. If

$$\lim_{t\to 0}\omega(t) = 0 \quad \text{and} \quad \lim_{t\to 0}\omega(t)\log\frac{1}{|t|} = +\infty,$$

then there is a function $f \in L(T)$ such that

$$h^{-1}[\omega(h)]^{-1} \int_0^h |f(x+h) - f(x)|\, dx \le A(f) \quad \text{for} \quad h \in]0, \pi],$$

almost everywhere, yet $\sigma[f]$ diverges almost everywhere.

2.4.10. Compare the Dini, Lebesgue and Marcinkiewicz tests for the convergence of Fourier series on a set of positive measure.

CHAPTER III

Convergence and Summability of Trigonometric Fourier Series and Their Conjugates in the Spaces $L^p(T)$, $p \in]0,+\infty[$

3.1. Convergence and Summability of the Series $\sigma[f]$ and $\overline{\sigma}[f]$ in the Spaces $L^p(T)$, $p \in]0,1]$

In 1925, Kolmogorov [3] showed that if $f \in L(T)$, then for any $p \in]0,1[$, it holds

$$\|S_n(f) - f\|_p \to 0, \quad \|\overline{S}_n(f) - \overline{f}\|_p \to 0 \quad \text{as} \quad n \to \infty. \tag{3.1.1}$$

Applying the corresponding results of Flett [2], Y. Chen [1], [3], and the author (see Sections 1.1 and 1.2 of this book), one can conclude that

$$\|[S_n(f) - f]\omega\|_p \to 0, \quad \|[\overline{S}_n(f) - \overline{f}]\,\omega\|_p \to 0 \quad \text{as} \quad n \to \infty$$

for some weight functions ω.

Analogously, applying a result of Lesniewicz [1], we have: if $f \in L(T)$, then

$$\int\limits_T \Phi\Big[|S_n x, f) - f(x)|\Big]\,dx \to 0, \quad \int\limits_T \Phi\Big[|\overline{S}_n x, f) - \overline{f}(x)|\Big]\,dx \to 0 \text{ as } n \to \infty$$

for any function Φ satisfying Lesniewicz's conditions.

If $\alpha \in]-\infty, 0[$ and $f \in L(T)$, then following the schemes of arguments of M. Riesz [1] and Zygmund [2], we can conclude that

$$\begin{aligned} &\|\sigma_n^{-\alpha}(f) - f\|_1 \to 0 \quad \text{as} \quad n \to \infty, \\ &\|t_n^{-\alpha}(f) - \overline{f}_n\|_1 \to 0 \quad \text{as} \quad n \to \infty. \end{aligned} \tag{3.1.2}$$

In view of (3.1.2) and the Kolmogorov inequality (1.1.21), it follows that if $f \in L(T)$, then for any $p \in]0,1[$, we have

$$\|t_n^{-\alpha}(f) - \overline{f}\|_p \to 0 \quad \text{as} \quad n \to \infty.$$

If $\alpha \in]0,+\infty[$ and is not natural, then the behavior of the means $\sigma_n^{-\alpha}(x,f)$ and $t_n^{-\alpha}(x,f)$ is essentially different from the above. For instance, there is a function $f \in C(T)$ such that for any $p \in]0,1[$,

$$\limsup_{n\to\infty} \|\sigma_n^{-\alpha}(f)\|_p = +\infty.$$

3.2. Convergence and Summability of $\sigma[f]$ and $\overline{\sigma}[f]$ in the Spaces $L^p(T)$, $p \in [1, +\infty]$

In 1927, M. Riesz [3] proved that if $f \in L^p(T)$ for some $p \in]1, +\infty[$, then

$$\|S_n(f) - f\|_p \to 0, \quad \|\overline{S}_n(f) - \overline{f}\|_p \to 0 \text{ as } n \to \infty. \tag{3.2.1}$$

In 1968, Hunt [2] showed that if $f \in L^p(T)$ for some $p \in]1, +\infty[$, then

$$\|S^*\|_p \le A(p)\|f\|_p. \tag{3.2.2}$$

We have used the function S^* and this result of Hunt while proving Theorem 2.2.4. It should be added that (3.2.2) implies the first assertion from (3.2.1).

From the corresponding results presented in Section 1.1, one can conclude that if $f \in L^p(T)$ for some $p \in]1, +\infty[$, then

$$\left\{ \int_T |S_n(x, f) - f(x)|^p \omega(x)\, dx \right\}^{1/p} \to 0 \quad \text{as} \quad n \to \infty.$$

In particular, the assertions of this type (for the corresponding weights ω) can be deduced from the results of Hardy and Littlewood [8], Hirschman [1], Flett [2], Gaposhkin [1], Y.Chen [1], Helson and Szegö [1], Forelli [1], Hunt, Muckenhoupt and Wheeden [1].

In 1974, Hunt and Young [1] proved that in the class $L^p(T, \omega)$ with $p \in]1, +\infty[$, the inequality

$$\left\{ \int_T |S^*(x)|^p \omega(x)\, dx \right\}^{1/p} \le A(p, \omega) \left\{ \int_T |f(x)|^p \omega(x)\, dx \right\}^{1/p}$$

holds iff the weight function $\omega \in (A_p)$. Hence it follows that if $f \in L^p(T, \omega)$, $p \in]1, +\infty[$, then the relation

$$\left\{ \int_T |S_n(x, f) - f(x)|^p \omega(x)\, dx \right\}^{1/p} \to 0 \text{ as } n \to \infty$$

is valid only if $\omega \in (A_p)$.

Let now $\alpha \in]-\infty, 0[$ and $p \in]1, +\infty[$. Using the results of M. Riesz [2], [3] (see (1.1.10) and (3.2.1)), we get

$$\|\sigma_n^{-\alpha}(f) - f\|_p \to 0, \quad \|t_n^{-\alpha}(f) - \overline{f}\|_p \to 0 \text{ as } n \to \infty. \tag{3.2.3}$$

Here too we should note that some results given in Section 1.1 imply the relations of the type (3.2.3) for some weight functions ω.

In 1962, Rosenblum [1] found out necessary and sufficient conditions for the weight ω to have the following property: if $f \in L^p(T, \omega)$ for some $p \in [1, +\infty[$, then

$$\left\{ \int_T |\sigma_n^1(x, f) - f(x)|^p \omega(x)\, dx \right\}^{1/p} \to 0 \text{ as } n \to \infty.$$

In 1975, Muckenhoupt [2] showed that for $p \in]1, +\infty[$, these conditions of Rosenblum are equivalent to the condition $\omega \in (A_p)$. The same range of questions is treated in another paper of Muckenhoupt [1], as well as in E. Sawyer [1].

In 1982, Nakhman and Osilenker [1] studied the problem of weighted estimates of weak and strong types of linear means of $\sigma[f]$, constructed by means of so called triangular matrices.

In 1929, Zygmund [4] proved that if $f \in L\log^+ L(T)$, then

$$\|S_n(f) - f\|_1 \to 0, \quad \|\overline{S}_n(f) - \overline{f}\|_1 \to 0 \quad \text{as } n \to \infty. \tag{3.2.4}$$

From the results of Flett [2] and Y. Chen [1] given in Section 1.1, one can obtain the assertions generalizing the above theorem of Zygmund for the case of some concrete weight functions ω. As far as we know, there are no definitive results in this direction.

In 1944, Lozinskiĭ [1] generalized the assertion of Zygmund [4] in the sense that he considered the convergence of $\sigma[f]$ in a nonreflexive Orlicz space. In the same paper he found some generalizations of M. Riesz's theorem [3] (see (3.2.1)). In this direction the paper of Gurielashvili [1] should also be mentioned.

Zygmund's theorem (see (3.2.4)) is nonimprovable in a certain sense (this is noted in Lozinskiĭ [1]). Namely, let $\Phi : \mathbb{R}_+ \to \mathbb{R}_+$, $\Phi \uparrow \infty$ as $t \uparrow \infty$, and

$$\Phi(t)(t \log t)^{-1} \to 0 \quad \text{as } t \to \infty.$$

Then there is $f \in L(T)$ such that

$$\int_T \Phi(|f(x)|)\, dx < +\infty,$$

yet

$$\limsup_{n\to\infty} \|S_n(f) - f\|_1 > 0.$$

It should be noted that from the results of Lozinskiĭ [1] and his method, one can deduce the convergence of the means $\sigma_n^{-\alpha}(x, f)$ with $\alpha \in]-\infty, 0[$ in certain Orlicz spaces. The case $\alpha = -1$ is treated in detail in this work.

A theorem by Ryan implies that if L_Φ is a reflexive Orlicz space and $f \in L_\Phi(T)$, then for any $\alpha \in]-\infty, 0[$, we have

$$\left.\begin{aligned} &\int_T \Phi\left[\left|\sigma_n^{-\alpha}(x, f) - f(x)\right|\right] dx \to 0, \\ &\int_T \Phi\left[\left|t_n^{-\alpha}(x, f) - \overline{f}(x)\right|\right] dx \to 0 \end{aligned}\right\} \quad \text{as } n \to \infty.$$

For weighted Orlicz spaces, similar statements follow from the results of Tkebuchava [1].

In 1925, Zygmund [3] showed that if $f \in Lip\,\alpha$ for some $\alpha \in]0, 1]$, then for any $\beta \in [0, \alpha[$ we have

$$\|\sigma_n^{-\beta}(f) - f\|_C \to 0 \quad \text{as } n \to \infty. \tag{3.2.5}$$

In 1928, Hardy and Littlewood [6] generalized this result. In particular, they proved that if $p \in]1,+\infty[$, $\alpha \in]0,1[$, and

$$f \in C(T) \cap Lip(\alpha, p) \quad \text{as} \quad \alpha p \in]1,+\infty[,$$

then for any $\beta \in [0,\alpha[$, (3.2.5) holds.

In 1957, Yano [1] established that the assertion of Hardy and Littlewood remains true in the case $\alpha p = 1$.

Moreover, Hardy and Littlewood [6] showed that if $p \in]1,+\infty[$, $\alpha \in]0,1[$, $\alpha p \in]1,+\infty[$, and $f \in C(T) \cap Lip^*(\alpha,p)$, i.e., $f \in C(T)$ and

$$|h|^{-\alpha}\left\{\int_T |f(x+h) - f(x)|^p dx\right\}^{1/p} \to 0 \quad \text{as} \quad h \to 0,$$

then

$$\|\sigma_n^{-\alpha}(f) - f\|_C \to 0 \quad \text{as} \quad n \to \infty.$$

Using the corresponding argument of Ul'yanov [5], we can verify that the last assertion of Hardy and Littlewood is not valid in the case where

$$f \in C(T) \cap Lip^*(\alpha, p), \quad \alpha \in]0,1[, \quad \alpha p \in]0,1].$$

In the papers [19], [24] of the author some statements are given concerning the $(C,-\alpha)$-summability ($\alpha \in]0,+\infty[$ and is not natural) of series $\sigma[f]$ and $\overline{\sigma}[f]$ in the norm of $L^p(T)$ with $p \in [1,+\infty]$. Below we set forth these results. The results of [14] dealing with the divergence of $\sigma[f]$ and $\overline{\sigma}[f]$ in the norm of $L(T)$ are also presented. In the end of the section, some further statements are pointed out in this direction as well as their relation to the proven theorems.

Theorem 3.2.1. *Let $[a,b] \subset T$, $b-a>0$, $p \in [1,+\infty]$, and $\alpha \in [0,+\infty[$, $\alpha \neq n$ for $n \in \mathbb{N}$. Suppose that $f \in L(T)$ and $f \in L^p([a,b])$ for some $p \in]1,+\infty]$. If*

$$h^{-\alpha}\left\{\int_a^b \left[\int_h^\pi t^{\alpha-1}|\varphi(x,t) - \varphi(x,t+h)|\, dt\right]^p dx\right\}^{1/p} \to 0 \quad as \quad h \to 0+, \tag{3.2.6}$$

then

$$\lim_{n\to\infty}\left\{\int_c^d |\sigma_n^{-\alpha}(x,f) - f(x)|^p dx\right\}^{1/p} = 0$$

for any $[c,d]$ with $a<c<d<b$.

Proof. We will assume that $\alpha \in [0,1[$. As in the proof of Theorem 2.2.1, the scheme of our proof showes that such a restriction is unessential. The case $p=+\infty$ was treated earlier (see Theorem 2.2.2). However, the reader may include this case considering the corresponding estimates in the norm of $C(T)$.

We will follow the scheme of the proof of Theorem 2.2.1 and try not to repeat typical reasoning.

Since

$$\int_{\pi-h}^{\pi} \varphi(x,t)\,dt = \int_{h}^{\pi-h} \left[\varphi(x,t+h) - \varphi(x,t)\right] dt + \int_{h}^{2h} \varphi(x,t)\,dt \quad \text{for } h \in \left]0, \frac{\pi}{3}\right[,$$

Minkowski inequality and (3.2.6) imply

$$h^{-\alpha}\left\{\int_{c}^{d}\left|\int_{\pi-h}^{\pi} \varphi(x,t)\,dt\right|^{p} dx\right\}^{1/p} \to 0 \quad \text{as } h \to 0+.$$

Hence

$$h^{-\alpha}\left\{\int_{c}^{d}\left|\int_{0}^{h} \varphi(x,t+\pi)\,dt\right|^{p} dx\right\}^{1/p} \to 0 \quad \text{as } h \to 0+. \tag{3.2.7}$$

While proving Theorem 2.2.1, on the basis of the conditions of that theorem and (2.2.31) and (2.2.32) we have established (2.2.33) at a point $x \in T$. Using again (2.2.32), (3.2.7) and the conditions of the theorem, and applying Minkowski's inequality, we find

$$k^{\alpha}\left\{\int_{c}^{d} |A_k(x)|^{p} dx\right\}^{1/p} \to 0 \quad \text{as } k \to \infty,$$

whence (see (2.2.60))

$$k^{\alpha} a_k(f) \to 0, \quad k^{\alpha} b_k(f) \to 0 \quad \text{as } k \to \infty. \tag{3.2.8}$$

According to our earlier reasoning (see (2.2.34)), we have

$$\sigma_n^{-\alpha}(x,f) - f(x) = \frac{1}{\pi}\int_{0}^{\pi} \varphi(x,t) K_n^{-\alpha}(t)\,dt =$$

$$= \frac{1}{\pi}\int_{0}^{\pi/n} \varphi(x,t) K_n^{-\alpha}(t)\,dt + \frac{1}{\pi}\int_{\pi/n}^{\pi} \varphi(x,t) K_n^{-\alpha}(t)\,dt \equiv$$

$$\equiv I_1(x,f,\alpha,n) + I_2(x,f,\alpha,n), \quad n > 4. \tag{3.2.9}$$

Integrating by parts and taking into account that $f \in L^p([a,b])$ with $p \in]1, +\infty[$, by (2.2.13) and (2.2.17) we get

$$\left\{\int_{c}^{d} |I_1(x,f,\alpha,n)|^{p} dx\right\}^{1/p} \to 0 \quad \text{as } n \to \infty. \tag{3.2.10}$$

As while proving Theorem 2.2.1 (see (2.2.36)), using (2.2.9) we can write $I_2(x,f,\alpha,n)$ as

$$I_2(x,f,\alpha,n) = \frac{1}{\pi}\int_{\pi/n}^{\pi} \varphi(x,t)\varphi_n^{-\alpha}(t)\,dt +$$

$$+\frac{1}{\pi}\int_{\pi/n}^{\pi}\varphi(x,t)r_n^{-\alpha}(t)\,dt \equiv P_1(x,f,\alpha,n)+P_2(x,f,\alpha,n).$$

Again integrating by parts and using (2.2.15), (2.2.19) and the conditions of the theorem, we conclude

$$\left\{\int_c^d |P_2(x,f,\alpha,n)|^p dx\right\}^{1/p} \to 0 \quad \text{as} \quad n\to\infty. \tag{3.2.11}$$

To evaluate $P_1(x,f,\alpha,n)$, it suffices to estimate

$$H(x,f,\alpha,n)\equiv n^\alpha\int_{\pi/n}^{\pi}\varphi(x,t)\omega(-\alpha,t)\cos nt\,dt.$$

According to (2.2.38), we have

$$2H(x,f,\alpha,n)=n^\alpha\left\{\int_{\pi/n}^{\pi}\left[\varphi(x,t)-\varphi\left(x,t+\frac{\pi}{n}\right)\right]\omega(-\alpha,t)\cos nt\,dt+\right.$$

$$+\int_{\pi/n}^{\pi-\pi/n}\varphi\left(x,t+\frac{\pi}{n}\right)\left[\omega(-\alpha,t)-\omega\left(-\alpha,t+\frac{\pi}{n}\right)\right]\cos nt\,dt-$$

$$-\int_0^{\pi/n}\varphi\left(x,t+\frac{\pi}{n}\right)\omega\left(-\alpha,t+\frac{\pi}{n}\right)\cos nt\,dt+$$

$$\left.+\int_{\pi-\pi/n}^{\pi}\varphi(x,t)\omega(-\alpha,t)\cos nt\,dt\right\}\equiv\sum_{i=1}^{4}B_i(x,f,\alpha,n).$$

On the basis of (2.2.21) and (3.2.6), we find

$$\left\{\int_c^d |B_1(x,f,\alpha,n)|^p dx\right\}^{1/p} \to 0 \quad \text{as} \quad n\to\infty.$$

Using (2.2.22), we get

$$\left\{\int_c^d |B_2(x,f,\alpha,n)|^p dx\right\}^{1/p} \to 0 \quad \text{as} \quad n\to\infty.$$

It is easy to check that it also holds

$$\left\{\int_c^d |B_3(x,f,\alpha,n)|^p dx\right\}^{1/p} \to 0 \quad \text{as} \quad n\to\infty.$$

If $\alpha = 0$, then it is clear that

$$\left\{\int_c^d |B_4(x,f,\alpha,n)|^p dx\right\}^{1/p} \to 0 \quad \text{as} \quad n \to \infty,$$

and if $\alpha \in]0,1[$, then using (3.2.8) and (2.2.49), quite analogously to (2.2.50) we have

$$\left\{\int_c^d |B_4(x,f,\alpha,n)|^p dx\right\}^{1/p} \to 0 \quad \text{as} \quad n \to \infty.$$

Therefore we can say that

$$\left\{\int_c^d |P_1(x,f,\alpha,n)|^p dx\right\}^{1/p} \to 0 \quad \text{as} \quad n \to \infty.$$

But then by (3.2.11),

$$\left\{\int_c^d |I_2(x,f,\alpha,n)|^p dx\right\}^{1/p} \to 0 \quad \text{as} \quad n \to \infty.$$

The last relation along with (3.2.9) and (3.2.10) implies

$$\lim_{n\to\infty}\left\{\int_c^d |\sigma_n^{-\alpha}(x,f) - f(x)|^p dx\right\}^{1/p} \to 0 \quad \text{as} \quad n \to \infty. \qquad \blacksquare$$

Remark 3.2.1. In the proof of Theorem 3.2.1, we have for several times made use of integration by parts. In some cases, the corresponding estimates can also be obtained by another method. However, we have used integration by parts for two reasons: in the first place, for the sake of unity of method, and in the second (essential) place, this method has been completely applicable when we studied the $(C,-\alpha)$-summability of $\sigma[f]$ at single points.

The method of Theorem 3.2.1 can be used to verify that it holds

Theorem 3.2.2. *Let $[a,b] \subset T$, $b-a>0$, $p \in [1,+\infty]$, and $\alpha \in [0,+\infty[$ $(\alpha \neq n,\ n \in \mathbb{N})$. Suppose that $f \in L(T)$ and $f \in L^p([a,b])$ for some $p \in]1,+\infty]$. If*

$$h^{-\alpha}\left\{\int_a^b \left[\int_h^\pi t^{\alpha-1}|\psi(x,t)-\psi(x,t+h)|\,dt\right]^p dx\right\}^{1/p} \to 0 \quad \textit{as} \quad h \to 0+,$$

then

$$\lim_{n\to\infty}\left\{\int_c^d |t_n^{-\alpha}(x,f) - \overline{f}_n(x)|^p dx\right\}^{1/p} = 0$$

for any segment $[c,d]$ with $a<c<d<b$.

Remark 3.2.2. If in the conditions of Theorem 3.2.1 (the same is true for Theorem 3.2.2) we replace (3.2.6) by

$$h^{-\alpha}\left\{\int_a^b\left[\int_h^\pi t^{\alpha-1}|\varphi(x,t)-\varphi(x,t+h)|\,dt\right]^p dx\right\}^{1/p} \le A(f,p) \quad \text{for } h\in]0,\pi[,$$

then $(C,-\alpha)$-means of $\sigma[f]$ will be uniformly bounded in the norm of $L^p([c,d])$.

Remark 3.2.3. Theorem 3.2.1 cannot be considered as similar to the localization principle. It should be noted that for (C,α)-means, the localization principle, generally speaking, is not valid for $\alpha\in]-\infty,0[$.

It is not difficult to verify that if $p\in[1,+\infty[$, $\alpha\in]0,1[$ and $f\in Lip^*(\alpha,p)$, then all the conditions of Theorem 3.2.1 (with $[a,b]=[-\pi,\pi]$) are fulfilled. So it holds

Corollary 3.2.1. *If $f\in Lip^*(\alpha,p)$, $p\in[1,+\infty[$ and $\alpha\in]0,1[$, then $\sigma[f]$ is $(C,-\alpha)$-summable to f in the $L^p(T)$-norm.*

Thus if $f\in Lip^*(\alpha,p)$, $p\in[1,+\infty[$ and $\alpha\in]0,1[$, then the restriction $\alpha p\in]1,+\infty[$ is unnecessary for the $(C,-\alpha)$-summability of $\sigma[f]$ in the $L^p(T)$-norm.

Various consequences can be obtained from Theorems 3.2.1 and 3.2.2. For instance, if $\alpha=0$ and $[a,b]=[-\pi,\pi]$, we have

Corollary 3.2.2. *If $f\in L(T)$ and*

$$\int_T\left[\int_h^\pi t^{-1}|\varphi(x,t)-\varphi(x,t+h)|\,dt\right]dx\to 0 \quad as \quad h\to 0+, \tag{3.2.12}$$

then $\sigma[f]$ converges to f in the $L(T)$-norm.

In particular, if $f\in L(T)$ and its integral modulus of continuity satisfies

$$\omega(\delta,f)_1\log\frac{1}{\delta}\to 0 \quad as \quad \delta\to 0+, \tag{3.2.13}$$

then $\sigma[f]$ converges to f in the norm of $L(T)$.

Remark 3.2.4. Condition (3.2.6) (therefore conditions (3.2.12) and (3.2.13) as well) is not necessary for the $(C,-\alpha)$-summability of series $\sigma[f]$ of individual functions with $\alpha\in[0,1[$. However, for a certain class of functions and for fixed $p\in[1,+\infty]$ and $\alpha\in[0,1[$, it is necessary. The same is true for the boundedness in $L^p([c,d])$ of the norm of $(C,-\alpha)$-means of $\sigma[f]$ with $\alpha\in[0,1[$. Consider several cases:

a) $\alpha=0$, $[a,b]=[-\pi,\pi]$, $p=+\infty$. In this case, the validity of the remark follows from a well-known result of Faber [1] (see also Zygmund [15, Ch. VIII], Fejér [2]) on the nonimprovability of the Dini–Lipschitz test (see, e.g., Bari [2, Ch. IV], Zygmund [15, Ch. II]). If $\alpha=0$, $[a,b]=[-\pi,\pi]$, and $p\in]1,+\infty[$, then, as already pointed out (see (3.2.1)), according to M. Riesz's theorem [3]

$$\|\sigma_n^0(f)-f\|_p\to 0 \quad \text{as} \quad n\to\infty.$$

Therefore it is appropriate to assume that $\alpha=0$, $[a,b]=[-\pi,\pi]$, and $p=1$. Let $a_k\downarrow 0$ as $k\uparrow\infty$ and the sequence $(a_k)_{k\geq 0}$ be convex, i.e., $\Delta^2 a_k\geq 0$ for all

$k \in \mathbb{N}_0$, where $\mathbb{N}_0$ is the set of all nonnegative integers. Then, according to results by W. Young [1], [2] and Kolmogorov [2] (see also Zygmund [15, Ch. V]), the series

$$\frac{a_0}{2} + \sum_{k=1}^{\infty} a_k \cos kx \tag{3.2.14}$$

converges almost everywhere to a function $f \in L(T_+)$ and is $\sigma[f]$. Moreover,

$$\|\sigma_n^0(f) - f\|_1 \geq Aa_n \log n \quad \text{for} \quad n \geq n_0 > 1. \tag{3.2.15}$$

On the other hand, in 1973 Telyakovskiĭ [1] established certain results which imply that for series of the type (3.2.14) with the above conditions imposed on $(a_k)_{k \geq 0}$, the relation

$$\omega\left(\frac{1}{n}, f\right)_1 \leq \frac{A}{n} \sum_{k=0}^{n} a_k \tag{3.2.16}$$

holds. Denote by M_0 the set af all $f \in L(T_+)$ for which $\sigma[f]$ is of the form (3.2.14) with $a_k \downarrow 0$ as $k \uparrow \infty$, $\Delta^2 a_k \geq 0$ for $k \in \mathbb{N}_0$, and

$$\sum_{k=0}^{n} a_k \leq Ana_n.$$

By (3.2.15) and (3.2.16), we can conclude that in the class M_0, the condition

$$\omega\left(\frac{1}{n}, f\right)_1 \log n \to 0 \quad \text{as} \quad n \to \infty$$

is necessary for the convergence of $\sigma[f]$ in the norm of $L(T)$. Therefore condition (3.2.12) is necessary as well (see also (3.2.13) in the class M_0).

b) $\alpha \in]0, 1[$, $[a, b] = [-\pi, \pi]$, $p = 1$. It can be easily verified that if

$$\|\sigma_n^{-\alpha}(f) - f\|_1 \to 0 \quad \text{as} \quad n \to \infty,$$

then (3.2.8) holds. Then, according to (3.2.16), we conclude: the condition

$$n^\alpha \omega\left(\frac{1}{n}, f\right)_1 \to 0 \quad \text{as} \quad n \to \infty,$$

is necessary in the class M_0 for the $(C, -\alpha)$-summability of $\sigma[f]$ in the norm of $L(T)$. Hence it follows that the condition (3.2.6) of Theorem 3.2.1 (with $\alpha \in]0, 1[$ and $p = 1$) is necessary in the class M_0 for the corresponding statement to be true.

c) $\alpha \in]0, 1[$, $[a, b] = [-\pi, \pi]$, $p \in]1, +\infty]$. If $p = +\infty$, then using the scheme of the proof of Theorem 2.2.5 (other arguments can also be used, including those based on the results of Lorentz [1]), we can conclude that for a certain subclass $M_1 \subset C(T)$, the condition (3.2.6) of Theorem 3.2.1 is necessary. On the other hand, if $p \in]1, +\infty[$, then using the corresponding estimates for $\omega(\frac{1}{n}, f)_p$ established by M. and S. Izumi [2], or by T. Tevzadze, we get sure that in this case, the condition (3.2.6) of Theorem 3.2.1 is necessary for a certain subclass $M_2 \subset C(T)$.

It should be noted that a remark of the type 3.2.4 (with appropriate modifications) is true for Theorem 3.2.2 as well.

In 1932, Zygmund [5] set forth with complete proof the following statement of F. Riesz.

There is a function $f \in L(T)$ such that $\overline{f} \in L(T)$, yet

$$\limsup_{n\to\infty} \|S_n(f) - f\|_1 > 0.$$

In the papers [14] and [21], we gave a somewhat improved version of this theorem which we present below.

Theorem 3.2.3. *There is a function $f \in L(T)$ such that $\overline{f} \in L(T)$ and*

a) $\omega(\delta, f)_1 \log \dfrac{2\pi}{\delta} \le A(f),\ \ \omega(\delta, \overline{f})_1 \log \dfrac{2\pi}{\delta} \le A(f)$ *for* $\delta \in]0, \pi]$;

b) $\limsup_{n\to\infty} \|S_n(f) - f\|_1 > 0.$

Proof. Let $n \equiv n(j)$ $(j = 1, 2, \dots)$ be defined by (2.3.9), and the expressions $V(n-1, n, x)$ and $\overline{V}(n-1, n, x)$ by (2.3.1) and (2.3.2), respectively. According to Lemma 2.3.1 (see (2.3.3)),

$$\sum_{j=0}^{\infty} 2^{-j} \left\{ \int_T \left[|V(n-1, n, x)| + |\overline{V}(n-1, n, x)| \right] dx \right\} < +\infty.$$

Therefore the functions $f : T \to [-\infty, +\infty]$ and $\varphi : T \to [-\infty, +\infty]$ defined by

$$f(x) = \sum_{j=1}^{\infty} 2^{-j} V(n-1, n, x), \tag{3.2.17}$$

$$\varphi(x) = \sum_{j=1}^{\infty} 2^{-j} \overline{V}(n-1, n, x) \tag{3.2.18}$$

are summable on T. By (2.3.9), the different polynomials $V(n-1, n, x)$ and $\overline{V}(n-1, n, x)$ do not contain similar terms. Therefore the series in right sides of (3.2.17) and (3.2.18) can be represented as trigonometric series of the forms

$$\sum_{k=0}^{\infty} a_k \cos kx \tag{3.2.19}$$

and

$$\sum_{k=1}^{\infty} b_k \sin kx, \tag{3.2.20}$$

respectively. It is not difficult to see that the partial sums of (3.2.19) and (3.2.20) with the subscript $n(j) + 2n(j-2)$ converge in the norm of $L(T)$ to f and φ, respectively. Then (3.2.19) and (3.2.20) are the Fourier series of the functions f and φ, respectively. Moreover, $\varphi = \overline{f}$ almost everywhere.

To prove the assertion a) of the theorem, we use the method we have used in establishing the validity of the assertion b) of Theorem 2.3.1.

Let $h \in]0, 10^{-2}[$ and $j_0(h)$ be the greatest natural number for which

$$\left(\exp_2 \exp_2 j\right)^2 \le h^{-1} \quad (j = 1, 2, \dots, j_0). \tag{3.2.21}$$

Therefore

$$\left(\exp_2 \exp_2(j_0 + 1)\right)^2 > h^{-1}.$$

Hence we can conclude that

$$2^{-j_0-1} \log \frac{1}{h} \le A. \tag{3.2.22}$$

According to (2.3.1) and (2.3.2), we have

$$\|V(n,k,\cdot)\|_C \le An, \quad \|\overline{V}(n,k,\cdot)\|_C \le An.$$

Therefore, using the Lagrange formula and an inequality of Bernstein [2], we find

$$\|V(n-1,n,\cdot+h) - V(n-1,n,\cdot)\|_C \le An^2 h, \tag{3.2.23}$$

$$\|\overline{V}(n-1,n,\cdot+h) - \overline{V}(n-1,n,\cdot)\|_C \le An^2 h. \tag{3.2.24}$$

Now by (3.2.17), (2.3.3), (3.2.21), (3.2.22), and (3.2.23), we get

$$\int_T |f(x+h) - f(x)|\,dx \le$$

$$\le \sum_{j=1}^{j_0} 2^{-j} \int_T |V(n-1,n,x+h) - V(n-1,n,x)|\,dx +$$

$$+ \sum_{j=j_0+1}^{\infty} 2^{-j} \int_T |V(n-1,n,x+h) - V(n-1,n,x)|\,dx \le A(f)\left(\log\frac{1}{h}\right)^{-1}.$$

Quite analogously, taking into account (3.2.18), (2.3.3), (3.2.21), (3.2.22), and (3.2.24), we obtain

$$\int_T |\overline{f}(x+h) - \overline{f}(x)|\,dx \le A(f)\left(\log\frac{1}{h}\right)^{-1}.$$

Thus the validity of a) is established. Let us now prove b).

Let $n \equiv n(j) = 2^{2^j}$ $(j = 1, 2, \dots)$ and

$$V_2(n,x) = \frac{1}{n}\sum_{k=1}^{n-1} k\cos(k+n)x. \tag{3.2.25}$$

Consider the trigonometric series

$$\cos nx + \cos(n+1)x + \cdots + \cos(n+k)x + \cdots$$

and denote by $S_m(x)$ and $\sigma_m(x)$ its partial sums and $(C,1)$-means, respectively. We have

$$S_{n-1}(x) - \sigma_{n-1}(x) = V_2(n,x). \tag{3.2.26}$$

But

$$S_{n-1}(x) = D_{2n-1}(x) - D_{n-1}(x) = \frac{\sin(2n-\frac{1}{2})x - \sin(n-\frac{1}{2})x}{2\sin\frac{x}{2}} =$$
$$= \frac{\sin 2nx - \sin nx}{2\operatorname{tg}\frac{x}{2}} + g(n,x), \tag{3.2.27}$$

where

$$\sup_{x,n} |g(n,x)| \leq 1.$$

According to the definition,

$$\sigma_{n-1}(x) \sum_{k=0}^{n-1} \left(1 - \frac{k}{n}\right) \cos(n+k)x =$$
$$= \cos nx \sum_{k=0}^{n-1} \left(1 - \frac{k}{n}\right) \cos kx - \sin nx \sum_{k=0}^{n-1} \left(1 - \frac{k}{n}\right) \sin kx =$$
$$= \cos nx K_{n-1}(x) - \sin nx \overline{K}_{n-1}(x) + \frac{1}{2} \cos nx, \tag{3.2.28}$$

where $K_n(x)$ and $\overline{K}_{n-1}(x)$ are the Fejér's kernel and its conjugate, respectively. Therefore (3.2.25)–(3.2.28) imply

$$|V_2(n,x)| > \left|\frac{\sin 2nx}{2\operatorname{tg}\frac{x}{2}}\right| - 2K_{n-1}(x) - 2 \quad \text{for} \quad |x| \in]0,\pi].$$

Hence it follows that

$$\int_T |V_2(n,x)|\, dx \geq A \log n,$$

i.e.,

$$\int_T |V_2(n,x)|\, dx \geq A \cdot 2^j, \quad j \geq j_1 > 1 \tag{3.2.29}$$

since (see, e.g., Bari [2, Ch. I])

$$\int_T |D_n(x)|\, dx \approx \frac{4}{\pi^2} \log n \quad \text{as} \quad n \to \infty.$$

It is easy to see that if f is defined by (3.2.17), then

$$S_m(x,f) = \sum_{j=1}^{\lambda} 2^{-j} V(n-1,n,x) + 2^{-\lambda-1} g[V(n-1,n,x)],$$

where λ is a number depending on m, and g denotes a certain part (in some cases it is absent) of the trigonometric polynomial $V(n-1,n,x)$. Put

$$p(j) = 2n(j) - 1, \quad q(j) = 3n(j-1) - 2.$$

According to (3.2.25), we have

$$S_{p(j)}(x,f) - S_{q(j)}(x,f) = 2^{-j} V_2[n(j),x].$$

Hence, by (3.2.25), we find

$$\|S_{p(j)}(f) - S_{q(j)}(f)\|_1 > A,$$

so that

$$\limsup_{m\to\infty} \|S_m(f) - f\|_1 > 0. \qquad \blacksquare$$

It is well-known (see, e.g., Bari [2, Ch. VIII]) that if f and $\overline{f}$ are summable on T and one of the series $\sigma[f]$ and $\sigma[\overline{f}]$ converges in the norm of $L(T)$, then the another one also converges in the same norm. Therefore, in Theorem 3.2.3, for $\sigma[\overline{f}]$ we have the following relation

$$\limsup_{m\to\infty} \|\overline{S}_m(f) - \overline{f}\|_1 > 0.$$

Analyzing the proof of Theorem 3.2.3, we can conclude that it holds

Theorem 3.2.4. *There is a function f for which*

$$\omega(\delta, f) \log \frac{2\pi}{\delta} \le A(f) \quad \text{for } \delta \in]0, \pi],$$

yet

$$\limsup_{n\to\infty} \|S_{2^n}(f) - f\|_1 > 0.$$

It should be noted that for the space $C(T)$, the validity of a similar theorem follows from the corresponding results of Faber [1] and Fejér [2].

Remark 3.2.5. For several times we have made use of the fact that if the functions f and $\overline{f}$ are summable on T, then $\overline{\sigma}[f] = \sigma[\overline{f}]$. This result was established by Kolmogorov [3] in 1925. Three-four years later, using different methods, it was proved by Titchmarsh [2] and Smirnov [1].

In 1973, Nevai [1] proved that if $f \in C(T)$ and uniformly by $x \in T$

$$f(x+h) - f(x) > -\varepsilon(h)\Big(\log\frac{1}{h}\Big)^{-1} \quad \text{for } h \in]0, \pi],$$

where $\varepsilon(h) > 0$ and $\varepsilon(h) \to 0$ as $h \to 0+$, then $\sigma[f]$ is uniformly convergent. Therefore the well-known Dini–Lipschitz test for the uniform convergence of $\sigma[f]$ can be improved.

In [20] and [25], the author showed that the following assertion holds: let $f \in C(T)$, $\varepsilon(h) \ge 0$, and $\varepsilon(h) \to 0$ as $h \to 0$. If $\alpha \in]0, 1[$ and

$$f(x+h) - f(x) > -\varepsilon(h)h^\alpha \quad \text{as } h \in]0, \pi]$$

uniformly by $x \in T$, then $\sigma[f]$ is summable by $(C, -\alpha)$-method in the $C(T)$-norm.

The scheme of the proof of this result which is presented in [25] showes that Nevai's theorem and the just formulated assertion are consequences of Theorem 2.2.2.

Indeed, put

$$g(t, \alpha) = \begin{cases} t^{-1} & \text{for } \alpha = 0,\ t \in]0, \pi], \\ t^{\alpha-1} & \text{for } \alpha \in]0, 1[,\ t \in]0, \pi]; \end{cases}$$

$$\beta(h,\alpha)=\begin{cases}\left(\log\dfrac{2\pi}{h}\right)^{-1} & \text{for } \alpha=0,\ h\in]0,\pi],\\ h^{\alpha} & \text{for } \alpha\in]0,1[,\ h\in]0,\pi].\end{cases}$$

Consider the expression

$$B_0(x,f,\alpha,n)\equiv h^{-\alpha}\int_h^{\pi}|f(x+t+h)-f(x+t)|g(t,\alpha)\,dt.$$

We have

$$\begin{aligned}B_0(x,f,\alpha,n)&=h^{-\alpha}\int_h^{\pi}|f(x+t+h)-f(x+t)|\,g(t,\alpha)\,dt\le\\&\le h^{-\alpha}\int_h^{\pi}|f(x+t+h)-f(x+t)+\varepsilon(h)\beta(h,\alpha)|g(t,\alpha)\,dt+\\&+h^{-\alpha}\varepsilon(h)\beta(h,\alpha)\int_h^{\pi}g(t,\alpha)\,dt.\end{aligned}\tag{3.2.30}$$

It is clear that

$$h^{-\alpha}\varepsilon(h)\beta(h,\alpha)\int_h^{\pi}g(t,\alpha)\,dt\to 0\quad\text{as}\quad h\to 0+.\tag{3.2.31}$$

For any $\varepsilon_0\in]0,1[$, we choose $h_0(\varepsilon_0)\in]0,\pi]$ such that

$$\|f\|_C h^{1-\alpha}<(1-\alpha)\varepsilon_0\quad\text{for } h\in]0,h_0].\tag{3.2.32}$$

According to (3.2.30) and (3.2.31), it suffices to evaluate the expression

$$\begin{aligned}B_0^*(x,f,\alpha,n)&\equiv h^{-\alpha}\int_h^{\pi}[f(x+t+h)-f(x+t)]g(t,\alpha)\,dt=\\&=h^{-\alpha}\int_h^{h/h_0}[f(x+t+h)-f(x+t)]g(t,\alpha)\,dt+\\&+h^{-\alpha}\int_{h/h_0}^{\pi}[f(x+t+h)-f(x+t)]g(t,\alpha)\,dt=\\&=h^{-\alpha}\int_h^{h/h_0}[f(x+t+h)-f(x+t)]g(t,\alpha)\,dt+\\&+h^{-\alpha}\int_{h/h_0+h}^{\pi}f(x+t)[g(t-h,\alpha)-g(t,\alpha)]\,dt+\end{aligned}$$

$$+h^{-\alpha}\int_{\pi}^{\pi+h} f(x+t)g(t-h,\alpha)\,dt -$$

$$-h^{-\alpha}\int_{h/h_0}^{h/h_0+h} f(x+t)g(t,\alpha)\,dt.$$

Hence, using (3.2.32) and the fact that $f\in C(T)$, we find

$$B_0^*(x,f,h,\alpha)\to 0 \quad \text{as} \quad h\to 0+.$$

So, by (3.2.31) and (3.2.30), we have

$$B_0(x,f,h,\alpha)\to 0 \quad \text{as} \quad h\to 0+.$$

A series of corollaries can be obtained from Theorems 3.2.1 and 3.2.2, but we do not dwell on this (some of them have been mentioned above).

3.3. On the Series $\sigma[f]$ and $\sigma[\bar{f}]$ for Even and Odd f

First of all, we set forth some well-known results we will need in the sequel.

In 1906, Fatou [1] proved that if $f\in L(T)$ and

$$ka_k(f)\to 0,\quad kb_k(f)\to 0 \quad \text{as} \quad k\to\infty, \tag{3.3.1}$$

then $\sigma[f]$ converges almost everywhere. If, however, $f\in C(T)$ and (3.3.1) is fulfilled, then

$$\|S_n(f)-f\|_C\to 0 \quad \text{as} \quad n\to\infty. \tag{3.3.2}$$

In 1919, Hardy and Littlewood [3] showed that if $f\in C(T)$ and

$$k|a_k(f)|\le A(f),\quad k|b_k(f)|\le A(f) \quad \text{for} \quad k\in\mathbb{N}, \tag{3.3.3}$$

then (3.3.2) holds.

Some further works of Hardy and Littlewood [4], [5] are also dedicated to the same range of problems.

In 1932, Paley [1] established that if $f\in C(T)$ and

$$a_k(f)\ge 0,\quad b_k(f)\ge 0,\quad k\in\mathbb{N},$$

then (3.3.2) is valid.

In 1933, Szász [1] proved that if $f\in C(T)$ and for some $A\in]0,+\infty[$

$$ka_k(f)>-A,\quad kb_k(f)>-A \quad \text{for} \quad k\in\mathbb{N}, \tag{3.3.4}$$

then (3.3.2) is true.

It should be noted that this result improves Paley's theorem. The same is true for the statement of Hardy and Littlewood (see(3.3.3)) since instead of (3.3.3) only one-sided inequalities are required.

From the above mentioned result of Hardy and Littlewood and a theorem of the Tauber type established by them in [1], it follows that if $f\in C(T)$ and its Fourier coefficients satisfy (3.3.3), then $\sigma[f]$ is $(C,-\alpha)$-summable for any $\alpha\in]0,1[$ in the

$C(T)$-norm. If, however, the conditions of either Paley's or Szasz's theorem are fulfilled, then for $\alpha \in]0,1[$ the series $\sigma[f]$ may be nonsummable by $(C,-\alpha)$-method at every point $x \in T$.

Thus, for the uniform $(C,-\alpha)$-summability of $\sigma[f]$ with $\alpha \in]0,1[$, two-sided inequalities of the type (3.3.3) are essential.

Now consider the functions F and G defined by (2.1.10).

In 1961, Kinukava and Igari [1] proved that if $f \in B(T_+)$, then $F \in B(T_-)$ and $G \in B(T_-)$.

In 1964 and 1965, Wang [1], [2] investigated the problem on finding nonimprovable conditions for (2.1.6)–(2.1.9) to be the Fourier series of bounded or continuous functions. He, in particular, repaired some incorrectnesses in the paper by M. and S. Izumi [1]. Naturally, Wang studied the continuity of the functions F and G.

In the works [6], [7], and [13, Ch. II], the author gave some statements concerning the functions F, G and the series $\sigma[F]$, $\sigma[G]$. Below we present some of them.

Theorem 3.3.1. a) *Let* $f \in B(T_+)$ *and*

$$|S_n(0,f)| \le A(f) \quad \text{for} \quad n \in \mathbb{N}_0. \tag{3.3.5}$$

Then

$$\|S_n(F)\|_C \le A(f) \quad \text{for} \quad n \in \mathbb{N}; \tag{3.3.6}$$

b) *If* $f \in C(T_+)$ *and* (3.3.5) *is fulfilled, then*

$$\|S_n(F) - F\|_C \to 0 \quad \text{as} \quad n \to \infty.$$

Proof. According to the conditions of the theorem, we have

$$\sigma[f] = \frac{a_0(f)}{2} + \sum_{k=1}^{\infty} a_k(f)\cos kx$$

and

$$\sigma[\overline{f}] = \sum_{k=1}^{\infty} a_k(f)\sin kx.$$

Since f and $\overline{f}$ are summable on T (see, e.g., Bari [2, p. 545]), we have

$$\sum_{k=1}^{\infty} k^{-1}|a_k(f)| < +\infty.$$

Without loss of generality we can assume that

$$\sum_{k=1}^{\infty} k^{-1}a_k(f) = 0.$$

Then, taking $x \in]0,\pi]$, we get

$$F(x) = \operatorname{ctg}\frac{x}{2}\int_0^x \overline{f}(t)\,dt = \operatorname{ctg}\frac{x}{2}\sum_{k=1}^{\infty} a_k(f)\int_0^x \sin kt\,dt =$$

$$= -\operatorname{ctg}\frac{x}{2}\sum_{k=1}^{\infty} k^{-1}a_k(f)(\cos kx - 1) =$$

$$= -\sum_{k=1}^{\infty} k^{-1}a_k(f)\operatorname{ctg}\frac{x}{2}\cos kx.$$

Hence, putting

$$V_k \equiv V_k(f) = \sum_{i=k}^{\infty} i^{-1}a_i(f), \tag{3.3.7}$$

we find

$$F(x) = -\sum_{k=1}^{\infty}(V_k - V_{k+1})\operatorname{ctg}\frac{x}{2}\cos kx =$$

$$= \lim_{q\to\infty}\left\{-\sum_{k=1}^{q}(V_k - V_{k+1})\operatorname{ctg}\frac{x}{2}\cos kx\right\} =$$

$$= \sum_{k=1}^{\infty} V_k \operatorname{ctg}\frac{x}{2}[\cos(k-1)x - \cos kx] =$$

$$= 2\sum_{k=1}^{\infty} V_k \cos\frac{x}{2}\sin\left(k - \frac{1}{2}\right)x =$$

$$= \sum_{k=1}^{\infty} V_k[\sin kx + \sin(k-1)x] =$$

$$= \sum_{k=1}^{\infty}(V_k + V_{k+1})\sin kx =$$

$$= \sum_{k=1}^{\infty}[k^{-1}a_k(f) + 2V_{k+1}(f)]\sin kx.$$

If we denote

$$r_k \equiv r_k(f) = (2k)^{-1}a_k(f) + V_{k+1}(f), \tag{3.3.8}$$

then the last relation can be written as

$$F(x) = 2\sum_{k=1}^{\infty} r_k \sin kx, \tag{3.3.9}$$

this equality being obviously true also for $x = 0$ and $x = \pi$. Therefore $\sigma[F]$ everywhere converges to F.

By (3.3.7) and (3.2.8), we have

$$r_k = (2k)^{-1}a_k(f) + \lim_{q\to\infty}\left\{\sum_{i=k+1}^{\infty} i^{-1}[S_i(0,f) - S_{i-1}(0,f)]\right\} =$$

$$= (2k)^{-1}a_k(f) - (k+1)^{-1}S_k(0,f) + \sum_{i=k}^{\infty}[i(i+1)]^{-1}S_i(0,f),$$

whence, according to (3.3.5), it follows that

$$k|r_k| \leq A(f) \quad \text{for} \quad k \in \mathbb{N}. \tag{3.3.10}$$

On the other hand, it is well-known that

$$S_n(x, F) - \sigma_n^1(x, F) = (n+1)^{-1} \sum_{k=1}^{n} k r_k \sin kx. \tag{3.3.11}$$

According to a theorem of Kinukava and Igari [1], if $f \in B(T_+)$, then $F \in B(T_-)$, so that

$$\|\sigma_n^1(F)\|_C \leq A(f), \quad n \in \mathbb{N}.$$

But then (3.3.9)–(3.3.11) imply (3.2.6). Therefore the assertion a) of the theorem is proved.

Suppose now that $f \in C(T_+)$. By Wang [1], $F \in C(T_-)$. In view of (3.3.10), the Fourier coefficients of F satisfy the conditions (3.3.3) of the theorem of Hardy and Littlewood. Therefore

$$\|S_n(F) - F\|_C \to 0 \quad \text{as} \quad n \to \infty. \quad \blacksquare$$

Theorem 3.3.2. a) *Let $f \in B(T_+)$ and* (3.3.5) *be fulfilled. Then*

$$\|S_n(G)\|_C \leq A(f) \quad \textit{for} \quad n \in \mathbb{N};$$

b) *If $f \in C(T_+)$, $f(0) = a_0(f) = 0$ and* (3.3.5) *holds, then*

$$\|S_n(G) - G\|_C \to 0 \quad \textit{as} \quad n \to \infty.$$

Proof. According to a theorem of Kinukava and Igari [1], if $f \in B(T_+)$, then $G \in B(T_-)$, and the results of Wang [1], [2] imply that if $f \in C(T_+)$ and $f(0) = a_0(f) = 0$, then $G \in C(T_-)$. Further, if $x \in]0, \pi[$, then the function $\operatorname{ctg} \frac{x}{2}$ has a bounded variation on $[x, \pi]$. Therefore, by a theorem of W. Young [2], we have

$$G(x) = \int_x^\pi \overline{f}(t) \operatorname{ctg} \frac{t}{2}\, dt = \sum_{k=1}^{\infty} a_k(f) \int_x^\pi \operatorname{ctg} \frac{t}{2} \sin kt\, dt.$$

Since $S_0(0, f) = 0$, then $G(x)$ can be represented as

$$G(x) = \sum_{k=1}^{\infty} [S_k(0, f) - S_{k-1}(0, f)] \int_x^\pi \operatorname{ctg} \frac{t}{2} \sin kt\, dt =$$

$$= \lim_{q \to \infty} \left\{ \sum_{k=1}^{q} [S_k(0, f) - S_{k-1}(0, f)] \int_x^\pi \operatorname{ctg} \frac{t}{2} \sin kt\, dt \right\} =$$

$$= \sum_{k=1}^{\infty} S_k(0, f) \int_x^\pi [\sin kt - \sin(k+1)t] \operatorname{ctg} \frac{t}{2}\, dt =$$

$$= - \sum_{k=1}^{\infty} S_k(0, f) \int_x^\pi [\cos kt + \cos(k+1)t]\, dt =$$

$$= \sum_{k=1}^{\infty} S_k(0,f)[k^{-1}\sin kx + (k+1)^{-1}\sin(k+1)x] =$$

$$= \sum_{k=1}^{\infty} k^{-1}[S_k(0,f) + S_{k-1}(0,f)]\sin kx, \tag{3.3.12}$$

(3.3.12) being true for any $x \in T$. This proves the validity of the theorem. ∎

Remark 3.3.1. To establish (3.3.9), there is no need of everywhere convergence of $\sigma[f]$. Therefore, for a certain subclass of $L(T)$, the series $\sigma[F]$ and $\sigma[G]$ behave "better" – they are everywhere convergent. Moreover, the following can be claimed.

1) For any $\alpha \in]0,1[$, the series $\sigma[F]$ and $\sigma[G]$ are everywhere $(C,-\alpha)$-summable to F and G, respectively.

2) Under the conditions of the assertions a) of Theorems 3.3.1 and 3.3.2, for any $\alpha \in]0,1[$ we have

$$\|\sigma_n^{-\alpha}(F)\|_C \le A(f), \quad \|\sigma_n^{-\alpha}(G)\|_C \le A(f) \quad \text{for} \quad n \in \mathbb{N}.$$

3) Under the conditions of the assertions b) of Theorems 3.3.1 and 3.3.2, for any $\alpha \in]0,1[$ we have

$$\|\sigma_n^{-\alpha}(F) - F\|_C \to 0, \quad \|\sigma_n^{-\alpha}(G) - G\|_C \to 0 \quad \text{as} \quad n \to \infty.$$

Let now $f \in B(T_-)$. Then

$$\sigma[f] = \sum_{k=1}^{\infty} b_k(f)\sin kx$$

and

$$\sigma[\overline{f}] = -\sum_{k=1}^{\infty} b_k(f)\cos kx.$$

Taking an arbitrary $x \in]0,\pi[$, we have

$$F(x) = \operatorname{ctg}\frac{x}{2}\int_0^x \overline{f}(t)\,dt = -\operatorname{ctg}\frac{x}{2}\sum_{k=1}^{\infty} b_k(f)\int_0^x \cos kt\,dt =$$

$$= -\sum_{k=1}^{\infty} k^{-1}b_k(f)\operatorname{ctg}\frac{x}{2}\sin kx.$$

Hence, defining

$$\beta_k \equiv \beta_k(f) = -\sum_{i=k}^{\infty} i^{-1}b_i(f), \qquad \alpha_k = (2k)^{-1}b_k(f) + \beta_k(f), \tag{3.3.13}$$

we find

$$\operatorname{ctg}\frac{x}{2}\int_0^x \overline{f}(t)\,dt = \sum_{i=k}^{\infty}(\beta_k - \beta_{k+1})\operatorname{ctg}\frac{x}{2}\sin kx =$$

$$= \operatorname{ctg}\frac{x}{2}\sin x\sum_{k=t}^{\infty}k^{-1}b_k(f)+\sum_{k=2}^{\infty}\beta_k\operatorname{ctg}\frac{x}{2}[\sin kx-\sin(k-1)x]=$$

$$=-2\cos^2\frac{x}{2}+\sum_{k=1}^{\infty}k^{-1}b_k(f)+2\sum_{k=2}^{\infty}\beta_k\cos\left(k-\frac{1}{2}\right)x\,\cos\frac{x}{2}=$$

$$=-(1+\cos x)\sum_{k=1}^{\infty}k^{-1}b_k(f)+\sum_{k=2}^{\infty}\beta_k[\cos kx+\cos(k-1)x]=$$

$$=-\sum_{k=1}^{\infty}k^{-1}b_k(f)+2\sum_{k=1}^{\infty}\alpha_k\cos kx.$$

Thus

$$F(x)=-\sum_{k=1}^{\infty}k^{-1}b_k(f)+2\sum_{k=1}^{\infty}\alpha_k\cos kx,\qquad x\in]0,\pi[. \tag{3.3.14}$$

Furthermore,

$$G(x)=\int_0^x\overline{f}(t)\operatorname{ctg}\frac{t}{2}\,dt=-\sum_{k=1}^{\infty}b_k(f)\int_x^{\pi}\operatorname{ctg}\frac{t}{2}\cos kt\,dt=$$

$$=-\sum_{k=1}^{\infty}k^{-1}[S_k(0,f)-S_{k-1}(0,f)]\int_x^{\pi}\operatorname{ctg}\frac{t}{2}\,\cos kt\,dt.$$

Putting

$$S_k^*(f)=\sum_{i=1}^{k}b_i(f),\quad \delta_k\equiv\delta_k(f)=-k^{-1}[S_k^*(f)+S_{k-1}^*(f)], \tag{3.3.15}$$

we get

$$G(x)=\sum_{k=1}^{\infty}S_k^*(f)\int_x^{\pi}[\cos(k+1)t-\cos kt]\operatorname{ctg}\frac{t}{2}\,dt=$$

$$=-\sum_{k=1}^{\infty}S_k^*(f)\int_x^{\pi}2\sin\left(k+\frac{1}{2}\right)t\,\cos\frac{t}{2}\,dt=$$

$$=-\sum_{k=1}^{\infty}S_k^*(f)\int_x^{\pi}[\sin(k+1)t+\sin kt]\,dt=$$

$$=\sum_{k=1}^{\infty}S_k^*(f)[k(k+1)]^{-1}(-1)^k-$$

$$-\sum_{k=1}^{\infty}S_k^*(f)[(k+1)^{-1}\cos(k+1)x+k^{-1}\cos kx]=$$

$$=\sum_{k=1}^{\infty}S_k^*(f)[k(k+1)]^{-1}(-1)^k-\cos x+$$

$$+\sum_{k=1}^{\infty} \delta_k \cos kx, \quad x \in]0, \pi[. \tag{3.3.16}$$

Define the function $f : T \to \mathbb{R}$ by

$$f(x) = -\sum_{k=1}^{\infty} \frac{\sin kx}{k \ln k}.$$

It is well-known (see, e.g., Zygmund [15, Ch. V]) that f is absolutely continuous. If in (3.3.13) and (3.3.15) we put $b_i(f) = -(i \ln i)^{-1}$ for $i \geq 2$, then we obtain

$$\alpha_k \geq \frac{A}{k \ln k}, \quad \delta_k \geq \frac{A}{k \ln k} \quad \text{for} \quad k \geq 2.$$

This means that according to (3.3.14) and (4.4.16), we have

Theorem 3.3.3. *There is an absolutely continuous odd function $f : T \to \mathbb{R}$ satisfying*

$$\|S_n(f) - f\|_C \to 0 \quad \textit{as} \quad n \to \infty,$$

for which

$$F \notin C(T_+), \quad G \notin C(T_+)$$

and

$$\lim_{n\to\infty} \|S_n(F)\|_C = +\infty, \quad \lim_{n\to\infty} \|S_n(G)\|_C = +\infty.$$

On the other hand, on the basis of (3.3.14)–(3.3.16) we can conclude that it holds

Theorem 3.3.4. *Let $f \in B(T_-)$. If F and G belong to $B(T_+)$, and*

$$\|S_k(f)\|_C \leq A(f) \quad \textit{for} \quad k \in \mathbb{N}, \tag{3.3.17}$$

then

$$\|S_n(F)\|_C \leq A(f), \quad \|S_n(G)\|_C \leq A(f) \quad \textit{for} \quad n \in \mathbb{N}.$$

Thus naturally the question arises: what nontrivial conditions should be imposed on $f \in L(T_-)$ for F and G to belong to $B(T_+)$? This question is partly answered by

Theorem 3.3.5. *Let $f \in L(T_-)$ and $\overline{f} \in L(T_+)$. If*

$$\frac{1}{n} \sum_{k=1}^{n} k|b_k(f)| \leq A(f) \quad \textit{for} \quad n \in \mathbb{N}, \tag{3.3.18}$$

then (3.3.17) *is necessary and sufficient for $F \in B(T_+)$.*

This theorem follows from the argument of Fatou [1] concerning the summability of $\sigma[f]$ by Lebesgue's method (see, e.g., Zygmund [15, Ch. IX]).

In particular, if $b_k(f) \geq 0$ for $k \in \mathbb{N}$ and (3.3.18) holds, then $F \in B(T_-)$ iff the series

$$\sum_{k=1}^{\infty} b_k(f)$$

is convergent.

3.4. Unsolved Problems

3.4.1. Let a function $f : T \to [-\infty, +\infty]$ be measurable and almost everywhere finite. Let, moreover, $\alpha \in]0,1[$, $p \in]1,+\infty[$, and $\varepsilon \in]0,1[$. Is there a function $g \in C(T)$ such that

$$\mu_1\Big[\{x : f(x) \neq g(x),\ \ x \in T\}\Big] < \varepsilon$$

and $\sigma[g]$ is $(C,-\alpha)$-summable in the norm of $L^p(T)$?

3.4.2. Let $\alpha \in]0,1[$. Is there a function $f \in L(\log^+ L)^\alpha(T)$ such that $\sigma[f]$ diverges on a set $E \subset T$ with $\mu_1(E) > 0$ (almost everywhere, everywhere) and converges in the norm of $L(T)$?

The converse problem has a trivial solution.

Indeed, let $f : T \to [-\infty, +\infty]$ be defined as

$$f(x) = \sum_{k=0}^{\infty} \frac{\cos kx}{\ln(k+2)}.$$

According to results of W. Young [1], [2] and Kolmogorov [2] (see also Zygmund [15, Ch. V]), $f \in L(T)$ and

$$\limsup_{n\to\infty} \|S_n(f) - f\|_1 > 0.$$

Using a relevant result of Telyakovskiĭ [1] (see also (3.2.16)), we get

$$\omega\left(\frac{1}{n}, f\right)_1 \log(n+1) \leq A(f) \quad \text{for} \quad n \in N.$$

Applying now the imbedding Theorem 2 of Ul'yanov [6], we can conclude that $f \in L(\log^+ L)^\alpha(T)$ for any $\alpha \in]0,1[$.

3.4.3. Let $\alpha \in]0,1]$. Is there a function $f \in L(T)$ such that

$$\omega(\delta, f)_1 \left(\log \frac{2\pi}{\delta}\right)^\alpha \leq A(f) \qquad \text{for} \quad \delta \in]0,\pi],$$

$\sigma[f]$ diverges on a set $E \subset T$ with $\mu_1(E) > 0$ (almost everywhere, everywhere), and

$$\limsup_{n\to\infty} \|S_n(f) - f\|_1 > 0\ ?$$

3.4.4. Let $\alpha \in]0,1[$. Is there a function $f \in C(T)$ such that $\sigma[f]$ is $(C,-\alpha)$-summable on a set $E \subset T$ with $\mu_1(E) > 0$ (almost everywhere, everywhere), yet for some (any) $p \in]0,+\infty[$,

$$\limsup_{n\to\infty} \|\sigma_n^{-\alpha}(f) - f\|_p > 0\,?$$

CHAPTER IV

Some Approximating Properties of Cesàro Means of the Series σ[f] and $\overline{\sigma}$[f]

4.1. Approximating Properties of Cesàro Means of $\sigma[f]$

By now, a great deal of fundamental and profound results has been established concerning approximating properties of means of Cesàro, Poisson–Abel, Hardy, and of general linear means of the series $\sigma[f]$ and $\overline{\sigma}[f]$. Most of them are set forth in monographs and proceedings of international conferences, schools and symposia. Even several monographs would not suffice for a complete account of the results obtained in this direction. Therefore below we will survey only those of them which are necessary in the study of the questions treated in this chapter. We note, however, that the monographs by Alexits [2], Akhiezer [1], Bari [2], Vallée–Poussin [1], Jackson [3], Dzyadyk [1], Zygmund [15], Korneĭchuk [1], Leindler [1], Lorentz [2], Natanson [1], Nikol'skiĭ [5], Stepanets [1], Timan [1], as well as the papers by Korneĭchuk [2] and Tikhomirov [1], give the reader an idea of the most important investigations in this field.

In 1885, Weierstrass [1] proved that if $f \in C(T)$, then

$$E_n(f)_C \to 0 \quad \text{as} \quad n \to \infty,$$

where $E_n(f)_C$ is the best uniform approximation of f by trigonometric polynomials of degree $\leq n$.

Lebesgue [3] was appearently the first to apply in 1906 this statement to the theory of trigonometric series. He proved that if $f \in C(T)$, then

$$\|S_n(f) - f\|_C \leq AE_n(f)_C \log(n+2), \quad n \in \mathbb{N}, \tag{4.1.1}$$

Later, in 1910, he in [4] established that if $f \in Lip\,\alpha$ with $\alpha \in]0,1]$, then

$$\|S_n(f) - f\|_C \leq A(f)n^{-\alpha} \log(n+2) \quad \text{for} \quad n \in \mathbb{N}, \tag{4.1.2}$$

the order in (4.1.2) being nonimprovable.

Remark 4.1.1. In 1911, Jackson [1] (see also his paper [2]) proved that if $f \in C(T)$, then

$$E_n(f)_C \leq 12\omega\Big(\frac{1}{n}, f\Big)_C \quad \text{for} \quad n \in \mathbb{N}. \tag{4.1.3}$$

Therefore, by (4.1.3), the inequality (4.1.2) is a consequence of (4.1.1). But, as we have seen, the paper of Jackson was published a year later than the work of Lebesgue [3].

Using (4.1.2), we can easily verify that if $f \in Lip\,\alpha$ with $\alpha \in]0,1]$, then for any $n \in \mathbb{N}$,

$$\|\sigma_n^1(f) - f\|_C \le \begin{cases} A(f)n^{-\alpha}\log(n+2) & \text{if } \alpha \in]0,1[, \\ A(f)n^{-\alpha}\log^2(n+2) & \text{if } \alpha = 1. \end{cases}$$

In 1912, Bernstein [1] improved these estimates. In particular, he showed that if $f \in Lip\,\alpha$ with $\alpha \in]0,1]$, then for any $n \in \mathbb{N}$,

$$\|\sigma_n^1(f) - f\|_C \le \begin{cases} A(f)n^{-\alpha} & \text{if } \alpha \in]0,1[, \\ A(f)n^{-\alpha}\log(n+2) & \text{if } \alpha = 1, \end{cases}$$

the order being unimprovable.

In 1940, Nikol'skiĭ [1] proved that if $f \in Lip\,\alpha$ with $\alpha \in]0,1]$, then for any $n \ge n_0 > 1$,

$$\sup_{f \in Lip\,\alpha} \|\sigma_n^1(f) - f\|_C =$$

$$= \begin{cases} \dfrac{2}{\pi n}\log n + \gamma_n & \text{if } \alpha = 1, \\ \dfrac{2\Gamma(\alpha)}{\pi(1-\alpha)n^\alpha}\sin\dfrac{\alpha\pi}{2} + \gamma_n(\alpha) & \text{if } \alpha \in]0,1[, \end{cases} \tag{4.1.4}$$

where $n|\gamma_n| \le A$, $n^\alpha|\gamma_n(\alpha)| \le A(\alpha)$.

In 1950, Natanson [2] found that if $f \in C(T)$, then

$$\|\sigma_n^1(f) - f\|_C \le 3\omega\Big[\frac{\pi(1+2\log n)}{4n}, f\Big] \quad \text{for} \quad n \in \mathbb{N}.$$

In 1954, Izumi [1] showed that if $f \in Lip\,\alpha$ with $\alpha \in]0,1]$, then

$$\|\sigma_n^\alpha(f) - f\|_C \le A(f,\alpha)n^{-\alpha}\log(n+2), \quad n \in \mathbb{N}.$$

In 1956, Flett [1] established a series of theorems concerning approximating properties of (C,α)-means of $\sigma[f]$ at single points and in the norm of $C(T)$, where $\alpha \in [0,+\infty[$. In particular, he generalized some results of Bernstein and others.

In 1960, Taberski [1] showed that:

a) if $f \in C(T)$, then for any $n \in \mathbb{N}$,

$$\|\sigma_n^\alpha(f) - f\|_C \le \begin{cases} A(f,\alpha)\omega(n^{-\alpha}, f) & \text{if } \alpha \in]0,1[, \\ A(f,\alpha)\omega[n^{-1}\log(n+2), f] & \text{if } \alpha \in [1,\infty[; \end{cases}$$

b) if $f \in Lip\,\alpha$ with $\alpha \in]0,1]$, then for any $n \in \mathbb{N}$,

$$\|\sigma_n^\beta(f) - f\|_C \le \begin{cases} A(f,\beta)n^{-\alpha} & \text{if } \alpha \in]0,\beta[, \\ A(f,\beta)n^{-\alpha}\log(n+2) & \text{if } \alpha = \beta, \\ A(f,\beta)n^{-\beta} & \text{if } \alpha \in]\beta,1[, \end{cases}$$

$$\|\sigma_n^\beta(f)-f\|_C \le \begin{cases} A(f,\beta)n^{-\alpha} & \text{if } \alpha\in]0,1[,\ \beta\in[1,+\infty[,\\ A(f,\beta)n^{-1}\log(n+2) & \text{if } \alpha=1,\ \beta\in[1,+\infty[.\end{cases}$$

Similar estimates can also be found in the monograph by Alexits [2, Ch. IV].

Note that Taberski's estimates can be obtained from the above paper of Flett [1].

In 1961, Stechkin [5] proved that if $f\in C(T)$, then

$$\|\sigma_{n-1}^1(f)-f\|_C \le An^{-1}\sum_{k=1}^{\infty}E_k(f)_C \quad \text{for} \quad n\in\mathbb{N}, \tag{4.1.5}$$

this inequality being true (with appropriate changes) also for $f\in L^p(T)$ with $p\in[1,+\infty[$. We should add that for $p\in]1,+\infty[$, the analogue of (4.1.5), i.e.,

$$\|\sigma_{n-1}^1(f)-f\|_p \le A(p)n^{-1}\sum_{k=1}^{n}E_k(f)_p \quad \text{for} \quad n\in\mathbb{N},$$

follows from the inequality (1.1.10) of M. Riesz.

In 1949, Zamansky [1] showed that if $f\in C(T)$ and

$$\omega^{(2)}(\delta,f)\delta^{-1}\le A(f) \quad \text{for} \quad \delta\in]0,\pi],$$

then

$$\sigma_n^1(x,f)-f(x)=$$

$$=\frac{1}{2\pi}\int_a^\infty t^{-2}\Big[f\Big(x+\frac{2t}{n}\Big)+f\Big(x-\frac{2t}{n}\Big)-2f(x)\Big]\,dt+B_0^{**}(x,f,n) \text{ with } \alpha\in]0,+\infty[,$$

where

$$n\|B_0^{**}(\cdot,f,n)\|_C\le A(f) \quad \text{for} \quad n\in\mathbb{N}.$$

In 1958, Efimov [1] proved that if $f\in C(T)$, then

$$\sigma_n^1(x,f)-f(x)=$$

$$=\frac{1}{2\pi}\int_a^\infty t^{-2}\Big[f\Big(x+\frac{2t}{n}\Big)+f\Big(x-\frac{2t}{n}\Big)-2f(x)\Big]\,dt+B_0^{**}(x,f,n),\ \ \alpha\in]0,+\infty[,$$

where

$$\|B_0^{**}(\cdot,f,n)\|_C\Big[\omega^{(2)}\Big(\frac{1}{n},f\Big)_C\Big]^{-1}\le A(f) \quad \text{for} \quad n\in\mathbb{N}.$$

In 1962, Gui [1] established that if $f\in C(T)$ and $\alpha\in]0,+\infty[$, then

$$\sigma_n^\alpha(x,f)-f(x)=\frac{1}{2\pi}\int_a^\infty t^{-2}\Big[f\Big(x+\frac{2t}{n}\Big)+f\Big(x-\frac{2t}{n}\Big)-2f(x)\Big]\,dt+$$
$$+B_0^{**}(x,f,n,\alpha) \text{ with } \alpha\in]0,+\infty[, \tag{4.1.6}$$

where

$$\|B_0^{**}(\cdot,f,n,\alpha)\|_C\Big[\omega^{(2)}\Big(\frac{1}{n},f\Big)_C\Big]^{-1}\le A(f) \quad \text{for} \quad n\in\mathbb{N}.$$

In 1964, Ul'yanov [3] proved that:

a) If $f \in L^p(T)$, where either $p = 1$ or $p = +\infty$, and $\alpha \in]0, +\infty[$, then

$$\|\sigma_n^\alpha(f) - f\|_p \le A(\alpha)\omega\Big[\frac{\log(n+2)}{n+1}, f\Big]_p \quad \text{for} \quad n \in \mathbb{N}. \tag{4.1.7}$$

b) If $f \in L^p(T)$ for $p \in]1, +\infty[$, and $\alpha \in [1, +\infty[$, then

$$\|\sigma_n^\alpha(f) - f\|_p \le A(p,\alpha)\omega\Big(\frac{1}{n}, f\Big)_p \quad \text{for} \quad n \in \mathbb{N}. \tag{4.1.8}$$

Moreover, he showed that for any fixed α and p, the orders in estimates (4.1.7) and (4.1.8) cannot be improved. In the same paper, Ul'yanov [3] established that if $\alpha \in]0,1[$ and $\beta \in]0, +\infty[$, then the order of approximation of functions f from $Lip\,\alpha$ and $Lip(\alpha, 1)$ by the means $\sigma_n^\beta(x, f)$ in the norms of $C(T)$ and $L(T)$, respectively, does not depend on β. It should be added that a similar conclusion in the case $p = +\infty$ follows from (4.1.6).

In 1965, Buadze [1] showed that (4.1.8) is true for $\alpha \in]0, 1[$ as well. The work of Trigub [1] should also be mentioned here.

Thus the approximating properties of Cesàro means of positive orders were sufficiently clarified.

In 1925, Zygmund [3] actually proved that if $f \in Lip\,\alpha$ with $\alpha \in]0, 1]$, then for any $\beta \in]0, \alpha[$,

$$\|\sigma_n^{-\beta}(f) - f\|_C \le A(f,\beta)n^{\beta-\alpha} \quad \text{for} \quad n \in \mathbb{N}. \tag{4.1.9}$$

Later, in 1928, Hardy and Littlewood [6] showed that if $f \in C(T) \cap Lip(\alpha, p)$ with $\alpha \in]0, 1]$ and $\alpha p \in]1, +\infty[$, then $f \in Lip(\alpha - \frac{1}{p})$. Therefore, in this case, by (4.2.1) we have

$$\|S_n(f) - f\|_C \le A(p, f)n^{1/p-\alpha}\log(n+2) \quad \text{for} \quad n \in \mathbb{N}.$$

In 1955, Izumi [2] found out that in fact a more precise estimate

$$\|S_n(f) - f\|_C \le A(p, f)n^{1/p-\alpha} \quad \text{for} \quad n \in \mathbb{N} \tag{4.1.10}$$

is valid.

In 1955, Satô [2] (see also her work [1]) proved that if $f \in C(T)$ and $\alpha \in]0, 1[$, then

$$\|\sigma_n^{-\alpha}(f) - f\|_C \le$$
$$\le A(f,\alpha)\Big\{\omega\Big(\frac{1}{n}, f\Big)_C\Big[n\omega\Big(\frac{1}{n}, f\Big)\Big]_1\Big[\omega\Big(\frac{1}{n}, f\Big)_C^{-1}\Big]^\alpha + n^{-1}\int_{1/n}^{\pi} t^{-2}\omega(t, f)_C dt\Big\}.$$

The problems arose:

1) Let $f \in L^p(T)$ for some $p \in [1, +\infty]$ and let $\alpha \in]0, 1[$. Is it possible to establish precise in the sense of order estimates for

$$\|\sigma_n^{-\alpha}(f) - f\|_p$$

in the terms of $L^p(T)$ smoothness moduli of f?

2) In the statement of Izumi [2], is it possible to replace the partial sums $S_n(x, f)$ by $\sigma_n^{-\alpha}(x, f)$ for some $\alpha \in]0, 1[$?

3) Investigate similar questions for $\overline{\sigma}[f]$.

In the works [3], [4], [5], [13, Ch. II], and [30], we set down some results connected with the formulated problems.

In this section we present our results concerning approximating properties of the means $\sigma_n^\alpha(x,f)$ in the spaces $L^p(T)$ with $p \in [1,+\infty]$.

In the sequel, for $p \in [1,+\infty]$, $\alpha \in]-\infty,+\infty[$, and a function $f \in L^p(T)$, we will denote

$$\lambda_n^{(0)} \equiv \lambda_n^{(0)}(f,\alpha,p) =$$
$$= (n^\alpha+1)\omega^{(2)}\left(\frac{1}{n},f\right)_p + n^{-1}\int\limits_{1/n}^{\pi} t^{-2}\omega^{(2)}(t,f)_p dt \quad \text{for } n \in \mathbb{N}. \tag{4.1.11}$$

Theorem 4.1.1. *Let $f \in L^p(T)$ for some $p \in [1,+\infty]$ and let $\alpha \in]0,1[$. Then*

$$\|\sigma_n^{-\alpha}(f) - f\|_p \le A(p,\alpha)\lambda_n^{(0)}. \tag{4.1.12}$$

Proof. Without loss of generality, we can assume that f is not equivalent to a constant. Let T_n be the best approximating trigonometric polynomial (of order $\le n$) for f in $L^p(T)$. Put

$$Z_n(x,t) \equiv Z_n(x,t,f) = T_n(x+t) + T_n(x-t) - 2T_n(x),$$
$$g_n(x,t) \equiv g_n(x,t,f) = \varphi(x,t) - Z_n(x,t).$$

We have

$$\sigma_n^{-\alpha}(x,f) - f(x) = \frac{1}{\pi}\int\limits_0^{\pi/n} \varphi(x,t)K_n^{-\alpha}(t)\,dt + \frac{1}{\pi}\int\limits_{\pi/n}^{\pi} \varphi(x,t)K_n^{-\alpha}(t)\,dt,$$

whence, by (2.2.9), we get

$$\sigma_n^{-\alpha}(x,f)-f(x)=\frac{1}{\pi}\int\limits_0^{\pi/n}\varphi(x,t)K_n^{-\alpha}+\frac{1}{\pi}\int\limits_{\pi/n}^{\pi}\varphi(x,t)\varphi_n^{-\alpha}(t)\,dt+$$
$$+\int\limits_{\pi/n}^{\pi}\varphi(x,t)r_n^{-\alpha}(t)\,dt \equiv \sum_{k=1}^{3} M_k(x,f,\alpha,n). \tag{4.1.13}$$

Using (2.2.13) and (2.2.15), respectively, we find

$$\|M_1(\cdot,f,\alpha,n)\|_p \le A(\alpha)\omega^{(2)}\left(\frac{1}{n},f\right)_p, \tag{4.1.14}$$

$$\|M_3(\cdot,f,\alpha,n)\|_p \le A(\alpha)n^{-1}\int\limits_{1/n}^{\pi} t^{-2}\omega^{(2)}(t,f)_p dt. \tag{4.1.15}$$

Consider now the expression $M_2(x,f,\alpha,n)$ in (4.1.13). We have

$$M_2(x,f,\alpha,n)=\frac{1}{\pi}\int_{\pi/n}^{\pi} g_n(x,t)\varphi_n^{-\alpha}(t)\,dt+$$

$$+\frac{1}{\pi}\int_{\pi/n}^{\pi} Z_n(x,t)\varphi_n^{-\alpha}(t)\,dt\equiv R_1(x,f,\alpha,n)+R_2(x,f,\alpha,n). \tag{4.1.16}$$

In view of (2.2.5) and (2.2.11), we obtain

$$\|R_1(\cdot,f,\alpha,n)\|_p\le A(\alpha)(n^\alpha+1)E_n(f)\int_{\pi/n}^{\pi} t^{\alpha-1}dt\le$$

$$\le\begin{cases}A(\alpha)(n^\alpha+1)E_n(f)_p & \text{if } \alpha\in]0,1[,\\ A(\alpha)E_n(f)_p & \text{if } \alpha\in]-1,0[.\end{cases} \tag{4.1.17}$$

Put

$$\gamma(t)\equiv\gamma_\alpha(t)=\left(2\sin\frac{t}{2}\right)^{\alpha-1},$$

$$\beta(t)\equiv\beta(t,\alpha,n)=\left(n+\frac{1}{2}-\frac{\alpha}{2}\right)t+\frac{\alpha\pi}{2}.$$

Then, by (4.1.16), we get

$$R_2(x,f,\alpha,n)=\frac{1}{\pi A_n^{-\alpha}}\int_{\pi/n}^{\pi} Z_n(x,t)\gamma(t)\sin\beta\,dt.$$

Integrating by parts and taking into account the fact that $\cos\beta(\pi)=0$, we find

$$\begin{aligned}R_2(x,f,\alpha,n)&=\left[\pi A_n^{-\alpha}\left(n+\frac{1-\alpha}{2}\right)\right]^{-1}Z_n\left(x,\frac{\pi}{n}\right)\gamma\left(\frac{\pi}{n}\right)\cos\beta\left(\frac{\pi}{n}\right)+\\ &+\left[\pi A_n^{-\alpha}\left(n+\frac{1-\alpha}{2}\right)^2\right]^{-1}Z_n(x,\pi)\gamma'(\pi)\sin\beta(\pi)-\\ &-\left[\pi A_n^{-\alpha}\left(n+\frac{1-\alpha}{2}\right)^2\right]^{-1}\left[Z_n'\left(x,\frac{\pi}{n}\right)\gamma\left(\frac{\pi}{n}\right)+\right.\\ &\left.+Z_n\left(x,\frac{\pi}{n}\right)\gamma'\left(\frac{\pi}{n}\right)\right]\sin\beta\left(\frac{\pi}{n}\right)-\\ &-\left[\pi A_n^{-\alpha}\left(n+\frac{1-\alpha}{2}\right)^2\right]^{-1}\int_{\pi/n}^{\pi} Z_n''(x,t)\gamma(t)\sin\beta(t)\,dt-\\ &-\left[\pi A_n^{-\alpha}\left(n+\frac{1-\alpha}{2}\right)^2\right]^{-1}\int_{\pi/n}^{\pi} Z_n'(x,t)\gamma'(t)\sin\beta(t)\,dt-\end{aligned}$$

$$-\left[\pi A_n^{-\alpha}\left(n+\frac{1-\alpha}{2}\right)^2\right]^{-1}\int_{\pi/n}^{\pi} Z_n(x,t)\gamma''(t)\sin\beta(t)\,dt \equiv$$

$$\equiv \sum_{k=1}^{6} L_k(x,f,\alpha,n). \tag{4.1.18}$$

But since

$$\omega^{(2)}(\delta,T_n)_p \le A\omega^{(2)}(\delta,f)_p,$$

we have

$$\|L_1(\cdot,f,\alpha,n)\|_p \le A(\alpha)\omega^{(2)}\left(\frac{1}{n},f\right)_p. \tag{4.1.19}$$

On the other hand,

$$n^2\omega\left(\frac{1}{n},f\right)_p \ge A.$$

Therefore the estimate

$$\|L_2(\cdot,f,\alpha,n)\|_p \le A(\alpha)(n^\alpha+1)\omega^{(2)}\left(\frac{1}{n},f\right)_p \tag{4.1.20}$$

holds. Furthermore, if T_n is an arbitrary trigonometric polynomial of order $\le n$ and

$$\Delta_h^k T(x) = \sum_{i=0}^{k}(-1)^i\binom{i}{k}T_n[x+(k-2i)h],$$

then, as Nikol'skiĭ [3] showed in 1948, for $h=\frac{\pi}{2n}$ and any $p\in[1,+\infty]$ there holds

$$\|T_n^{(k)}\|_p \le A(k)n^k\|\Delta_h^k T_n\|_p \quad \text{for} \quad k\in\mathbb{N}. \tag{4.1.21}$$

Using (4.1.18) and (4.1.21), we find

$$\begin{aligned}
\|L_3(\cdot,f,\alpha,n)\|_p \le{}& A(\alpha)n^{\alpha-2}\Bigg[\int_{-\pi/n}^{\pi/n}\|T''(\cdot+s)\|_p ds\left|\gamma\left(\frac{\pi}{n}\right)\right| + \\
&+\omega^{(2)}\left(\frac{\pi}{n},T_n\right)_p\left|\gamma'\left(\frac{\pi}{n}\right)\right|\Bigg] \le \\
&\le A(p,\alpha)n^{\alpha-2}\Bigg[n^{-1}\|T_n^1\|_p\left|\gamma\left(\frac{\pi}{n}\right)\right| + \\
&+\omega^{(2)}\left(\frac{\pi}{n},f\right)_p\left|\gamma'\left(\frac{\pi}{n}\right)\right|\Bigg] \le \\
&\le A(p,\alpha)n^{\alpha-2}\Bigg[n\omega^{(2)}\left(\frac{1}{n},f\right)_p\left|\gamma\left(\frac{\pi}{n}\right)\right| + \\
&+\omega^{(2)}\left(\frac{\pi}{n},f\right)_p\left|\gamma'\left(\frac{\pi}{n}\right)\right|\Bigg] \le \\
&\le A(p,\alpha)\omega^{(2)}\left(\frac{1}{n},f\right)_p,
\end{aligned} \tag{4.1.22}$$

$$\|L_4(\cdot,f,\alpha,n)\|_p \le A(\alpha)n^{\alpha-2}\int_{\pi/n}^{\pi}\|T_n''\|_p|\gamma(t)|\,dt \le$$
$$\le A(\alpha)(n^\alpha+1)\omega^{(2)}\Big(\frac{1}{n},f\Big)_p, \tag{4.1.23}$$
$$\|L_5(\cdot,f,\alpha,n)\|_p \le A(\alpha)n^{\alpha-2}\int_{\pi/n}^{\pi}\Big\|\int_{-t}^{t}T_n''(\cdot+s)\,ds\Big\|_p|\gamma'(t)|\,dt \le$$
$$\le A(p,\alpha)n^{\alpha-2}\|T_n''\|_p\int_{\pi/n}^{\pi}t|\gamma(t)|\,dt \le$$
$$\le A(p,\alpha)(n^\alpha+1)\omega^{(2)}\Big(\frac{1}{n},f\Big)_p. \tag{4.1.24}$$

It is easy to see that

$$\|L_6(\cdot,f,\alpha,n)\|_p \le A(\alpha)n^{\alpha-2}\int_{\pi/n}^{\pi}t^{\alpha-3}\omega^{(2)}(t,T_n)_p\,dt \le$$
$$\le A(\alpha)n^{\alpha-2}\int_{\pi/n}^{\pi}t^{\alpha-3}\omega^{(2)}(t,f)_p\,dt \le$$
$$\le A(\alpha)n^{\alpha}\omega^{(2)}\Big(\frac{1}{n},f\Big)_p\int_{\pi/n}^{\pi}t^{\alpha-1}\,dt \le$$
$$\le A(\alpha)(n^\alpha+1)\omega^{(2)}\Big(\frac{1}{n},f\Big)_p. \tag{4.1.25}$$

Thus, on the basis of (4.1.13)–(4.1.20) and (4.1.22)–(4.1.25), we get

$$\|\sigma_n^{-\alpha}(f)-f\|_p \le A(p,\alpha)\lambda_n^{(0)} \qquad \blacksquare$$

Remark 4.1.2. It is not difficult to see that Theorem 4.1.1 is also true in the case where $\alpha \in]-\infty,-1]$, or where $\alpha \in]1,+\infty[\setminus\mathbb{N}$. But for $\alpha \in]1,+\infty[$, this theorem is not so meaningfull for, if f is not equivalent to a constant and $\alpha \in [2,+\infty[\setminus\mathbb{N}$, then

$$n^\alpha\omega^{(2)}\Big(\frac{1}{n},f\Big)_p \to 0 \quad \text{as} \quad n\to\infty.$$

Remark 4.1.3. If $\alpha \in]-\infty,0[$, then in (4.1.11) the term

$$n^{-1}\int_{1/n}^{\pi}t^{-2}\omega^{(2)}(t,f)_p\,dt$$

is dominant, i.e.,

$$\|\sigma_n^{-\alpha}(f) - f\|_p \le A(p,\alpha)n^{-1} \int_{1/n}^{\pi} t^{-2}\omega^{(2)}(t,f)_p\, dt. \tag{4.1.26}$$

If $\alpha \in]0,1[$, then in the expression of λ_n^0, both terms are essential. In the case $\alpha \in]1,+\infty[\setminus \mathbb{N}$, in λ_n^0 the term

$$(n^\alpha + 1)\omega^{(2)}\Big(\frac{1}{n}, f\Big)_p$$

is dominant since

$$n^{-1} \int_{1/n}^{\pi} t^{-2}\omega^{(2)}(t,f)_p\, dt \le An\omega^{(2)}\Big(\frac{1}{n}, f\Big)_p \quad \text{for} \quad n \in \mathbb{N}.$$

Various corollaries can be obtained from Theorem 4.4.1. We present

Corollary 4.1.1. *Let $f \in L^p(T)$ for some $p \in [1,+\infty]$ and let $\alpha \in]0,1[$. Then*

$$\|\sigma_n^{-\alpha}(f) - f\|_p \le A(p,\alpha)n^\alpha\omega\Big(\frac{1}{n}, f\Big)_p \quad \textit{for} \quad n \in \mathbb{N}. \tag{4.1.27}$$

Indeed, according to (4.1.11), we have

$$\|\sigma_n^{-\alpha}(f) - f\|_p \le A(p,\alpha)\Big[n^\alpha\omega\Big(\frac{1}{n}, f\Big)_p + n^{-1}\int_{1/n}^{\pi} t^{-2}\omega^{(2)}(t,f)_p\, dt\Big].$$

On the other hand,

$$n^{-1}\int_{1/n}^{\pi} t^{-2}\omega^{(2)}(t,f)_p\, dt \le A\omega\Big(\frac{1}{n}, f\Big)_p \log(n+1) \quad \text{for} \quad n \in \mathbb{N}.$$

Therefore (4.1.27) holds. ■

In the case $p \in]1,+\infty[$, Theorem 4.4.1 can be improved. To be more precise, it holds

Theorem 4.1.2. *Let $f \in L^p(T)$ for some $p \in]1,+\infty[$ and let $\alpha \in]0,+\infty[\setminus \mathbb{N}$. Then*

$$\|\sigma_n^{-\alpha}(f) - f\|_p \le A(p,\alpha)\Big[(n^\alpha + 1)E_{[\frac{n}{2}]}(f)_p + n^{-1}\sum_{k=1}^{[\frac{n}{2}]} E_k(f)_p\Big] \quad \textit{for} \ n \ge 2.$$

The validity of this theorem can easily be verified using M. Riesz's inequality (1.1.10).

Naturally, the problem arises whether the order in the conclusion of Theorem 4.4.1 can be improved. Below, in terms of the classes H_p^ω, we formulate some related statements.

According to Theorem 4.1.2, for functions $f \in L^p(T)$ with $p \in]1,+\infty[$, Theorem 4.4.1 can be made more precise. Therefore we will consider the case where either $f \in C(T)$ or $f \in L(T)$.

In 1955, Stechkin [2] (see also his work [5] and Dzyadyk [1, Ch. III]) showed that for any continuity modulus ω, there is a least convex majorant – a continuity modulus ω^{**} such that

$$\omega(t) \le \omega^{**}(t) \le 2\omega(t)$$

for $t \in [0,1]$. Therefore, in the classes H_p^ω we consider below, we will assume that ω is a convex continuity modulus.

I. Let $p = +\infty$ and $\alpha \in [-\infty, 0[$. Then, as already noted, the conclusion of Theorem 4.4.1 can be expressed by (4.1.26). Consider the function $f : T \to \mathbb{R}$ defined by

$$f(x) = \sum_{k=1}^{\infty} \left[\omega\left(\frac{1}{k}\right) - \omega\left(\frac{1}{k+1}\right)\right] \cos kx, \tag{4.1.28}$$

and assume that

$$\delta^{-1}\omega(\delta) \uparrow \infty \quad \text{for} \quad \delta \downarrow 0. \tag{4.1.29}$$

From the results of Totik [1] (see also his paper [2]), it follows that $f \in H_p^\omega$ and

$$\|\sigma_n^{-\alpha}(f) - f\|_C \ge A(\alpha) n^{-1} \int_{1/n}^{\pi} t^{-2}\omega(t)\,dt. \tag{4.1.30}$$

Assume now that

$$\delta^{-1}\omega(\delta) \le A \quad \big(\omega(\delta) = A\delta, \ \ \delta \in [0,\pi]\big). \tag{4.1.31}$$

Applying the appropriate assertions from Ul'yanov [3], we conclude that if a function $f : T \to \mathbb{R}$ is defined by

$$f(x) = \left|\sin\frac{x}{2}\right| \quad \text{for} \quad x \in T,$$

then $f \in H_p^\omega$ and

$$\|\sigma_n^{-\alpha}(f) - f\|_C \ge A(\alpha) n^{-1} \int_{1/n}^{\pi} t^{-2}\omega(t)\,dt. \tag{4.1.32}$$

II. Let now $p = 1$ and $\alpha \in]-\infty, 0[$. Then, as already noted, the conclusion of Theorem 4.4.1 can be expressed by (4.1.26). Suppose that (4.1.29) holds and

$$\sum_{k=1}^{\infty} k^{-1}\omega\left(\frac{1}{k}\right) < +\infty. \tag{4.1.33}$$

Then, according to W. Young [1], [2] (see also Sidon [1], Hille and Tamarkin [1], [2], Zygmund [15, Ch. V]),

$$\sum_{k=1}^{\infty} \omega\left(\frac{1}{k}\right) \sin kx \quad \text{for} \quad x \in T \tag{4.1.34}$$

is the Fourier series of a function $f \in L(T)$. In view of a result by Telyakovskiĭ [1],

$$\omega\left(\frac{1}{n}, f\right)_1 \le A\left[n^{-1}\sum_{k=1}^{n}\omega\left(\frac{1}{k}\right) + \sum_{k=1}^{\infty}k^{-1}\omega\left(\frac{1}{k}\right)\right]. \tag{4.1.35}$$

Using a scheme of Ul'yanov [3], we find

$$\|\sigma_n^{-\alpha}(f) - f\|_1 \ge \int_0^{\pi}[\sigma_n^{-\alpha}(x, f) - f(x)]\, dx =$$

$$= (A_n^{-\alpha})^{-1}\sum_{k=0}^{n}A_{n-k}^{-\alpha-1}\int_0^{\pi}\left[\sum_{i=k+1}^{\infty}\omega\left(\frac{1}{i}\right)\sin ix\right]dx =$$

$$= (A_n^{-\alpha})^{-1}\sum_{k=0}^{n}A_{n-k}^{-\alpha-1}\left[\sum_{i=k+1}^{\infty}i^{-1}\omega\left(\frac{1}{i}\right)[1-(-1)^i]\right].$$

But

$$\sum_{i=k+1}^{\infty}i^{-1}\omega\left(\frac{1}{i}\right)[1-(-1)i] \ge \sum_{i=k+1}^{\infty}(2i+1)^{-2}(2i+1)\omega\left(\frac{1}{2i+1}\right) \ge$$

$$\ge (2k+3)\omega\left(\frac{1}{2k+3}\right)\sum_{i=k+1}^{\infty}(2i+1)^{-2} \ge A\omega\left(\frac{1}{k+1}\right) \quad \text{for } k \in \mathbb{N}.$$

Therefore, when $n > 1$, we get

$$\|\sigma_n^{-\alpha}(f) - f\|_1 \ge A(A_n^{-\alpha})^{-1}\sum_{k=0}^{[\frac{n}{2}]}A_{n-k}^{-\alpha-1}\omega\left(\frac{1}{k+1}\right) \ge$$

$$\ge A(\alpha)n^{-1}\int_{1/n}^{\pi}t^{-2}\omega(t)\, dt. \tag{4.1.36}$$

Suppose now that (4.1.31) holds. Consider the trigonometric series

$$\sum_{k=1}^{\infty}k^{-1}\sin kx \qquad \text{for} \quad x \in T. \tag{4.1.37}$$

It is well-known (see, e.g., Zygmund [15, Ch. I]) that (4.1.37) is everywhere convergent and its sum has a bounded variation. Therefore a well-known theorem of Hardy and Littlewood [5] (see also their work [6]) implies that $f \in H_1^{\omega}$. Using the appropriate assertion from Ul'yanov [3], we get

$$\|\sigma_n^{-\alpha}(f) - f\|_1 \ge A(\alpha)n^{-1}\int_{1/n}^{\pi}t^{-2}\omega(t)\, dt \qquad \text{for } n \in \mathbb{N}. \tag{4.1.38}$$

It should be noted that if (4.1.33) fails to be fulfilled, i.e.,

$$\sum_{k=1}^{\infty}\frac{1}{k}\omega\left(\frac{1}{k}\right) = +\infty,$$

then (4.1.31), in general, is not a Fourier series. Therefore, in this case ($p=1$), the above presented method does not work.

Let now $p=+\infty$ and $\alpha \in]0,1[$. Then, as already pointed out, the conclusion of Theorem 4.4.1 reads as

$$\|\sigma_n^{-\alpha}(f)-f\|_C \geq A(\alpha)n^\alpha \omega^{(2)}\Big(\frac{1}{n},f\Big)_C \quad \text{for} \quad n\in \mathbf{N}.$$

Using an appropriate assertion by Akhobadze [4], [5], we conclude that the order in the last estimate cannot be improved for the whole space $C(T)$. Note that the question whether Theorem 4.1.1 is the best possible (for $p=+\infty$) could be investigated using some other reasons as well. The same question for $p=1$ and $\alpha \in]0,1[$ still remains open.

Thus the problem whether Theorem 4.4.1 is the best possible is solved only in part.

Theorem 4.1.3. *Let $f\in C(T)\cap Lip(\alpha,p)$ for some $p\in]1,+\infty[$ and $\alpha\in]0,1[$. If $\alpha p\in]1,+\infty[$, then for any $n\in \mathbf{N}$,*

$$\|\sigma_n^{-\beta}(f)-f\|\|_C \leq \begin{cases} A(f,p,\beta)n^{1/p-\alpha} & \text{if } \beta\in \Big]0,\dfrac{1}{p}\Big[,\\ A(f,p,\beta)n^{1/p-\alpha}[\log(n+2)]^{1-1/p} & \text{if } \beta=\dfrac{1}{p},\\ A(f,p,\beta)n^{\beta-\alpha} & \text{if } \beta\in \Big]\dfrac{1}{p},\alpha\Big[. \end{cases}$$

Proof. By (2.2.34), we have

$$\begin{aligned} &\sigma_n^{-\beta}(x,f)-f(x)=\\ &=\frac{1}{\pi}\int\limits_0^{\pi/n}\varphi(x,t)K_n^{-\beta}(t)\,dt+\frac{1}{\pi}\int\limits_{\pi/n}^{\pi}\varphi(x,t)K_n^{-\beta}(t)\,dt\equiv\\ &\equiv I_1(x,f,\beta,n)+I_2(x,f,\beta,n). \end{aligned} \tag{4.1.39}$$

According to (2.2.13), we find

$$\|I_1(\cdot,f,\beta,n)\|_C\leq A(\beta)\omega^{(2)}\Big(\frac{1}{n},f\Big)_C. \tag{4.1.40}$$

Then, using (2.2.9), we can write $I_2(x,f,\beta,n)$ as

$$\begin{aligned} &I_2(x,f,\beta,n)=\\ &=\frac{1}{\pi}\int\limits_{\pi/n}^{\pi}\varphi(x,t)\varphi_n^{-\beta}(t)\,dt+\frac{1}{\pi}\int\limits_{\pi/n}^{\pi}\varphi(x,t)r_n^{-\beta}(t)\,dt\equiv\\ &\equiv P_1(x,f,\beta,n)+P_2(x,f,\beta,n). \end{aligned} \tag{4.1.41}$$

Applying (2.2.15), we get

$$\|P_2(\cdot,f,\beta,n)\|_C \le A(\beta)n^{-1}\int_{1/n}^{\pi} t^{-2}\omega^{(2)}(t,f)_C\,dt. \tag{4.1.42}$$

As we have seen, to evaluate $P_4(x,f,\beta,n)$ it suffices to estimate the expression

$$H(x,f,\beta,n) = n^{\beta}\int_{\pi/n}^{\pi}\varphi(x,t)\omega(-\beta,t)\cos nt\,dt.$$

We have (see (2.2.38))

$$\begin{aligned}
&2H(x,f,\beta,n) = \\
&= n^{\beta}\Big\{\int_{\pi/n}^{\pi-\pi/n}\Big[\varphi(x,t)-\varphi\Big(x,t+\frac{\pi}{n}\Big)\Big]\omega(-\beta,t)\cos nt\,dt + \\
&+\int_{\pi/n}^{\pi-\pi/n}\varphi\Big(x,t+\frac{\pi}{n}\Big)\Big[\omega(-\beta,t)-\omega\Big(-\beta,t+\frac{\pi}{n}\Big)\Big]\cos nt\,dt - \\
&-\int_0^{\pi/n}\varphi\Big(x,t+\frac{\pi}{n}\Big)\omega\Big(-\beta,t+\frac{\pi}{n}\Big)\cos nt\,dt + \\
&+\int_{\pi-\pi/n}^{\pi}\varphi(x,t)\omega(-\beta,t)\cos nt\,dt\Big\} \equiv \\
&\equiv \sum_{k=1}^{4} B_k(x,f,\beta,n), \quad n>1.
\end{aligned} \tag{4.1.43}$$

Using Hölder's inequality, according to the condition of the theorem we find

$$\begin{aligned}
&\|B_1(\cdot,f,\beta,n)\|_C \le \\
&\le A(\alpha)\Big\{\int_T\Big|\varphi(x,t)-\varphi\Big(x,t+\frac{\pi}{n}\Big)\Big|^p dt\Big\}^{1/p}\Big\{\int_{\pi/n}^{\pi} t^{-(1-\beta)q}dt\Big\}^{1/q},
\end{aligned}$$

whence

$$\|B_1(\cdot,f,\beta,n)\|_C \le$$

$$\le \begin{cases} A(f,p,\beta)n^{1/p-\alpha}, & \beta\in\Big]0,\dfrac{1}{p}\Big[, \\ A(f,p,\beta)n^{1/p-\alpha}[\log(n+2)]^{1-1/p}, & \beta=\dfrac{1}{p}, \\ A(f,p,\beta)n^{\beta-\alpha}, & \beta\in\Big]\dfrac{1}{p},\alpha\Big[. \end{cases} \tag{4.1.44}$$

On the other hand, by (2.2.22) we have

$$\|B_2(\cdot,f,\beta,n)\|_C \le A(\beta)n^{\beta-1}\int\limits_{1/n}^{\pi} t^{\beta-2}\omega^{(2)}(t,f)_C\,dt. \tag{4.1.45}$$

It is not difficult to see that

$$\|B_3(\cdot,f,\beta,n)\|_C \le A(\beta)\omega^{(2)}\Big(\frac{1}{n},f\Big)_C. \tag{4.1.46}$$

We can assume that f differs from a constant (otherwise the theorem is obvious), so

$$n\omega\Big(\frac{1}{n},f\Big)_C \ge A.$$

Hence

$$\|B_4(\cdot,f,\beta,n)\|_C \le A(\beta)\omega\Big(\frac{1}{n},f\Big)_C. \tag{4.1.47}$$

Therefore, using (4.1.39)–(4.1.47), we get convinced that the conclusion of the theorem is true. ■

We should note that for $p=+\infty$, Theorem 4.1.3 coincides with an assertion of Lebesgue [4].

Let $f\in C(T)$ and $\alpha\in]0,1]$. For any $\beta\in[-1,\alpha[$ (we could assume that $\beta\in]-\infty,\alpha[$ but, as we have already seen not once, this demand is not essential), we put

$$\nu(f,\alpha,\beta,n)=\Big[\omega\Big(\frac{1}{n},f\Big)_C\Big]^{\alpha-\beta}$$

if $\alpha\in]0,1]$, $\beta\in]-1,0[$, $\beta\in]0,\alpha[$, and $\alpha-\beta\le 1$;

$$\nu(f,\alpha,\beta,n)=\omega\Big(\frac{1}{n},f\Big)_C$$

if $\alpha\in]0,1]$, $\beta\in]-1,0[$, and $\alpha-\beta\ge 1$;

$$\nu(f,\alpha,\beta,n)=\Big[\omega\Big(\frac{1}{n},f\Big)_C\Big]^{\alpha}$$

if $\beta=0$, $\alpha\in]0,1[$. Furthermore, we assume

$$\lambda_n^{(1)}\equiv\lambda_n^{(1)}(f,\alpha,\beta,n)=$$

$$=\nu(f,\alpha,\beta,n)+n^{-1}\int\limits_{1/n}^{\pi} t^{-2}\omega(t,f)_C\,dt \quad \text{for } n>1. \tag{4.1.48}$$

Theorem 4.1.4. *Let $\alpha\in]0,1]$ and $\beta\in[-1,\alpha[$. If $f\in C(T)\cap Lip(\alpha,\frac{1}{\alpha})$, then*

$$\|\sigma_n^{-\beta}(f)-f\|_C \le A(f,\alpha,\beta)\lambda_n^{(1)}.$$

Proof. We will not give here a detailed proof since we should like to avoid the repetition of our above arguments. It is assumed that f is nonconstant. If $\beta=-1$, then the theorem is obvious. Let therefore $\beta\in]-1,\alpha[$ and $\beta\neq 0$. Using (4.1.39) and analyzing the proof of Theorem 4.1.3, we get convinced that the essential point is to obtain the needed estimates for $B_1(x,f,\beta,n)$ and $B_2(x,f,\beta,n)$ in (4.1.43). Denoting

$$b(n)\equiv b(n,f)=\left[\omega\Big(\frac{1}{n},f\Big)_C\right]^{-1}\quad\text{for}\quad n>1,$$

we have

$$\|B_1(\cdot,f,\beta,n)\|_C=$$

$$=n^\beta\Bigg\{\int\limits_{\pi/n}^{b(n)/n}\Big[\varphi(x,t)-\varphi\Big(x,t+\frac{\pi}{n}\Big)\Big]\omega(-\beta,t)\cos nt\,dt+$$

$$+\int\limits_{b(n)/n}^{\pi-\pi/n}\Big[\varphi(x,t)-\varphi\Big(x,t+\frac{\pi}{n}\Big)\Big]\omega(-\beta,t)\cos nt\,dt\equiv$$

$$\equiv L_1^{(0)}(x,f,\beta,n)+L_2^{(0)}(x,f,\beta,n).$$

Plainly

$$\|L_1^{(0)}(\cdot,f,\beta,n)\|_C\le$$

$$\le A(\beta)\omega\Big(\frac{1}{n},f\Big)_C n^\beta\int\limits_{1/n}^{b(n)/n}t^{\beta-1}dt\le A(\beta)\left[\omega\Big(\frac{1}{n},f\Big)_C\right]^{1-\beta}.$$

Furthermore, if $\alpha=1$, then

$$\|L_2^{(0)}(\cdot,f,\beta,n)\|_C\le$$

$$\le A(\beta)n\left[\omega\Big(\frac{1}{n},f\Big)_C\right]^{1-\beta}\Big\|\int\limits_T\Big|\varphi(\cdot,t)-\varphi\Big(\cdot,t+\frac{\pi}{n}\Big)\Big|\,dt\Big\|_C\le$$

$$\le A(f,\beta)\left[\omega\Big(\frac{1}{n},f\Big)_C\right]^{1-\beta},$$

and if $\alpha\in]0,1[$, then using Hölder's inequality ($p^{-1}=\alpha$, $q^{-1}=1-\alpha$), we get

$$\|L_2^{(0)}(\cdot,f,\beta,n)\|_C\le$$

$$\le A(\beta)n^\beta\Bigg\|\Bigg\{\int\limits_T\Big|\varphi(\cdot,t)-\varphi\Big(\cdot,t+\frac{\pi}{n}\Big)\Big|^{\frac{1}{\alpha}}dt\Bigg\}^\alpha\Bigg\|_C\Bigg\{\int\limits_{b(n)/n}^{\pi}t^{-\frac{1-\beta}{1-\alpha}}dt\Bigg\}^{1-\alpha}\le$$

$$\le A(f,\alpha,\beta)\left[\omega\Big(\frac{1}{n},f\Big)_C\right]^{\alpha-\beta}.$$

Thus

$$\|B_1(\cdot,f,\beta,n)\|_C\le A(f,\alpha,\beta)\nu(f,\alpha,\beta,n).$$

Using (2.2.22), we find

$$\|B_2(\cdot,f,\beta,n)\|_C \le$$
$$\le A(\beta)n^{\beta-1}\left[\int\limits_{1/n}^{b(n)/n} t^{\beta-2}\omega(t,f)_C\,dt + \int\limits_{b(n)/n}^{\pi} t^{\beta-2}\omega(t,f)_C\,dt\right] \le$$
$$\le A(f,\beta)\left[\omega\left(\frac{1}{n},f\right)_C\right]^{1-\beta}.$$

Therefore

$$\|B_2(\cdot,f,\beta,n)\|_C \le A(f,\alpha,\beta)\nu(f,\alpha,\beta,n).$$

Thus we can consider Theorem 4.1.4 as proved since the case $\beta=0$ and $\alpha\in]0,1[$ can be verified easier. ■

We should add that Theorem 4.1.4 remains also true in the case $\beta=0$ and $\alpha=1$. This result is due to Kominar [1]. To be more precise, in this case the conditions of Theorem 4.1.4 imply

$$\|S_n(f)-f\|_C \le A(f)\omega\left(\frac{1}{n},f\right)_C \log\left[\omega\left(\frac{1}{n},f\right)_C\right]^{-1}.$$

In connection with this estimate of Kominar, the paper of Stechkin [1] and a note of Telyakovskiĭ should also be mentioned (RZh Mat, 1962, 7Б89).

Remark 4.1.4. Let $f\in L^p(T)$ for some $p\in[1,+\infty]$. Put

$$a(n)\equiv a(n,f,p)=\left[\omega\left(\frac{1}{n},f\right)_C\right]^{-1}.$$

The expression

$$n^{-1}\int\limits_{1/n}^{\pi} t^{-2}\omega(t,f)_p dt$$

which appears in the above assertions, can be estimated as:

1) $$n^{-1}\int\limits_{1/n}^{\pi} t^{-2}\omega(t,f)_p dt = n^{-1}\int\limits_{1/n}^{a(n)/n} t^{-2}\omega(t,f)_p dt +$$
$$+n^{-1}\int\limits_{a(n)/n}^{\pi} t^{-2}\omega(t,f)_p dt \le A\omega\left(\frac{1}{n},f\right)_p \log\left[\omega\left(\frac{1}{n},f\right)_p\right]^{-1} \quad \text{for } n>1;$$

2) $$n^{-1}\int\limits_{1/n}^{\pi} t^{-2}\omega(t,f)_p dt \le A\omega\left[\frac{\log(n+1)}{n},f\right]_p \quad \text{for } n\in\mathbb{N}.$$

Various corollaries can be deduced from Theorem 4.1.4. We formulate one of them

Corollary 4.1.2. *Let $\beta \in]0,1[$. If $f \in C(T)$ has a bounded variation, then*

$$\|\sigma_n^{-\beta} - f\|_C \leq A(f,\beta)\left[\omega\left(\frac{1}{n}, f\right)_C\right]^{1-\beta}. \tag{4.1.49}$$

Finally, we make some remarks.

In 1977, Storozhenko [1], [2] established some approximating properties of the means $\sigma_n^\alpha(x,f)$, $\alpha > -1$, $\alpha \neq 0$, in $L^p(T)$, $p \in]0,1[$. She pointed out that for $p \in]1,+\infty[$ and $\alpha \in]-1,0[$, the estimate (4.1.27) can be obtained.

In 1982, Akhobadze [3] established an estimate of the type (4.1.49) in terms of the modulus of continuity of $f \in C(T)$.

In 1983, Lekishvili [1] established a pointwise asymptotic equality (of the Zamansky type, see (4.1.6)) for $\sigma_n^\alpha(x,f) - f(x)$ in the case where $f \in L^p(T)$, $p \in [1,+\infty]$ and either $\alpha \in]-1,0[$ or $\alpha \in]0,+\infty[$. In the case $\alpha = 1$, a similar result was obtained by Lebed' and Avdeenko [1].

In the works of many authors mentioned in this chapter, the approximating properties in $L^p(T)$, $p \in]0,+\infty]$, of other linear means are also investigated, among them of Poisson–Abel means. Important results in this connection can be found in the monograph of Stepanets [1, Ch. II, IV, VI].

4.2. Approximating Properties of Cesàro Means of the Series $\overline{\sigma}[\overline{f}]$

By virtue of results of Bernstein [1] and Privalov [1], one can conclude that if $f \in Lip\,\alpha$ with $\alpha \in]0,1[$, then

$$\|t_n^1(f) - \overline{f}\|_C \leq A(f)n^{-\alpha} \quad \text{for} \quad n \in \mathbb{N}. \tag{4.2.1}$$

On the other hand, using Privalov's theorem and appropriate results of Ul'yanov [3], we make sure that if $f \in Lip\,\alpha$ with $\alpha \in]0,1[$ and $\beta \in]0,+\infty[$, then

$$\|t_n^\beta(f) - \overline{f}\|_C \leq A(f,\beta)n^{-\alpha} \quad \text{for} \quad n \in \mathbb{N}. \tag{4.2.2}$$

This way, however, it is impossible to claim that (4.2.1) and (4.2.2) hold also in the case $f \in Lip\,1$.

In 1941, Alexits [1] proved that if $f \in Lip\,1$, then

$$\|t_n^1(f) - \overline{f}\|_C \leq A(f)n^{-1} \quad \text{for} \quad n \in \mathbb{N}. \tag{4.2.3}$$

Thus Alexits found out that although in some cases the function $\overline{f}$ does not have the properties of f (summability, continuity, and others), in the sense of approximability properties it behaves "better". The point is that, as already noted, Bernstein's results [1] imply that in the class $Lip\,1$, for the quantity

$$\|\sigma_n^1(f) - f\|_C$$

the order $n^{-1}\log(n+1)$ can be attained.

In 1945, Zygmund [8], using another method, proved (4.2.3) anew, and Flett [1] improved this inequality.

In 1959, Efimov [2] (see also his work [3]) established that if $f \in C(T)$ and for any $k = 2, 3, \dots$

$$\left\| \sum_{i=0}^{k} (-1)^{k-1} \binom{k}{i} f\left[\cdot + (k-2i)\frac{h}{2} \right] \right\|_C \le |h| \quad \text{for} \quad |h| \in]0, \pi],$$

then

$$\overline{f}(x) - t_{n-1}^1(x, f) = f\left(x + \frac{1}{2n}\right) - f\left(x - \frac{1}{2n}\right) + B(x, n),$$

where

$$n\|B(\cdot, n)\|_C \le A(f).$$

As already noted, in 1940 Nikol'skiĭ [1] began an extensive investigation of the asymptotic behavior of the expressions

$$\sup_{f \in M} \|\sigma_n^\beta(f) - f\|_C \tag{4.2.4}$$

for various subclasses M of $C(T)$.

In 1946, he [2] studied analogous problems for $L(T)$. In these works, Nikol'skiĭ considered, among others, the cases $\beta = 0$ and $\beta = 1$.

Let $r \in \mathbb{N}$ and ω be an arbitrary modulus of continuity. For any $p \in [1, +\infty]$, the symbol $W^r H_p^\omega(T)$ is used to denote the class of all the functions $f \in C(T)$ having the $(r-1)$-th order absolutely continuous derivative and satisfying

$$f^{(r)} \in H_p^\omega(T).$$

Moreover, we define

$$W^0 H_p^\omega(T) \equiv H_p^\omega(T), \quad W^r H_p^\omega(T) \equiv W^r H_p^\alpha \text{ with } \omega(\delta) = \delta^\alpha, \quad \alpha \in]0, 1].$$

Let now either $p = +\infty$ or $p = 1$. The class $\widetilde{W}^r H_p^\omega(T)$ is defined as the set of all conjugates to functions from $W^r H_p^\omega(T)$, where ω is an arbitrary modulus of continuity if $r \in \mathbb{N}$, and ω satisfies Dini's condition if $r = 0$ (see also (4.1.33)), i.e.,

$$\int_0^\pi t^{-1}\omega(t)\, dt < +\infty.$$

In 1945, Nikol'skiĭ [2] found precise orders of decrease for the quantities

$$\sup_{f \in \widetilde{W}^r H_\infty^\alpha(T)} \|t_n^1(f) - \overline{f}\|_C \quad \text{with} \quad r \in \mathbb{N}, \quad \alpha \in]0, 1].$$

In 1946, Nagy [2] (see also his work [1]) established an asymptotic equality for

$$\sup_{f \in \widetilde{W}^r H_\infty^1(T)} \|t_n^1(f) - \overline{f}\|_C \quad \text{with} \quad r \in \mathbb{N}.$$

In 1961, Stechkin [5] obtained a precise in the sense of order estimate for

$$\sup_{f \in H_\infty^\omega(T)} \|t_n^1(f) - \overline{f}\|_C.$$

In 1962, Taberski [2] proved an asymptotic equality for

$$\sup_{f\in Lip\,\alpha} \|t_n^{-\beta}(f)-\overline{f}\|_C \quad \text{with} \quad \beta=-3,-4,\ldots, \quad \alpha\in]0,1].$$

In 1979, Zadereĭ and Stepanets [1] got (see also Stepanets [1, Ch. IV]) an asymptotic equality for

$$\sup_{f\in \widetilde{W}^r H_\infty^\omega(T)} \|t_n^1(f)-\overline{f}\|_C \quad \text{with} \quad r\in\mathbb{N}.$$

In 1980, Falaleev [1] established an asymptotic equality for

$$\sup_{f\in Lip\,\alpha} \|t_n^{-\beta}(f)-\overline{f}\|_C$$

with $\beta\in]-\infty,-3]$ and $\alpha\in]0,1]$. The analogous question is investigated by the same author in [2] in the case $\beta\in]-3,-2]$.

In connection with the above works of Nikol'skiĭ [1], [2], we should refer to Telyakovskiĭ [2], and Baskakov and Telyakovskiĭ [1].

In the works [3], [4], [5], [13, Ch. II], and [30], we proved assertions concerning approximating properties of the means

$$t_n^\alpha(x,f) \quad \text{with} \quad \alpha\in]-1,+\infty[$$

in $L^p(T)$, where $p\in[1,+\infty]$. Some of them are presented below.

Theorem 4.2.1. *Let $f\in L^p(T)$ for some $p\in[1,+\infty]$. If $\alpha\in]-\infty,0[\cup]0,1[$, then*

$$\|t_n^{-\alpha}(f)-\overline{f}_n\|_p \le A(p,\alpha)(n^\alpha+1)\omega\Big(\frac{1}{n},f\Big)_p \qquad \text{for} \quad n\in\mathbb{N}. \tag{4.2.5}$$

Proof. As we have seen not once above, it suffices to consider the case $|\alpha|\in]0,1[$. By (2.2.2), (2.2.6), (2.2.8), and (2.2.10), we have

$$\begin{aligned} t_n^{-\alpha}(x,f)-\overline{f}_n(x) &= -\frac{1}{\pi}\int_0^{\pi/n}\psi(x,t)\tau_n^\alpha(t)\,dt - \frac{1}{\pi}\int_{\pi/n}^{\pi}\psi(x,t)\psi_n^{-\alpha}(t)\,dt - \\ &\quad -\frac{1}{\pi}\int_{\pi/n}^{\eta}\psi(x,t)\gamma_n^{-\alpha}(t)\,dt \equiv \sum_{i=1}^{3} q_i(x,f,\alpha,n). \end{aligned} \tag{4.2.6}$$

In view of (2.2.14), we get

$$\|q_1(\cdot,f,\alpha,n)\|_p \le A(p,\alpha)\omega\Big(\frac{1}{n},f\Big)_p. \tag{4.2.7}$$

Using (2.2.16), we obtain

$$\|q_3(\cdot,f,\alpha,n)\|_p \le A(p,\alpha)n^{-2}\int_{1/n}^{\pi} t^{-3}\omega(t,f)_p\,dt \le$$
$$\le A(p,\alpha)\omega\Big(\frac{1}{n},f\Big)_p n^{-2}\int_{1/n}^{\pi} t^{-3}(nt+1)\,dt \le$$
$$\le A(p,\alpha)\omega\Big(\frac{1}{n},f\Big)_p. \tag{4.2.8}$$

To evaluate $q_2(x,f,\alpha,n)$, it suffices to find an estimate for

$$P(x,f,\alpha,n) \equiv n^{\alpha}\int_{\pi/n}^{\pi} \psi(x,t)\omega(-\alpha,t)\cos nt\,dt,$$

where $\omega(-\alpha,t)$ is defined by (2.2.21). But $P(x,f,\alpha,n)$ can be represented as

$$2P(x,f,\alpha,n) =$$
$$= n^{\alpha}\Bigg\{\int_{\pi/n}^{\pi-\pi/n}\Big[\psi(x,t)-\psi\Big(x,t+\frac{\pi}{n}\Big)\Big]\omega(-\alpha,t)\cos nt\,dt +$$
$$+\int_{\pi/n}^{\pi-\pi/n}\psi\Big(x,t+\frac{\pi}{n}\Big)\Big[\omega(-\alpha,t)-\omega\Big(-\alpha,t+\frac{\pi}{n}\Big)\Big]\cos nt\,dt +$$
$$+\int_{\pi-\pi/n}^{\pi}\psi(x,t)\omega(-\alpha,t)\cos nt\,dt -$$
$$-\int_{0}^{\pi/n}\psi\Big(x,t+\frac{\pi}{n}\Big)\omega\Big(-\alpha,t+\frac{\pi}{n}\Big)\cos nt\,dt \equiv$$
$$\equiv \sum_{k=1}^{4} U_k(x,f,\alpha,n). \tag{4.2.9}$$

We have

$$\|U_1(\cdot,f,\alpha,n) \le A(p,\alpha)(n^{\alpha}+1)\omega\Big(\frac{1}{n},f\Big)_p. \tag{4.2.10}$$

Moreover, using (2.2.22), we get

$$\|U_2(\cdot,f,\alpha,n)\|_p \le A(p,\alpha)n^{\alpha-1}\int_{1/n}^{\pi} t^{\alpha-2}\omega(t,f)_p\,dt \le$$
$$\le A(p,\alpha)\omega\Big(\frac{1}{n},f\Big)_p. \tag{4.2.11}$$

On the other hand,

$$\begin{aligned}\|U_3(\cdot,f,\alpha,n)\|_p &\le A(p,\alpha)(n^\alpha+1)\int_{\pi-\pi/n}^{\pi}\omega(t,f)_p\,dt \le \\ &\le A(p,\alpha)(n^\alpha+1)\omega\Big(\frac{1}{n},f\Big)_p\int_{\pi-\pi/n}^{\pi}(nt+1)\,dt \le \\ &\le A(p,\alpha)(n^\alpha+1)\omega\Big(\frac{1}{n},f\Big)_p . \end{aligned} \tag{4.2.12}$$

It is easy to see that

$$\|U_4(\cdot,f,\alpha,n)\|_p \le A(p,\alpha)\omega\Big(\frac{1}{n},f\Big)_p . \tag{4.2.13}$$

Thus, based on (4.2.6)–(4.2.13), we get (4.2.5). ■

A number of corollaries can be obtained from Theorem 4.2.1. We formulate some of them.

Corollary 4.2.1. *Let $f \in L^p(T)$ for some $p \in [1,+\infty]$.*

a) *If $\alpha \in]0,1[$, then*

$$\|t_n^{-\alpha}(f) - \overline{f}_n\|_p \le A(p,\alpha)n^\alpha\omega\Big(\frac{1}{n},f\Big)_p .$$

b) *If $\alpha \in]-\infty,0[$, then*

$$\|t_n^{-\alpha}(f) - \overline{f}_n\|_p \le A(p,\alpha)\omega\Big(\frac{1}{n},f\Big)_p .$$

Corollary 4.2.2. *Let $f \in L^p(T)$ for some $p \in [1,+\infty]$ and*

$$\omega(\delta,f)_p \le A(f)\delta \quad \textit{for} \quad \delta \in]0,\pi].$$

a) *If $\alpha \in]0,1[$, then*

$$\|t_n^{-\alpha}(f) - \overline{f}\|_p \le A(f,p,\alpha)n^{\alpha-1};$$

b) *If $\alpha \in]-\infty,0[$, then*

$$\|t_n^{-\alpha}(f) - \overline{f}\|_p \le A(f,p,\alpha)n^{-1}.$$

Corollary 4.2.3. *Let $f \in L^p(T)$ for some $p \in]1,+\infty[$. Then*

$$\Big\|\int_0^{\pi/n}\psi(\cdot,t)\operatorname{ctg}\frac{t}{2}\,dt\Big\|_p \le A(p,\alpha)\omega\Big(\frac{1}{n},f\Big)_p \quad \textit{for } n>1.$$

Indeed, as Ul'yanov [3] has shown,

$$\|t_n^1(f) - \overline{f}\|_p \le A(p)\omega\Big(\frac{1}{n}, f\Big)_p.$$

Then, according to Theorem 4.2.1, we get

$$\Big\| \int_0^{\pi/n} \psi(\cdot, t)\operatorname{ctg}\frac{t}{2}\,dt \Big\|_p \le \|t_n^1(f) - \overline{f}_n\|_p + \|t_n^1(f) - \overline{f}\|_p \le A(p)\omega\Big(\frac{1}{n}, f\Big).$$

Thus it holds

Corollary 4.2.4. *Let $f \in L^p(T)$ for some $p \in]1, +\infty[$ and either $\alpha \in]0,1[$ or $\alpha \in]-\infty, 0[$. Then*

$$\|t_n^{-\alpha}(f) - f\|_p \le A(p,\alpha)(n^\alpha + 1)\omega\Big(\frac{1}{n}, f\Big)_p. \tag{4.2.14}$$

On the basis of this statement, one gets

Corollary 4.2.5. *Let $f \in L^p(T)$ for some $p \in]1, +\infty[$, and either $\alpha \in]0,1[$ or $\alpha \in]-\infty, 0[$. Then*

$$\|\sigma_n^{-\alpha}(f) - f\|_p \le A(p,\alpha)(n^\alpha + 1)\omega\Big(\frac{1}{n}, f\Big)_p. \tag{4.2.15}$$

If $f \in L^p(T)$ with either $p = 1$ or $p = +\infty$, then Theorem 4.2.1 can be used to estimate the expressions

$$\|\sigma_n^{-\alpha}(f) - f\|_p$$

from above by

$$(n^\alpha + 1)\omega\Big(\frac{1}{n}, f\Big)_p$$

in the case where $\sigma[f]$ has the type of power series (for $\alpha \in]-\infty, 0[$ see Zygmund [8]).

Corollary 4.2.6. *Let a modulus of continuity ω satisfy Dini's condition (see (4.1.33)). If $f \in H_p^\omega(T)$ with either $p = 1$ or $p = +\infty$, and $\alpha \in]-\infty, 0[$, then*

$$\|t_n^{-\alpha}(f) - \overline{f}\|_p \le A(\alpha) \int_0^{1/n} t^{-1}\omega(t)\,dt. \tag{4.2.16}$$

Indeed, by the condition, $\overline{f} \in L^p(T)$. The conclusion of Theorem 4.2.1 then implies

$$\|t_n^{-\alpha}(f) - \overline{f}\|_p \le A(\alpha)\omega\Big(\frac{1}{n}\Big) + A\int_0^{\pi/n} t^{-1}\omega(t)\,dt \le A(\alpha)\int_0^{1/n} t^{-1}\omega(t)\,dt.$$

For $p = +\infty$, (4.2.16) was established by Totik [1]. Moreover, he also proved the exactness in the class $H_\infty^\omega(T)$ of the order of the estimate (4.2.16) .

Now consider the problem of definitiveness in $H_1^\omega(T)$ of the order in the estimate (4.2.16).

Let ω satisfy (4.1.31). Put

$$f(x)=\frac{1}{2\log^2 2}+\sum_{k=1}^{\infty}\frac{\cos kx}{(k+1)\log^2(k+2)} \quad \text{for} \quad x\in T. \tag{4.2.17}$$

The functions f and $\overline{f}$ are obviously continuous. According to Telyakovskiĭ's inequality (3.2.16), we find

$$\omega\Big(\frac{1}{n},f\Big)_1\le A\frac{1}{n}\sum_{k=1}^{\infty}\frac{1}{(k+1)\log^2(k+1)}.$$

Hence

$$\omega\Big(\frac{1}{n},f\Big)_1\le An^{-1},$$

i.e., $f\in H_1^{\omega}(T)$. The last conclusion also follows from a related result by Hardy and Littlewood [5].

In 1942, Hille [1] proved that if $f\in C(T)$ and

$$n\|\sigma_n^1(f)-f\|_C\to 0 \quad \text{as} \quad n\to\infty,$$

then $f=const$. This remains true for the means $\sigma_n^{-\alpha}(x,f)$ with $\alpha\in]-\infty,0[$. It can be verified that if $f\in L(T)$ and for some $\alpha\in]-\infty,0[$ the relation

$$n\|\sigma_n^{-\alpha}(f)-f\|_1\to 0 \quad \text{as} \quad n\to\infty$$

is fulfilled, then $f=const$ almost everywhere.

Thus, for the function f defined by (4.2.17), we have

$$\limsup_{n\to\infty} n\|t_n^{-\alpha}(f)-\overline{f}\|_1>0.$$

But then also

$$\limsup_{n\to\infty}\Big(\int_0^{1/n}t^{-1}\omega(t)\,dt\Big)^{-1}\|t_n^{-\alpha}(f)-\overline{f}\|_1>0$$

since

$$\int_0^{1/n}t^{-1}\omega(t)\,dt\le An^{-1}.$$

Suppose now that ω satisfies (4.1.29) and assume

$$f(x)=\frac{1}{2}\omega(1)+\sum_{k=1}^{\infty}\omega\Big(\frac{1}{k+1}\Big)\cos kx \quad \text{for} \quad x\in T.$$

As above, using (3.2.16) we get

$$\omega\Big(\frac{1}{n},f\Big)_1\le An^{-1}\sum_{k=0}^{n}\omega\Big(\frac{1}{k+1}\Big). \tag{4.2.18}$$

On the other hand,

$$\|t_n^{-\alpha}(f)-\overline{f}\|_1\ge\int_0^{\pi}\big[t_n^{-\alpha}x,(f)-\overline{f}(x)\big]\,dx=$$

$$= (A_n^{-\alpha})^{-1} \sum_{k=0}^{n} A_{n-k}^{-\alpha-1} \int_0^{\pi} \Big[\sum_{i=k+1}^{\infty} i^{-1} \omega\Big(\frac{1}{i+1}\Big) \sin x \Big] dx =$$

$$= (A_n^{-\alpha})^{-1} \sum_{k=0}^{n} A_{n-k}^{-\alpha-1} \sum_{i=k+1}^{\infty} i^{-1} \omega\Big(\frac{1}{i+1}\Big) [1 - (-1)^i] \geq$$

$$\geq (A_n^{-\alpha})^{-1} \sum_{k=0}^{n} A_{n-k}^{-\alpha-1} \sum_{i=k+1}^{\infty} (2i+1)^{-1} \omega\Big(\frac{1}{i+1}\Big) \geq$$

$$\geq A(\alpha) n^{-1} \sum_{k=0}^{[\frac{n}{2}]} \sum_{i=k+1}^{\infty} (2i+1)^{-1} \omega\Big(\frac{1}{i+1}\Big) \geq$$

$$\geq A(\alpha) \sum_{i=n+1}^{\infty} (i+1)^{-1} \omega\Big(\frac{1}{i+1}\Big) \geq A(\alpha) \int_0^{1/n} t^{-1} \omega(t)\, dt$$

since

$$\int_0^{1/n} t^{-1} \omega(t)\, dt > \omega\Big(\frac{1}{n}\Big).$$

Analyzing the schemes of the proofs of Theorems 4.1.4 and 4.2.1, we conclude that it holds

Theorem 4.2.2. *Let* $\alpha \in]0,1[$ *and* $\beta \in [-1, \alpha[$. *If* $f \in C(T) \cap Lip(\alpha, \frac{1}{\alpha})$ *and* $\lambda_n^{(1)}$ *is defined by* (4.1.48), *then*

$$\|t_n^{-\beta}(f) - \overline{f}_n\|_C \leq A(f, \alpha, \beta) \lambda_n^{(1)}.$$

In Theorems 4.1.4 and 4.2.2, the restriction $\beta \in]-1, \alpha[$ is unessential since it can be replaced by $\beta \in]-\infty, \alpha[$.

4.3. Unsolved Problems

First of all, we note that we have already mentioned some open problems within the text of this chapter. As we have pointed out, they are not completely solved.

4.3.1. Let ω be an arbitrary modulus of continuity and either $p = +\infty$ or $p = 1$. Find asymptotic equalities for

$$\sup_{f \in W^r H_p^{\omega}(T)} \|\sigma_n^{-\alpha}(f) - f\|_p \quad \text{with} \quad r \in \mathbb{N}_0,$$

where either $\alpha \in]-\infty, 0[$ or $\alpha \in]0,1[$.

4.3.2. Let ω be an arbitrary modulus of continuity if $r \in \mathbb{N}$, and let ω satisfy the Dini condition if $r = 0$. Let, moreover, either $p = +\infty$ or $p = 1$. Find asymptotic equalities for

$$\sup_{f \in \widetilde{W}^r H_p^{\omega}(T)} \|t_n^{-\alpha}(f) - \overline{f}\|_p \quad \text{with} \quad r \in \mathbb{N}_0,$$

where either $\alpha \in [-\infty, 0[$ or $\alpha \in]0,1[$.

We should add that similar problems make also sense in the case $p \in]1, +\infty[$.

Part 2

MULTIPLE TRIGONOMETRIC SERIES

CHAPTER I

Conjugate Functions and Hilbert Transforms of Functions of Several Variables

1.1. Some Definitions and Auxiliary Statements

The theory of multiple trigonometric and orthogonal series is rapidly developping today. The results are set down in a number of monographs and surveys. We mention only a part of them: Alimov, Il'in, Nikishin [1], [2], Alimov, Ashurov and Pulatov [1], Golubov [1], D'yachenko [3], Dyn'kin [1], [2], Dyn'kin and Osilenker [1], Zhizhiashvili [16], [35], [46], Zygmund [16], Kislyakov [2], Stein [3], Stein and Weiss [1], Suetin [1], Topuria [1], Torchinsky [1], Khavin [2], Shapiro [1], Yanushauskas [1].

In [35], we presented results which deal with multiple trigonometric series and were established by various authors from 1903 to 1981. In [46], we added some other assertions, among them some new ones obtained since 1981.

In the second part of this book, we will mention only those results by other authors which are necessary for setting forth the studies of the author. We will make use of some notation from our works [16] and [35]. This notation, along with some other definitions, is given below.

As usual, $\mathbb{R}^n$ $(n \geq 1)$ denotes the n-dimentional euclidean space with standard linear operations. Points of $\mathbb{R}^n$ will be denoted by $x = (x_1, x_2, \dots, x_n)$, $y = (y_1, y_2, \dots, y_n)$, $s = (s_1, s_2, \dots, s_n), \dots$ (for $n = 1$, the letters x, y, s will denote points in $\mathbb{R}^1$, i.e., real numbers). In particular, $0 = (0, 0, \dots, 0)$, $1 = (1, 1, \dots, 1)$, $r = (r_1, r_2, \dots, r_n)$, $r_i \in]0, 1[$, $p = (p_1, p_2, \dots, p_n)$, $p_i \in \mathbb{Z}$, $m = (m_1, m_2, \dots, m_n)$, $m_i \in \mathbb{N}$, $1 \leq i \leq n$. Let $M \equiv \{1, 2, \dots, n\}$ $(n \in \mathbb{N}, n \geq 2)$ and B be an arbitrary subset of M. We will assume that $B' = M \setminus B$, and $k \equiv \overline{k}(B)$ is the number of elements of B (see, e.g., Halmos [1, Ch. I]). Plainly $\overline{k}(B) + \overline{k}(B') = n$. If $B \neq \varnothing$, then we assume that $B = \{i_1^B, i_2^B, \dots, i_k^B\} \equiv \{i_1, i_2, \dots, i_k\}$, where $i_1^B < i_2^B < \dots < i_k^B$ and $k = \overline{k}(B)$. For any $x \in \mathbb{R}^n$ and $B \subset M$, the symbol $x(B)$ denotes $(x_{i_1}, x_{i_2}, \dots, x_{i_k})$ and the symbol x_B will stand for the point of $\mathbb{R}^n$ whose coordinates with indices from B coincide with the corresponding coordinates of x, and those with indices from B' are zero. Besides, $x_M = x$, $x_\varnothing = 0$. For any $x \in \mathbb{R}^n$, by $-x_B \equiv x^1$ is denoted the point in $\mathbb{R}^n$ with $x_i^1 = -x_i$ for $i \in B$, and $x_i^1 = x_i$ for $i \in B'$. The symbol $\mathbb{R}^n(B)$ will mean the hyperplane spanned over the coordinate vectors with indices from B. The following notation is also used.

$$T^n = [-\pi,\pi]^n, \quad T^n(B) = T^n \cap \mathbb{R}^n(B),$$
$$K^n = [0,1]^n, \quad K^n(B) = K^n \cap \mathbb{R}^n(B),$$
$$T^n(\varepsilon) = \mathop{\mathsf{X}}_{i=1}^{n}\Big\{[-\pi,-\varepsilon_i]\cup[\varepsilon_i,\pi]\Big\}, \quad \varepsilon_i \in]0,\pi[, \quad 1\le i\le n,$$
$$T^n(B,\varepsilon) = T^n(\varepsilon)\cap \mathbb{R}^n(B),$$
$$xy = \sum_{k=1}^{n} x_k y_k, \quad \|x\| = \Big\{\sum_{k=1}^{n} x_k^2\Big\}^{1/2},$$
$$dx_B \equiv dx_{i_1}dx_{i_2}\cdots dx_{i_k},$$
$$\frac{1}{x} \equiv \Big(\frac{1}{x_1},\frac{1}{x_2},\dots,\frac{1}{x_n}\Big) \quad \text{for } x\in\mathbb{R}^n \text{ with non-zero coordinates},$$
$$\varepsilon(\mu) \equiv (\mu,\mu,\dots,\mu) = \mu\cdot 1 \quad \text{for} \quad \mu\in\mathbb{R}.$$

Everywhere below it is assumed that $\gamma \in]0,e^{-4n}]$ is a fixed number, and $I^n(\gamma) =]0,\gamma]^n$. If $\gamma_1 \in]0,\gamma]$ and $x_i \in]0,\gamma_1[$, $1\le i\le n$, then

$$E(x,\gamma_1,B) = \mathop{\mathsf{X}}_{i\in B}[0,x_i] \equiv [0,x_{i_1}]\times\cdots\times[0,x_{i_k}],$$
$$E_1(x,\gamma_1,B) = \mathop{\mathsf{X}}_{i\in B}]x_i,\gamma_1] \equiv]x_{i_1},\gamma_1]\times\cdots\times]x_{i_k},\gamma_1]$$

for $B\subset M$, $B\neq\varnothing$.

We will consider functions $f: T^n\to\mathbb{R}$ which are 2π-periodic in each of their arguments. If $h_{\{i\}} = (0,0,\dots,h_i,0,\dots,0)$ and

$$\Delta(f,x,h_{\{i\}}) = f(x+h_{\{i\}}) - f(x),$$
$$\overline{\Delta}(f,x,h_{\{i\}}) = f(x+h_{\{i\}}) - f(x-h_{\{i\}}),$$

then the symbols $\Delta(f,x,h_B)$ and $\overline{\Delta}(f,x,h_B)$ will denote the expressions which can be obtained by consecutive applying of Δ and $\overline{\Delta}$, respectively, to the arguments with indices belonging to B. We will also assume that

$$\Delta(f,x,h) = f(x+h) - f(x),$$
$$\Delta(f,x,h_B) = \Delta(f,x,h_{i_1},\dots,h_{i_k}).$$

The notation $h_B\to 0+$ means that $h_i\to 0+$, $i\in B$. If $m=(m_1,m_2,\dots,m_n)\in\mathbb{N}_0^n$, then $m\to\infty$ means that all $m_i\to\infty$, $1\le i\le n$, independently from each other. As to the notation $m_\lambda\to\infty$, it indicates that all $m_i\to\infty$ and for some $\lambda\in[1,+\infty[$, the relations

$$\lambda^{-1} \le \frac{m_i+1}{m_j+1} \le \lambda, \quad 1\le i,j\le n,$$

are fulfilled. The meaning of $\|m\|\to\infty$ is also clear.

The notation $x\ge y$ means that no coordinate of x is less then the corresponding one of y, and $x>y$ assumes that, in addition, for some index j, we have $x_j>y_j$.

Let $p \in]0, +\infty]$. If $f \in L^p(T^n)$, then

$$\|f\|_p = \left\{(2\pi)^{-n} \int_{T^n} |f(x)|^p \, dx\right\}^{1/p},$$

for $p = +\infty$ assuming, like in the one-dimentional case, that

$$L^\infty(T^n) \equiv C(T^n), \quad \|f\|_C = \sup_{x \in T^n} |f(x)|.$$

In the case $p \in [1, +\infty]$, $(L^p(T^n), \|\cdot\|_p) \equiv L^p(T^n)$ is a Banach space, and if $p \in]0, 1[$, then $L^p(T^n)$ is not normable. Putting

$$\rho(f_1, f_2) = \|f_1 - f_2\|_p^p, \quad p \in]0, 1[, \quad f_i \in L^p(T^n) \ (i = 1, 2),$$

we get the space $(L^p(T^n), \rho)$ which is complete with respect to the invariant metric ρ.

As usual, $\mathbb{C}$ will denote the field of complex numbers. If $n \in \mathbb{N}$ and $n \geq 2$, then $\mathbb{C}^n$ is an n-fold Cartesian product of $\mathbb{C}$, i.e., the points of $\mathbb{C}^n$ are ordered collections of n complex numbers $(z_1, z_2, \ldots, z_n)$, $z_i \in \mathbb{C}$, $1 \leq i \leq n$. In what follows, the points of $\mathbb{C}^n$ will be denoted by $z, v, \ldots$. It should be noted that algebraically we consider $\mathbb{C}^n$ to be an n-dimensional vector space over $\mathbb{C}$, and in the sense of topology, we regard $\mathbb{C}^n$ as the space $\mathbb{R}^{2n}$. If we introduce the scalar product in $\mathbb{C}^n$, i.e., for any z and v from $\mathbb{C}^n$ we put

$$(z, v) = \sum_{k=1}^{n} z_k \bar{v}_k$$

and

$$\|z\| = \sqrt{(z, z)},$$

then $(\mathbb{C}^n, \|\cdot\|) \equiv \mathbb{C}^n$ becomes an n-dimensional Hilbert space.

We denote by U^n (see, e.g., Zygmund [16, Ch. XVII], Rudin [3, Ch. I]) the open subset of $\mathbb{C}^n$ consisting of all $z \in \mathbb{C}^n$ satisfying

$$|z_k| < 1, \quad 1 \leq k \leq n.$$

It is clear that U^n is a Cartesian product of n copies of open unit discs. It is often called the unit n-cylinder or the unit polydisc in $\mathbb{C}^n$. The boundary of U^n is a union of all possible Cartesian products of unit circles $|z_k| = 1$, $1 \leq k \leq n$, and unit discs $|z_j| \leq 1$, $j \neq k$. Its subset defined by

$$|z_k| = 1, \quad 1 \leq k \leq n,$$

coincides with T^n and is often called the skeleton of the polydisc (n-cylinder) U^n. We will often use the notation

$$z = \xi + i\eta, \quad z_k = \xi_k + i\eta_k, \quad 1 \leq k \leq n,$$
$$z = re^{ix} = (r_1 e^{ix_1}, r_2 e^{ix_2}, \ldots, r_n e^{ix_n}).$$

Let now $D_n \subset U^n$ be a domain and let a function $f : D_n \to \mathbb{C}$ be continuous on D_n. The function f is said to be n-harmonic, if it is harmonic in every pair of arguments ξ_j and η_j, $1 \le j \le n$. A function $f : U^n \to \mathbb{C}$ is said to be regular on U^n, if it can be represented on U^n by an absolutely convergent n-fold power series, i.e.,

$$f(z) = \sum_{p \ge 0} c_p \prod_{k=1}^{n} z_k^{p_k} = \sum_{p \ge 0} c_p \prod_{k=1}^{n} r_k^{p_k} e^{ip_k x_k}, \quad z \in U^n.$$

Now suppose that $p \in]0, +\infty]$. The symbol $H^p(U^n) \equiv H^p$ stands for the set of all the functions $f : U^n \to \mathbb{C}$ which are regular on U^n and satisfy

$$\sup_{0 \le r < 1} \int_{T^n} \left| f\left(re^{ix}\right)\right|^p dx < +\infty,$$

for $p = +\infty$ assuming that

$$\sup_{z \in U^n} |f(z)| < +\infty.$$

If for $f \in H^p(T^n)$ we put

$$\|f\|_p = \sup_{0 \le r < 1} \left\{ (2\pi)^{-n} \int_{T^n} \left| f\left(re^{ix}\right)\right|^p dx \right\}^{1/p},$$

then for $p \in [1, +\infty]$, $H^p \equiv (H^p(T^n), \| \cdot \|)$ is a Banach space, and if for $p \in]0, 1[$ and $f_i \in H^p(T^n)$ $(i = 1, 2)$ we put

$$\rho(f_1, f_2) = \|f_1 - f_2\|_p^p,$$

then (H^p, ρ) is a complete metric space with invariant metric.

If $\varphi : \mathbb{R} \to [0, +\infty[$ is a convex function, then we denote by $H^\varphi(T^n)$ the set consisting of all the functions $f : U^n \to \mathbb{C}$ which are regular on U^n and satisfy

$$\sup_{0 \le r < 1} \int_{T^n} \varphi\left(|f(re^{ix})|\right) dx < +\infty.$$

Some results concerning the classes $H^\varphi(T^n)$ can be found in Shvedenko [1].

The theory of functions of several complex variables is intensively developing, being widely applicable in various fields of mathematics, mechanics, and physics. Various monographs present different sections of the modern theory. For example, we refer to Bochner and Martin [1], Fuks [1], [2], Gunning and Rossi [1], Hörmander [1], Hua [1], Rudin [1], [3], Shabat [1], Vladimirov [1], and others. In Khenkin and Chirka [1], results obtained before 1975 and dealing with boundary properties of holomorphic functions of several complex variables are set forth. Some problems from this field are also discussed in Shvedenko [1]. The modern state of various branches of the theory of functions of several complex variables is presented in Aĭzenberg [1], Akhiezer [1], Aleksandrov [1], Berenstein and Struppa [1], Chirka [1], Dektyarëv [1], Dolbeault [1], Khenkin [1], Leiterer [1], Morozov, Perelomov [1], Pinchuk [1], Poletskiĭ and Shabat [1], Ronkin [1], Roslyĭ, Khudaverdyan and Shvarts [1], Sadullaev [1], Tsikh [1], Tumanov [1], Vitushkin [1], Vladimirov and Sergeev [1], Yuzhakov [1], Zaĭdenberg and Lin [1].

In 1939, Bergman and Marcinkiewicz [1] investigated boundary properties of functions $f \in H^p(T^2)$, $p \in [1, +\infty[$. Later, in the works of Bers [1], Bochner [1], Zygmund

[11], [12], Calderón and Zygmund [1], fundamental results were obtained concerning boundary properties of functions $f \in H^p(T^n)$, $p \in]0,+\infty[$. From other results in this direction, we should mention Davis [1], Korányi [1], [2], Stein [3], [4], Stein and Weiss [1], Ch. Fefferman and Stein [1], Helson and Lowdenslager [1].

Let $f \in L^p(T^n)$ for some $p \in [1,+\infty[$. The expressions (see also Nikol'skiĭ [5, Ch. IV], Timan [1, Ch. III])

$$\omega_B(\delta,f)_p = \sup_{h\in T^n:|h_{i_m^B}|\le\delta_m} \|\Delta(f,\cdot,h_B)\|_p,\ \delta\in]0,\pi]^{\overline{k}(B)},\ \overline{k}(B)\in[2,n], \tag{1.1.1}$$

$$\omega_i(\delta,f)_p = \sup_{|h_i|\le\delta} \|\Delta(f,\cdot,h_{\{i\}})\|_p,\ \delta\in]0,\pi],\ 1\le i\le n, \tag{1.1.2}$$

$$\omega(\delta,f)_p = \sup_{|h|\le\delta} \|\Delta(f,\cdot,h)\|_p,\ \delta\in]0,\pi], \tag{1.1.3}$$

are called, respectively, mixed (in those variables whose indices belong to B), partial, and complete L^p-moduli of continuity of f. Mixed, partial and complete moduli of continuity of higher orders can also be considered.

Define $\Omega(m)$, $m \in \mathbb{N}$, as the class of all functions $\omega : [0,1]^m \to [0,+\infty[$ satisfying

1. $\omega(\delta) = 0$ when at least one coordinate of δ is zero;
2. ω is continuous on $[0,1]^m$;
3. ω is nondecresing on $[0,1]^m$ in the sense of Hardy [1], i.e., $\Delta(\omega,\delta,h_B) \ge 0$ for any $h \in [0,1]^m$ and $B \subset \{1,2,\dots,m\}$.
4. ω is semiadditive in each of its variables δ_i, $1 \le i \le m$.

A function from $\Omega(m)$ is called a modulus of continuity of the mixed type.

Let now $\omega_i \in \Omega(1)$, $1 \le i \le n$, and $p \in [1,+\infty[$. Put

$$H(\omega,B,L^p) \equiv H(\omega,B,L^p(T^n)) =$$
$$= \left\{f\in L^p(T^n) : \omega_B(\delta,f)_p \le A(f,p,n)\omega(\delta),\ \ \delta\in]0,\pi]^{\overline{k}(B)}\right\}$$
$$\text{for}\quad \varnothing \ne B\subset M \quad\text{and}\quad \omega\in\Omega(\overline{k}(B));$$
$$H(M,L^p) = \bigcap_{B\subset M} H(\omega,B,L^p);$$
$$H(\omega_i,L^p)\equiv H(\omega_i,L^p(T^n)) = H(\omega_i,\{i\},L^p),\quad 1\le i\le n;$$
$$H(n,L^p)\equiv H(n,L^p(T^n)) = \bigcap_{i=1}^{n} H(\omega_i,L^p);$$
$$H(\omega,L^p) = \left\{f\in L^p(T^n) : \omega_i(\delta,f)_p \le A(f,p,n)\omega(\delta),\ \ \delta\in]0,\pi],\ \ 1\le 1\le n\right\}$$
$$\text{for}\quad \omega\in\Omega(1).$$

Analogous definitions can be introduced for Orlicz spaces.

If

$$\omega(\delta(B)) = \prod_{j=1}^{\overline{k}(B)} \delta_j^{\alpha_j},\quad \alpha_j\in]0,1[^{\overline{k}(B)},$$

then we denote the corresponding class $H(\omega, B, L^p)$ by $Lip(B,p,\alpha)$, the symbol $Lip(B,\alpha)$ being used for $p=+\infty$. As to the class $H(M, L^p)$, we denote it in this case by $Lip(M,p,\alpha)$ (for $p=+\infty$, we mean that $Lip(M,p,\alpha) \equiv Lip(M,\alpha)$). If, however,

$$\omega_i(\delta)=\delta^{\alpha_i}, \quad \delta \in]0,\pi], \quad 1 \leq j \leq n,$$

then the class $H(n, L^p)$ is denoted by $Lip^{(n)}(\alpha,p)$, where for $p=+\infty$, we mean $Lip^{(n)}(\alpha,p) \equiv Lip^{(n)}\alpha$. In the case $\alpha_i=\alpha$, $\alpha \in]0,1]$, $1 \leq i \leq n$, the classes $Lip^{(n)}(\alpha,p)$ and $Lip^{(n)}\alpha$ are denoted by $Lip(\alpha,p)$ and $Lip\,\alpha$, respectively.

Let $\lambda(p)$ denote the number of zero coordinates of a point $p \in \mathbb{N}_0^n$. The series

$$S(n)=\sum_{p\geq 0} 2^{-\lambda(p)} \sum_{B \subset M} a_p^{(B)} \prod_{i\in B} \cos p_i x_i \prod_{j \in B'} \sin p_j x_j \tag{1.1.4}$$

is called the n-fold trigonometric series, where $(a_p^{(B)})_{p\geq 0}$ is an arbitrary n-fold sequence of real numbers.

Suppose now that $f \in L(T^n)$. Then the numbers

$$a_p^{(B)}(f)=\frac{1}{\pi^n}\int_{T^n} f(x) \prod_{i\in B} \cos p_i x_i \prod_{j\in B'} \sin p_j x_j \, dx \tag{1.1.5}$$

are called n-fold Fourier coefficients of f. The series (1.1.4) in which $(a_p^{(B)})_{p\geq 0} \equiv (a_p^{(B)}(f))_{p\geq 0}$ is called n-fold trigonometric Fourier series of the function f. In what follows, it is denoted by $\sigma_n[f]$ (for $n=1$, $\sigma_1[f] \equiv \sigma[f]$).

Remark 1.1.1. It is well-known that if $f \in L(T^n)$, then

$$\lim_{\|p\|\to\infty} a_p^{(B)}(f)=0.$$

Moreover, the same is true for any uniformly bounded orthonormal system in $L^2(T^n)$.

Consider the series (1.1.4). The conjugate n-fold trigonometric series in the variable x_l, $1 \leq l \leq n$, is defined as follows: in each addend of (1.1.4), $\cos p_l x_l$ is replaced by $\sin p_l x_l$ if $l \in B$, otherwise $\sin p_l x_l$ is replaced by $-\operatorname{sign}(p_l)\cos p_l x_l$. The conjugate trigonometric series of (1.1.4) in more than one variables is defined by consecutive application of the above operation to them. In what follows, the n-fold conjugate series to $\sigma_n[f]$ in the variables whose indices form a set B is denoted by $\overline{\sigma}_n[f,B]$ (for $n=1$, $\overline{\sigma}_1[f,B] \equiv \overline{\sigma}[f]$). It is clear that

$$\operatorname{Card}\left\{\overline{\sigma}_n[f,B] \,:\, B \subset M\right\}=2^n-1.$$

If $f \in L(T^n)$, then the expression

$$\overline{f}_B(x)=\text{V.P.}\Big(-\frac{1}{2\pi}\Big)^{\overline{k}(B)} \int_{T^n(B)} f(x+s_B) \prod_{i\in B} \operatorname{ctg}\frac{s_i}{2}\, ds_B \tag{1.1.6}$$

is called the conjugate function of n variables in those variables whose indices form the set B. We denote $\overline{f}_B(x) \equiv \overline{f}_i(x)$ for $B=\{i\}$. The equality (1.1.6) must be interpreted as

$$\overline{f}_B(x)=\lim_{\varepsilon\to 0+} \overline{f}_B(x,\varepsilon), \tag{1.1.7}$$

where

$$\overline{f}_B(x,\varepsilon) \equiv \left(-\frac{1}{2\pi}\right)^{\overline{k}(B)} \int\limits_{T^n(B,\varepsilon)} f(x+s_B)\prod_{i\in B}\operatorname{ctg}\frac{s_i}{2}\,ds_B \tag{1.1.8}$$

and $\varepsilon \to 0+$ means that $\varepsilon_i \to 0+$, $1 \le i \le n$, independently from each other, i.e., in the sense of Stolz–Pringsheim. Note that conjugate functions, for the time being, are defined formally.

We will need complex forms of the series $\sigma_n[f]$ and $\overline{\sigma}_n[f,B]$. If $f \in L(T^n)$ and

$$c_p(f) = \frac{1}{(2\pi)^n}\int\limits_{T^n} f(x)e^{-ipx}\,dx, \quad p \in \mathbb{Z}^n, \tag{1.1.9}$$

then

$$\sigma_n[f] = \sum_{p\in\mathbb{Z}^n} c_p(f)e^{ipx} \tag{1.1.10}$$

and

$$\overline{\sigma}_n[f,B] = \sum_{p\in\mathbb{Z}^n}(-1)^{\overline{k}(B)}\left(\prod_{j\in B}\operatorname{sign}p_j\right)c_p(f)e^{ipx}. \tag{1.1.11}$$

For any function $f \in L(\mathbb{R}^n)$, we say the expression

$$\widetilde{f}_B(x) = \text{V.P.}\frac{1}{\pi^k}\int\limits_{\mathbb{R}^n(B)} \frac{f(x_{B'}+s_B)\,ds_B}{\prod\limits_{j\in B}(s_j-x_j)} \tag{1.1.12}$$

to be the n-variable Hilbert transform in those variables whose indices form the set B (for $n=1$, $\widetilde{f}_B(x) \equiv \widetilde{f}(x)$, $x \in \mathbb{R}$). Add that in what follows, for any $f \in L^p(\mathbb{R}^n)$ with $p \in [1,+\infty[$ we put

$$\|f\|_p = \left\{\int\limits_{\mathbb{R}^n}|f(x)|^p dx\right\}^{1/p}.$$

Remark 1.1.2. As already noted in the first part of the book, the series

$$\frac{a_0}{2}+\sum_{k=1}^{\infty}a_k\cos kx+b_k\sin kx$$

and

$$-\sum_{k=1}^{\infty}b_k\cos kx-a_k\sin kx$$

are (formally) the real and imaginary parts of the power series

$$\frac{a_0}{2}+\sum_{k=1}^{\infty}(a_k-ib_k)z^k$$

on the unit circle ($z=e^{ix}$).

There is no similar connection in the multidimensional case. Consider this question in detail. Let

$$\Gamma_1 = \Big\{k \ : \ k = 2m-2, \ \ m \in \mathbb{N}\Big\},$$
$$\Gamma_2 = \Big\{k \ : \ k = 2m-1, \ \ m \in \mathbb{N}\Big\},$$

and

$$c_p = 2^{-n}\Bigg[\sum_{B\subset M:\overline{k}(B')\in\Gamma_1} (-1)^{\overline{k}(B)/2} a_p^{(B)} - - i\sum_{B\subset M:\overline{k}(B')\in\Gamma_2} (-1)^{(\overline{k}(B)-1)/2} a_p^{(B)}\Bigg]. \tag{1.1.13}$$

Consider the power series on U^n

$$s(n) \equiv \sum_{p\geq 0} 2^{-\lambda(p)} c_p \prod_{j=1}^{n} z_j^{p_j}. \tag{1.1.14}$$

On the skeleton of the polydisc U^n (i.e., when $z_j = e^{ix_j}$, $1 \leq i \leq n$), the series (1.1.4) and all the n-fold conjugate series in an even number of variables form (formally), to within the sign, the real part of (1.1.14), and all the n-fold conjugate series in an odd number of variables form the imaginary part of (1.1.14).

For a point $p = (p_1, p_2, \ldots, p_n)$, $n > 1$, denote by p^j the point which differs from p only in j-th coordinate, namely $p_j^j = -p_j$. We can define $a_p^{(B)}$, $p \in \mathbb{Z}\setminus\mathbb{N}_0^n$, such that

$$a_{p^j}^{(B)} = a_p^{(B)} \quad \text{for } \ 1\leq j\leq n \ \text{ and } \ \overline{k}(B) = n;$$
$$a_{p^j}^{(B)} = -a_p^{(B)} \quad \text{for } \ 1\leq j\leq n \ \text{ and } \ \overline{k}(B) \in [0, n[.$$

Then, according to (1.1.13), the series (1.1.14) can be represented as

$$s(n) = \sum_{p\in\mathbb{Z}^n} c_p e^{ipx}.$$

In what follows, we assume that

$$A_p(x,f) = 2^{-\lambda(p)} \sum_{B\subset M} a_p^{(B)}(f) \prod_{i\in B} \cos p_i x_i \prod_{j\in B'} \sin p_j x_j, \tag{1.1.15}$$

and $\overline{A}_p(x,f,B)$ is conjugate to $A(x,f)$ in those variables whose indices belong to B. We will also use the notation

$$\rho_p^2(f) = \sum_{B\subset M} [a_p^{(B)}(f)]^2. \tag{1.1.16}$$

Further, if $m \in \mathbb{N}_0^n$, then by $S_m(x,f)$ and $\overline{S}_m(x,f,B)$ we denote partial rectangular sums of the series $\sigma_n[f]$ and $\overline{\sigma}_n[f,B]$, respectively, i.e.,

$$S_m(x,f) = \sum_{p=0}^{m} A_p(x,f), \tag{1.1.17}$$

$$\overline{S}_m(x,f,B)=\sum_{p=0}^{m}\overline{A}_p(x,f,B). \tag{1.1.18}$$

We will call the expressions

$$\sigma_m^\alpha(x,f)=\left(\prod_{i=1}^{n}A_{m_i}^{\alpha_i}\right)^{-1}\sum_{p=0}^{m}\prod_{i=1}^{n}A_{m_i-p_i}^{\alpha_i-1}S_p(x,f), \tag{1.1.19}$$

$$t_m^\alpha(x,f,B)=\left(\prod_{i=1}^{n}A_{m_i}^{\alpha_i}\right)^{-1}\sum_{p=0}^{m}\prod_{i=1}^{n}A_{m_i-p_i}^{\alpha_i-1}\overline{S}_p(x,f,B) \tag{1.1.20}$$

n-fold Cesàro means of the series $\sigma_n[f]$ and $\overline{\sigma}_n[f,B]$, respectively, where $\alpha\in\,]-1,+\infty[^n$ and A_i^β are defined by (2.2.3) from Part 1.

By symbols $f(x,r)$ and $f(x,r,B)$ we denote n-fold Poisson–Abel means of the series $\sigma_n[f]$ and $\overline{\sigma}_n[f,B]$, respectively, i.e.,

$$f(x,r)=\sum_{p\geq 0}A_p(x,f)\prod_{i=1}^{m}r_i^{p_i}, \tag{1.1.21}$$

$$\overline{f}(x,r,B)=\sum_{p\geq 0}\overline{A}_p(x,f,B)\prod_{i=1}^{m}r_i^{p_i}, \tag{1.1.22}$$

where $r\in[0,1[^n$.

Using (1.1.5), (1.1.17), and (1.1.18), we have

$$S_m(x,f)=\frac{1}{\pi^n}\int_{T^n}f(x+s)\prod_{i=1}^{n}D_{m_i}(s_i)\,ds, \tag{1.1.23}$$

$$\overline{S}_m(x,f,B)=\frac{(-1)^{\overline{k}(B)}}{\pi^n}\int_{T^n}f(x+s)\prod_{i\in B'}D_{m_i}(s_i)\prod_{j\in B}\overline{D}_{m_j}(s_j)\,ds, \tag{1.1.24}$$

where D_{m_i} and $\overline{D}_{m_j}$ are, respectively, the Dirichlet kernel and its conjugate. If we use (2.2.7) and (2.2.8) from the first part, according to (1.1.19), (1.1.20), (1.1.23), and (1.1.24), we obtain

$$\sigma_m^\alpha(x,f)=\frac{1}{\pi^n}\int_{T^n}f(x+s)\prod_{i=1}^{n}K_{m_i}^{\alpha_i}(s_i)\,ds, \tag{1.1.25}$$

$$t_m^\alpha(x,f,B)=\frac{(-1)^{\overline{k}(B)}}{\pi^n}\int_{T^n}f(x+s)\prod_{i\in B'}K_{m_i}^{\alpha_i}(s_i)\prod_{j\in B}\tau_{m_j}^{\alpha_j}(s_j)\,ds. \tag{1.1.26}$$

Analogously, by (1.1.5), (1.1.21), and (1.1.22), we find

$$f(x,r)=\frac{1}{\pi^n}\int_{T^n}f(x+s)\prod_{i=1}^{n}P(r_i,s_i)\,ds, \tag{1.1.27}$$

$$\overline{f}(x,r,B)=\frac{(-1)^k}{\pi^n}\int_{T^n}f(x+s)\prod_{i\in B'}P(r_i,s_i)\prod_{j\in B}Q(r_j,s_j)\,ds, \tag{1.1.28}$$

where (see also (1.2.5) and (1.2.6) from the first part)

$$P(r_i, s_i) = \frac{1}{2} + \sum_{k=1}^{\infty} r_i^k \cos ks_i = \frac{1 - r_i^2}{2(1 - 2r_i \cos s_i + r_i^2)}, \tag{1.1.29}$$

$$Q(r_j, s_j) = \sum_{k=1}^{\infty} r_j^k \sin ks_j = \frac{r_j \sin s_j}{1 - 2r_j \cos s_j + r_j^2}. \tag{1.1.30}$$

We should add that

$$Q(r_j, s_j) = \frac{1}{2}\operatorname{ctg}\frac{s_j}{2} - \frac{(1 - r_j)^2 \operatorname{ctg}\frac{s_j}{2}}{2(1 - 2r_j \cos s_j + r_j^2)}, \quad |s_j| \in]0, \pi]. \tag{1.1.31}$$

In the relations (1.1.15) and (1.1.26), the coordinates of α vary independently from each other, so that some of them may be zero, others may be negative, and the rest may be positive. The case where $\alpha_i \in]-\infty, -1[\backslash\mathbb{Z}$ also can be considered.

In what follows, the following notation is used.

$$\sigma(f) \equiv \sigma(x) \equiv \sigma(x, f, \alpha) = \sup_m |\sigma_m^\alpha(x, f)|, \quad \alpha \in]0, +\infty[^n, \tag{1.1.32}$$

$$t_B(f) \equiv t_B(x) \equiv t_B(x, f, \alpha) = \sup_m |t_m^\alpha(x, f, B)|, \quad \alpha \in]0, +\infty[^n, \tag{1.1.33}$$

$$t_B(f) \equiv t(x) \quad \text{for} \quad n = 1,$$

$$F(f) \equiv F(x) \equiv F(x, f) = \sup_r |f(x, r)|, \tag{1.1.34}$$

$$F_B^*(f) \equiv F_B^*(x) \equiv F_B^*(x, f) = \sup_r |\overline{f}(x, r, B)|, \tag{1.1.35}$$

$$F_B^*(f) \equiv F^*(x) \quad \text{for} \quad n = 1,$$

$$t_B^*(f) \equiv t_B^*(x) \equiv t_B^*(x, f) = \sup_{\varepsilon_i, i \in B} |\overline{f}_B(x, \varepsilon)|, \tag{1.1.36}$$

$$t_B^*(f) \equiv t^*(x) \quad \text{for} \quad n = 1.$$

Further, if $B = \{i\}$, then

$$M_B(f) \equiv M_B(x, f) \equiv M_i(x, f) =$$

$$= \sup_{0 < |h_i| \le \pi} \int_{x_i}^{x_i + h_i} |f(x_{B'} + s_B)| \, ds_i, \quad 1 \le i \le n, \tag{1.1.37}$$

$$M_i(x, f) \equiv M(x, f) \quad \text{for} \quad n = 1.$$

If, however, $\overline{k}(B) \in [2, n]$, then

$$M_B(f) \equiv M_B(x, f) = M_{i_1} \cdots M_{i_k}(x, f). \tag{1.1.38}$$

We should note that in (1.1.32)–(1.1.38) it is not necessarily meant that all the expressions are almost everywhere finite. For $f \in L(T^n)$, $n \ge 2$ (if $n = 1$, they are almost everywhere finite), they are defined, for the time being, formally.

Consider a function defined by

$$\tau(f, z) \equiv \tau(z) = \sum_{p \ge 0} 2^{-\lambda(p)} \Bigg\{ \Bigg[\sum_{B \subset M : \overline{k}(B') \in \Gamma_1} (-1)^{\overline{k}(B)/2} a_p^B(f) \prod_{j=1}^{n} r_j^{p_j} e^{ix_j} \Bigg] +$$

$$+i\left[\sum_{B\subset M:\overline{k}(B')\in\Gamma_2}(-1)^{(\overline{k}(B)-1)/2}a_p^B(f)\prod_{j=1}^n r_j^{p_j}e^{ix_j}\right]\Bigg\},\tag{1.1.39}$$

$$r_j\in[0,1[,\quad 1\le j\le n.$$

Since

$$\lim_{\|p\|\to\infty}a_p^{(B)}(f)=0,$$

the function τ is, in fact, regular on U^n.

In what follows, we will need some results related to (1.1.32)–(1.1.39) in the one-dimensional case.

Let $a\in]0,1[$ and $x\in T$. Denote by $C_a(x)$ (see, e.g., Zygmund [16, Ch. XIV]) the domain limited by the two tangents to the circle $|z|=a$ drawn from the point e^{ix}, as well as by the largest of the arcs of this circle between the points of tangency.

In 1930, Hardy and Littlewood [7] proved the following assertions.

1. Let $f\in H^p(U)$ for some $p\in]1,+\infty[$. If

$$g(x)\equiv g(x,f,a)=\sup_{z\in C_a(x)}|f(z)|,\tag{1.1.40}$$

then $g\in L^p(T)$ and

$$\|g\|_p\le A(p)\|f\|_p.\tag{1.1.41}$$

2. a) Let $f\in L(T)$ and let $M(x,f)$ be defined (for $n=1$) by (1.1.37). Then for any $p\in]0,1[$,

$$\|M(\cdot,f)\|_p\le A(p)\|f\|_1;\tag{1.1.42}$$

b) if $f\in L\log^+L(T)$, then

$$\|M(\cdot,f)\|_1\le A\Big[1+\int_T|f(x)|\log^+|f(x)|\,dx\Big];\tag{1.1.43}$$

c) if $f\in L^p(T)$ for some $p\in]1,+\infty[$, then

$$\|M(\cdot,f)\|_p\le A(p)\|f\|_p.\tag{1.1.44}$$

In the same work, Hardy and Littlewood showed that (1.1.42)–(1.1.44) are also valid for the functions σ and F defined (for $n=1$) by (1.1.32) and (1.1.34), respectively.

It is well-known that (Zygmund [4], [15, Ch. VII], Titchmarsh [2]) if $\alpha\in]0,+\infty[$ and $f\in L(\log^+L)^\alpha(T)$, then

$$\int_T|\overline{f}(x)|\Big[\log^+[|\overline{f}(x)|+2]\Big]^{\alpha-1}dx\le$$

$$\le A(\alpha)\Big[1+\int_T|f(x)|\Big[\log^+|f(x)|\Big]^\alpha dx\Big].\tag{1.1.45}$$

The work of Magnaradze [1] should also be mentioned here.

In 1935, Jessen, Marcinkiewicz and Zygmund [1] (see also Zygmund [16, Ch. XVII]) proved that if $f \in L(\log^+ L)^\alpha(T)$ for some $\alpha \in \mathbb{N}$, then

$$\int_T M(x,f)|\big[\log^+ M(x,f)\big]^{\alpha-1} dx \le$$

$$\le A(\alpha)\Big[1 + \int_T |f(x)|\big[\log^+ |f(x)|\big]^\alpha dx\Big]. \tag{1.1.46}$$

As already noted, for $\alpha = 1$ this assertion was established by Hardy and Littlewood [7].

Everywhere below, we will assume $g(x) \equiv g(x, \tau(f), a)$ (see (1.1.40)), where the regular function $\tau \equiv \tau(f)$ is defined by (1.1.39) (for $n = 1$).

Using the above mentioned fundamental results by Kolmogorov [3], M. Riesz [3], Zygmund [4], Titchmarsh [2], Hardy and Littlewood [7] (see Zygmund [15, Ch. IV, VII] and (1.1.41)–(1.1.45)), we can conclude that the following statements are true.

1. Let $f \in L(T)$. Then for any $p \in]0,1[$, the inequalities

$$\|g(\tau(f))\|_p \le A(p)\|f\|_1, \tag{1.1.47}$$

$$\|\sigma(f)\|_p \le A(p,\alpha)\|f\|_1, \tag{1.1.48}$$

$$\|t(f)\|_p \le A(p,\alpha)\|f\|_1, \tag{1.1.49}$$

$$\|F(f)\|_p \le A(p)\|f\|_1, \tag{1.1.50}$$

$$\|F^*(f)\|_p \le A(p)\|f\|_1, \tag{1.1.51}$$

$$\|t^*(f)\|_p \le A(p)\|f\|_1 \tag{1.1.52}$$

hold.

2. Let $f \in L\log^+ L(T)$. Then

$$\|g(\tau(f))\|_1 \le A\Big[\int_T |f(x)|\log^+|f(x)|\,dx + 1\Big], \tag{1.1.53}$$

$$\|\sigma(f)\|_1 \le A(\alpha)\Big[\int_T |f(x)|\log^+|f(x)|\,dx + 1\Big], \tag{1.1.54}$$

$$\|t(f)\|_1 \le A(\alpha)\Big[\int_T |f(x)|\log^+|f(x)|\,dx + 1\Big], \tag{1.1.55}$$

$$\|F(f)\|_1 \le A\Big[\int_T |f(x)|\log^+|f(x)|\,dx + 1\Big], \tag{1.1.56}$$

$$\|F^*(f)\|_1 \le A\Big[\int_T |f(x)|\log^+|f(x)|\,dx + 1\Big], \tag{1.1.57}$$

$$\|t^*(f)\|_1 \le A\Big[\int_T |f(x)|\log^+|f(x)|\,dx + 1\Big]. \tag{1.1.58}$$

3. Let $f \in L^p(T)$ for some $p \in]1, +\infty[$. Then

$$\|g(\tau(f))\|_p \leq A(p)\|f\|_p, \tag{1.1.59}$$

$$\|\sigma(f)\|_p \leq A(p,\alpha)\|f\|_p, \tag{1.1.60}$$

$$\|t(f)\|_p \leq A(p,\alpha)\|f\|_p, \tag{1.1.61}$$

$$\|F(f)\|_p \leq A(p)\|f\|_p, \tag{1.1.62}$$

$$\|F^*(f)\|_p \leq A(p)\|f\|_p, \tag{1.1.63}$$

$$\|t^*(f)\|_p \leq A(p)\|f\|_p. \tag{1.1.64}$$

A number of more general theorems which are relevant to inequalities (1.1.41)-(1.1.64) have been established up today. But in what follows, the above formulated classical assertions will be sufficient. We should only note a result of Burkholder, Gundy and Silverstein [1], which provides the inequality inverse to (1.1.41), i.e.,

$$\sup_{0 \leq r < 1} \|\tau\|_p \leq A(p)\|g\|_p, \quad p \in]0, +\infty[,$$

only the case $p \in]0,1]$ being essential (according to M. Riesz's inequality (1.1.10) from the first part of the book).

Now we should say a few words about the types of convergence of multiple trigonometric series. We will confine ourselves to the series $\sigma_n[f]$ $(n \geq 2)$. This question is in detail discussed in the monograph by Zygmund [16, Ch. XVII]. Suppose that $S_m(x, f)$ are defined by (1.1.23). If at a point $x \in T$ there exists a finite limit

$$\lim_{m \to \infty} S_m(x, f) \equiv S(x),$$

then the series $\sigma_n[f]$ is said to be convergent to s at x in the sense of Stolz–Pringsheim (or by rectangles). If

$$\lim_{m_\lambda \to \infty} S_m(x, f) \equiv S(x)$$

for some $\lambda \in [1, +\infty[$, then the series $\sigma_n[f]$ is said to be λ-convergent (or convergent in the sense of λ-limit) to S at x. In the case where $\lambda = 1$ and $n \geq 3$, this convergence is called convergence by cubes, and for $n = 2$ - convergence by squares. It is easy to formulate these definitions in the cases where the convegence is considered in the metric of $L^p(T^n)$, $p \in]0, +\infty]$, or if almost everywhere convergence is considered, and so on. They likewise easily can be paraphrased for multiple methods of summation.

In what follows, when speaking about convergence or summability of a multiple trigonometric series (at a point, on a set of positive n-dimensional Lebesgue measure, almost everywhere, uniformly on T^n, and so on), we will omit the phrase "in the sense of Stolz–Pringsheim". Note also that for multiple trigonometric series, the phrases "almost everywhere", "everywhere", "uniformly" are relative to T^n. The same is true for the limit

$$\lim_{\varepsilon \to 0+} \overline{f}_B(x, \varepsilon).$$

1.2. On Existence of Conjugate Functions and of Hilbert Transforms

In 1938, Cesàri [1] showed that if $f: T^2 \to \mathbb{R}$ belongs to $Lip\,\alpha$ for some $\alpha \in]0,1]$, then all the conjugate functions $\overline{f}_B$ with $B \subset \{1,2\}$ exist everywhere and belong to $Lip\,\beta$ for any $\beta \in]0,\alpha[$.

In 1947, Sokol-Sokolowski [1] proved that if $f \in L(\log^+ L)^n(T^n)$, then all the conjugate functions of n variables exist almost everywhere.

In 1949, Zygmund [11] established that the same is true if $f \in L(\log^+ L)^{n-1}(T^n)$.

In 1957, in the work [14] he posed the question: if $f \in L(T^n)$, then does the function $\overline{f}_M$ exist almost everywhere in the sense of λ-limit?

In 1961, Stein [1] found out that if $n=2$, then in the class $L(\log^+ L)^{1-\varepsilon}(T^2)$ with $\varepsilon \in]0,1]$, the function $\overline{f}_M$ may fail to exist even in the case where in the definition of $\overline{f}_M$ the limit is meant in the sense of λ-convergence ($\lambda > 1$). However, as we showed in [8], the proof given by Stein contain some inaccuracies. The problem of existence of conjugate functions of several variables was thoroughly enough considered in our works [13, Ch. III], [16], [35, Ch. II], [46]. Some of our results in this direction were noted in [37]. In this section, the existence of conjugate functions and of Hilbert transform is studied in detail in the case of several variables.

As usual (see, e.g., Zygmund [16, Ch. XVII]), $C_a(x,n)$ denotes the Cartesian product of the domains $C_a(x_i)$, $1 \le i \le n$.

Let $f \in L(T^n)$ and let $\tau: U^n \to \mathbb{C}$ be defined by (1.1.39). Put

$$\begin{gathered} g_M(x) \equiv g_M(x,f,n,a) = \sup_{z \in C_a(x,n)} |\tau(f,z)| \quad \text{for } n \ge 2; \\ g_M(x) \equiv g(x,\tau,a) \quad \text{for } n=1, \end{gathered} \tag{1.2.1}$$

where g is defined by (1.1.40).

Lemma 1.2.1. *Suppose that $f \in L(\log^+ L)^{n-1}(T^n)$. If $\overline{f}(x,r,B)$ is defined by (1.1.28), then for any $p \in]0,1[$ and $B \subset M$, $B \ne \varnothing$, we have*

$$\|\overline{f}(\cdot,r,B)\|_p \le A(p,n)\Big\{\int_{T^n} |f(x)|\big[\log^+ |f(x)|\big]^{n-1} dx + 1\Big\}. \tag{1.2.2}$$

Proof. Let $B = M$. By virtue of (1.1.28), we find

$$\overline{f}(x,f,B) = \Big(-\frac{1}{\pi}\Big)^n \int_{T^n} f(x+s) \prod_{i=1}^n Q(r_i,s_i)\,ds.$$

Using the inequality (1.1.45) with respect to one of the arguments (for instance, for x_n), we get

$$\begin{aligned} &\int_T |\overline{f}(x,r,\{n\})|\big[\log^+ |\overline{f}(x,r,\{n\})|\big]^{n-2} dx_n \le \\ &\le A(n)\Big\{\int_T |f(x,r)|\big[\log^+ |f(x,r)|\big]^{n-1} dx_n + 1\Big\}. \end{aligned} \tag{1.2.3}$$

Repeating this process with respect to the arguments $x_{n-1}, x_{n-2}, \dots, x_2$, we obtain

$$\int_T |\overline{f}(x,r,\{n-1,n\})| \Big[\log^+ |\overline{f}(x,r\{n-1,n\})|\Big]^{n-3} dx_{n-1} \le$$

$$\le A(n)\bigg\{\int_T |\overline{f}(x,r,\{n\})| \Big[\log^+ |\overline{f}(x,r,\{n\})|\Big]^{n-2} dx_{n-1} + 1\bigg\}, \qquad (1.2.4)$$

$$\dots\dots\dots\dots\dots\dots\dots\dots\dots\dots\dots\dots$$

$$\int_T |\overline{f}(x,r,\{2,3,\dots,n\})| dx_2 \le$$

$$\le A(n)\bigg\{\int_T |\overline{f}(x,r,\{3,4,\dots,n\})| \log^+ |\overline{f}(x,r,\{3,4,\dots,n\})| \, dx_1 + 1\bigg\}. \qquad (1.2.5)$$

Using the inequality (1.1.21) from Part 1, we obtain

$$\bigg\{\int_T |\overline{f}(x,r,M)|^p dx_1\bigg\}^{1/p} \le A(p,n) \int_T |\overline{f}(x,r,\{2,3,\dots,n\})| \, dx_1. \qquad (1.2.6)$$

Furthermore, by (1.2.3)–(1.2.5) and (1.1.27), we have

$$\int_{T^n} |\overline{f}(x,r,\{2,3,\dots,n\})| \, dx \le$$

$$\le A(n)\bigg\{\int_{T^n} |f(x,r)| \Big[\log^+ |f(x,r)|\Big]^{n-1} dx + 1\bigg\}. \qquad (1.2.7)$$

In view of (1.2.6), we find

$$\int_{T^n} |\overline{f}(x,r,M)|^p dx \le$$

$$\le A(p,n) \int_{T^{n-1}} \Big[\int_T |\overline{f}(x,r,\{2,3,\dots,n\})| \, dx_1\Big]^p dx_2 \cdots dx_n.$$

Hence, using Hölder's inequality with

$$\frac{1}{p'} = p \quad \text{and} \quad \frac{1}{q'} = 1 - p,$$

we get

$$\int_{T^n} |\overline{f}(x,r,M)|^p dx \le A(p,n)\bigg\{\int_{T^n} \bigg|\int_{T^n} f(x+s) \prod_{i=2}^n Q(r_i,s_i) P(r_1,s_1) \, ds\bigg| dx\bigg\}^p,$$

whence, by (1.2.7), it follows the validity of (1.2.2) for $B = M$.

Suppose now that $\overline{k}(B) \in [1, n[$. Repeating the above scheme with respect to those arguments whose indices are in B, we get

$$\|\overline{f}(\cdot, r, B)\|_p \leq A(p,n)\left\{\int_{T^n} |f(x)|\Big[\log^+ |f(x)|\Big]^{\overline{k}(B)-1} dx + 1\right\}. \tag{1.2.8}$$

Thus we can say that the lemma is completely proved. ■

Let

$$f(x,r) \equiv \overline{f}(x,r,\varnothing).$$

Then, using (1.1.27), (1.1.28), and (1.1.39), the function $\tau : U^n \to \mathbb{C}$ can be represented as

$$\tau(z) = \sum_{B\subset M:\overline{k}(B)\in\Gamma_1} (-1)^{\overline{k}(B)/2}\overline{f}(x,r,B) + $$
$$+ i \sum_{B\subset M:\overline{k}(B)\in\Gamma_2} (-1)^{(\overline{k}(B)-1)/2}\overline{f}(x,r,B). \tag{1.2.9}$$

Thus, applying Lemma 1.2.1, we easily obtain

Corollary 1.2.1. *Let $f \in L(\log^+ L)^{n-1}(T^n)$ and $p \in]0,1[$. If the function τ is defined by* (1.1.39), *then $\tau \in H^p(T^n)$ and*

$$\|\tau\|_p \leq A(p,n)\left\{\int_{T^n} |f(x)|\Big[\log^+ |f(x)|\Big]^{n-1} dx + 1\right\}. \tag{1.2.10}$$

Lemma 1.2.2. *Suppose that $f \in L(\log^+ L)^{n-1}(T^n)$ and g_M is defined by* (1.2.1). *Then for any $p \in]0,1[$, we have $g_M \in L^p(T^n)$ and*

$$\|g_M\|_p \leq A(p,n,a)\left\{\int_{T^n} |f(x)|\Big[\log^+ |f(x)|\Big]^{n-1} dx + 1\right\}. \tag{1.2.11}$$

Proof. We will make use of the scheme of Zygmund [11]. Let $\rho \in]0,1[$ and

$$v(z,\rho) = \tau(\rho z). \tag{1.2.12}$$

It is clear that v is continuous in z on the closure of U^n. Put

$$\beta(z,\rho) = \log^+ |v(z,\rho)|, \tag{1.2.13}$$

$$\beta^*(x,\rho,a) = \max_{z\in C_a(x,n)} |\beta(z,\rho)|, \tag{1.2.14}$$

$$\gamma^*(x,\rho,a) = \max_{z\in C_a(x,n)} |v(z,\rho)|, \tag{1.2.15}$$

where $a \in]0,1[$ and $x \in T^n$.

Since β is a nonnegative continuous function which is subharmonic in each of the variables z_j, $1 \leq j \leq n$, then, using consecutively the appropriate assertion of Hardy and Littlewood [7], for any $q \in]1,+\infty[$ we get

$$\|\beta^*(\cdot,\rho,a)\|_q^q \leq A(n,a)\left(\frac{q}{q-1}\right)^{nq} \|\beta(\cdot,\rho)\|_q^q. \tag{1.2.16}$$

On the other hand, if $q \in [2, +\infty[$, then

$$\left(\frac{q}{q-1}\right)^{nq} \le 4^n.$$

Therefore, using (1.2.16) for $q = 2\lambda$ with $\lambda \in \mathbb{N}$, we obtain

$$\int_{T^n} \left| \sum_{\lambda=1}^{\infty} \frac{[\beta^*(x,\rho,a)]^{2\lambda} p^{2\lambda}}{(2\lambda)!} \right| dx \le A(n,a) \int_{T^n} \left[\sum_{\lambda=1}^{\infty} \frac{|\beta(z,\rho)|^{2\lambda} p^{2\lambda}}{(2\lambda)!} \right] dx.$$

Hence it easily follows that for any $p \in]0,1[$,

$$\int_{T^n} \operatorname{ch}\left[p\beta^*(x,\rho,a)\right] dx \le A(p,n,a) \int_{T^n} \operatorname{ch}\left[p|\beta(z,\rho)|\right] dx,$$

and because of $\frac{1}{2}e^t < \operatorname{ch} t < e^t$ for $t \in \mathbb{R}$,

$$\int_{T^n} \exp\left[p\beta^*(x,\rho,a)\right] dx \le A(p,n,a) \int_{T^n} \exp\left[p|\beta(z,\rho)|\right] dx.$$

Therefore, taking note of (1.2.13)–(1.2.15), we have

$$\int_{T^n} |\gamma^*(x,\rho,a)|^p dx \le A(p,n,a) \int_{T^n} \left[|\tau(\rho z)|^p dx + 1\right].$$

Using this relation for the function $L\tau$ with an arbitrary $L \in]0, +\infty[$, we get

$$\int_{T^n} |\gamma^*(x,\rho,a)|^p dx \le A(p,n,a) \int_{T^n} \left[|\tau(\rho z)|^p dx + L^{-p}\right],$$

which implies

$$\int_{T^n} |\gamma^*(x,\rho,a)|^p dx \le A(p,n,a) \int_{T^n} |\tau(\rho z)|^p dx. \tag{1.2.17}$$

On the other hand, by virtue of (1.2.1), (1.2.12), and (1.2.15), we can conclude that

$$\lim_{\rho \to 1-} \gamma^*(x,\rho,a) = g_M(x).$$

Then, using Fatou's theorem, according to (1.2.17) we get

$$\|g_M\|_p \le A(p,n,a) \sup_{\rho} \|\tau\|_p, \tag{1.2.18}$$

whence by Corollary 1.2.1 it follows (1.2.11). ■

Analyzing the proof of Lemma 1.2.2, we can conclude that if the function $\tau : U^n \to \mathbb{C}$ is regular and belongs to $H^p(T^n)$ for some $p \in]0, +\infty[$, then (1.2.18) holds.

Using the same argument as in the proof of Lemma 1.2.2, we make sure that it holds

Lemma 1.2.3. *Suppose that p and q are finite positive numbers, the function $\tau : U^n \to \mathbb{C}$ is regular, and*

$$\sup_r \int_{T^n} \left[|\tau(z)|^p + 1\right] \left[[\log^+ |\tau(z)|]^q + 1\right] dx < +\infty, \quad z = re^{ix}.$$

Then

$$\int_{T^n} [g_M(x)]^p [\log^+ g_M(x)]^q \, dx \leq$$

$$\leq A(p,q,n,a) \sup_r \int_{T^n} \left[|\tau(z)|^p + 1\right] \left[[\log^+ |\tau(z)|]^q + 1\right] dx. \tag{1.2.19}$$

It should be added that using the above presented method, a more general inequality can be proved.

Let now $B \subset M$ and $B \neq M$. As in the case of the symbol $\delta(B)$, we will assume that

$$r(B) = (r_{i_1}, r_{i_2}, \dots, r_{i_k}) \quad \text{and} \quad z(B) = (z_{i_1}, z_{i_2}, \dots, z_{i_k}).$$

For a function $f \in L(T^n)$, we put

$$f(x, r(B)) \equiv \overline{f}(x, r(B), \varnothing) = \frac{1}{\pi^k} \int_{T^n(B)} f(x + s_B) \prod_{i \in B} P(r_i, s_i) \, ds_B, \tag{1.2.20}$$

$$\overline{f}(x, r(B), B_1) =$$

$$= \frac{(-1)^{k_1}}{\pi^k} \int_{T^n(B)} f(x + s_B) \prod_{i \in B_1} Q(r_i, s_i) \prod_{j \in B \setminus B_1} P(r_j, s_j) \, ds_B, \tag{1.2.21}$$

$$\tau_1(f, z, B) \equiv \tau_1(z, B) = \sum_{B_1 \subset B : \overline{k}(B_1) \in \Gamma_1} (-1)^{\overline{k}(B_1)/2} \overline{f}(x, r(B), B_1) +$$

$$+ i \sum_{B_1 \subset B : \overline{k}(B_1) \in \Gamma_2} (-1)^{(\overline{k}(B_1) - 1)/2} \overline{f}(x, r(B), B_1). \tag{1.2.22}$$

We emphasize that in (1.2.20)–(1.2.22), the coordinates of x with indices from B' are fixed.

Analyzing the schemes of the proofs of Lemmas 1.2.1–1.2.3, we can conclude that the following statements are true.

Lemma 1.2.4. *Let $B \subset M$ and $f \in L(\log^+ L)^{\overline{k}(B)-1}(T^n)$. Then for any $p \in]0,1[$ and $B_1 \subset B$, we have*

$$\|\overline{f}(\cdot, r(B), B_1)\|_p \leq$$

$$\leq A(p,n) \left\{ \int_{T^n} |f(x)| \left[\log^+ |f(x)|\right]^{\overline{k}(B)-1} dx + 1 \right\}, \tag{1.2.23}$$

where $\overline{f}(x, r(B), B_1)$ is defined by (1.2.21).

Lemma 1.2.5. *Let $B \subset M$ and $f \in L(\log^+ L)^{\overline{k}(B)-1}(T^n)$ and τ_1 be defined by* (1.2.22). *Put*

$$g_B(x) \equiv g_B(x,\tau_1) = \sup_{z_i \in C_a(x_i),\, i\in B} |\tau_1(z,B)|. \tag{1.2.24}$$

Then for any $p \in]0,1[$, we have $g_B \in L^p(T^n)$ and

$$\|g_B\|_p \le A(p,n,a)\left\{\int_{T^n} |f(x)|\left[\log^+|f(x)|\right]^{\overline{k}(B)-1} dx + 1\right\}. \tag{1.2.25}$$

Lemma 1.2.6. *Let p and q be some finite positive numbers. If τ_1 is defined by* (1.2.22) *and*

$$\int_{T^n(B')} \sup_{r(B)}\left\{\int_{T^n(B)} \left[|\tau_1(z,B)|^p+1\right]\left[[\log^+|\tau_1(z,B)|]^q+1\right] dx_B\right\} dx_{B'} < +\infty,$$

then

$$\int_{T^n}[g_B(x)]^p[\log^+ g_B(x)]^q\, dx \le A(p,q,n,a)\int_{T^n(B')}\left\{\sup_{r(B)}\int_{T^n(B)} \left[|\tau_1(z,B)|^p+\right.\right.$$
$$\left.\left.+1\right]\left[[\log^+|\tau_1(z,B)\,|]^q+1\right] dx_B\right\} dx_{B'}. \tag{1.2.26}$$

Lemma 1.2.7. *Let $f \in L(\log^+ L)^n(T^n)$. Then for any $B \subset M$, it holds*

$$\|\overline{f}(\cdot,r,B)\|_1 \le A(n)\left\{\int_{T^n} |f(x)|\left[\log^+|f(x)|\right]^n dx + 1\right\}, \tag{1.2.27}$$

where $\overline{f}(x,r,B)$ is defined by (1.1.28).

Lemma 1.2.8. *Let $B \subset M$ and $f \in L(\log^+ L)^{\overline{k}(B)}(T^n)$. Then for any $B_1 \subset B$, it holds*

$$\|\overline{f}(\cdot,r(B),B_1)\|_1 \le A(n)\left\{\int_{T^n} |f(x)|\left[\log^+|f(x)|\right]^{\overline{k}(B)} dx + 1\right\}, \tag{1.2.28}$$

where $\overline{f}(x,r(B),B_1)$ is defined by (1.2.21).

Lemma 1.2.9. *Let $B \subset M$ and $f \in L(\log^+ L)^{\overline{k}(B)}(T^n)$. If g_B is given by* (1.2.24), *then*

$$\|g_B\|_1 \le A(n)\left\{\int_{T^n} |f(x)|\left[\log^+|f(x)|\right]^{\overline{k}(B)} dx + 1\right\}. \tag{1.2.29}$$

Let $B_1 \subset B \subset M$. We will often need the following functions

$$M_{B\setminus B_1}(x, g_{B_1}) \equiv M_{B\setminus B_1}(g_{B_1}), \tag{1.2.30}$$

where it is assumed

$$M_{B\setminus B_1}(g_{B_1}) \equiv \begin{cases} M_{B'}(g_{B_1}) & \text{if } B = M, \\ M_B(f) & \text{if } B_1 = \varnothing, \\ g_B & \text{if } B_1 = B, \end{cases} \tag{1.2.31}$$

and $M_B(f)$, g_M, g_B are defined by (1.1.37), (1.1.38), (1.2.1), and (1.2.24).

Using (1.1.27), (1.1.28), (1.1.37), (1.1.38), (1.2.1), (1.2.9), (1.2.30), and (1.2.31), we get

$$\sup_r |\overline{f}(x, r, B)| \le A(n)\Big\{ \sum_{B_1 \subset B} M_{B\setminus B_1}(x, g_{B_1}) \Big\}. \tag{1.2.32}$$

Analogously, on the basis of (1.1.27), (1.1.28), (1.2.20), (1.2.21), (1.2.22), (1.2.24), and (1.2.31), we find

$$\sup_{r(B_2)} |\overline{f}(x, r(B), B_2)| \le$$

$$\le A(n)\Big\{ \sum_{B_1 \subset B_2} M_{B\setminus B_1}(x, g_{B_1}) \Big\}, \quad B_2 \subset B. \tag{1.2.33}$$

Furthermore, using (1.1.8) with $\varepsilon_i = 1 - r_i$ for $1 \le i \le n$, (1.1.29), (1.1.30), (1.1.31), and (1.2.33), we can conclude that

$$\sup_{\varepsilon(B)} |\overline{f}_B(x, \varepsilon)| \le A(n)\Big\{ \sum_{B_1 \subset B} M_{B\setminus B_1}(x, g_{B_1}) \Big\}. \tag{1.2.34}$$

Lemma 1.2.10. *Let $f \in L(\log^+)^{n-1}(T^n)$. Then for any $p \in]0, 1[$, it holds*

$$\|M_{B'}(g_B)\|_p \le A(p, n, a)\Big\{ \int_{T^n} |f(x)| \big[\log^+ |f(x)|\big]^{n-1} dx + 1 \Big\}. \tag{1.2.35}$$

Proof. If $B = \varnothing$, then according to our notation, we have

$$M_M(x, g_B) = M_M(x, f).$$

Therefore, using the inequality (1.1.42) with respect to one of the variables (say x_1), we find

$$\Big\{ \int_T \big[M_M(x, f)\big]^p dx_1 \Big\}^{1/p} \le$$

$$\le A(p) \int_T \big[M_2 M_3 \cdots M_n(x, f)\big]\, dx_1. \tag{1.2.36}$$

Applying now (1.1.46) consecutively with respect to the variables $x_2, x_3, \dots, x_n$, we get

$$\int\limits_{T}^{n} \Big[M_2 M_3 \cdots M_n(x,f)\Big]\, dx \leq$$

$$\leq A(n)\Big\{ \int\limits_{T^n} |f(x)|\Big[\log^+ |f(x)|\Big]^{n-1} dx + 1\Big\}. \tag{1.2.37}$$

By (1.2.36), we have

$$\|M_M(f)\|_p^p \leq A(p,n)\Big\{ \int\limits_{T^n} \Big[M_2 M_3 \cdots M_n(x,f)\Big]\, dx\Big\}^p .$$

Hence, according to (1.2.37), the inequality (1.2.35) is proved in the case $B = \varnothing$.

Let $B = M$. In view of our notation,

$$M_{B'}(g_B) = g_M.$$

The inequality (1.2.35) then follows from Lemma 1.2.2 (see (1.2.11)).

Suppose now that $B \neq M$, $B \neq \varnothing$ and $B' = \{j_1, j_2, \dots, j_{k'}\}$ where $k' = \overline{k}(B')$. Using (1.1.42) with respect to the variable with the index j_1, we get

$$\int\limits_{T} \Big[M_{B'}(x, g_B)\Big]^p dx_{j_1} \leq$$

$$\leq A(p) \int\limits_{T} \Big[M_{j_2} M_{j_3} \cdots M_{j_{k'}}(x, g_B)\Big]\, dx_{j_1}. \tag{1.2.38}$$

If now we apply (1.1.46) consecutively with respect to the variables with indices $j_2, j_3, \dots, j_{k'}$, we find

$$\|M_{B'}(g_B)\|_p \leq A(p,n)\Big\{ \int\limits_{T^n} g_B(x)\Big[\log^+ g_B(x)\Big]^{k'-1} dx + 1\Big\},$$

whence, according to Lemma 1.2.6 (see (1.2.26) with $p = 1$ and $q = k' - 1$), we obtain

$$\|M_{B'}(g_B)\|_p \leq A(p,n,a)\Big\{ \int\limits_{T^n(B')} \Big[\sup_{r_i, i \in B} \int\limits_{T^n(B)} \Big[|\tau_1(z(B))| + 1\Big] \times$$

$$\times \Big[[\log^+ |\tau_1(z(B))|]^{k'-1} + 1\Big]\, dx_B\Big]\, dx_{B'}\Big\}. \tag{1.2.39}$$

On the other hand, taking note of (1.2.20)–(1.2.22), it is not difficult to check that

$$\Big[|\tau_1(z(B))| + 1\Big]\, \Big[[\log^+ |\tau_1(z(B))|]^{k'-1} + 1\Big] \leq$$

$$\leq A(n)\Big\{ \sum_{B_1 \subset B} |\overline{f}(x, r(B), B_1)|\Big[\log^+ |\overline{f}(x, r(B), B_1)|\Big]^{k'-1} +$$

$$+\sum_{B_1\subset B}|\overline{f}(x,r(B),B_1|+1\Big\}. \tag{1.2.40}$$

Now, using the method of the proof of Lemma 1.2.1, we find

$$\int\limits_{T^n(B)}|\overline{f}(x,r(B),B_1)|\Big[\log^+|\overline{f}(x,r(B),B_1)|\Big]^{k'-1}dx_B\le$$

$$\le A(n)\Big\{\int\limits_{T^n(B)}|f(x,r(B))|\Big[\log^+|f(x,r(B))|\Big]^{\overline{k}(B_1)+k'-1}dx_B+1\Big\}, \tag{1.2.41}$$

$$\int\limits_{T^n(B)}|\overline{f}(x,r(B),B_1)|dx_B\le$$

$$\le A(n)\Big\{\int\limits_{T^n(B)}|f(x,r(B))|\Big[\log^+|f(x,r(B))|\Big]^{\overline{k}(B_1)}dx_B+1\Big\}. \tag{1.2.42}$$

Thus, by (1.2.39)–(1.2.42), we have

$$\|M_{B'}(g_B)\|_p\le A(p,n,a)\Big\{\int\limits_{T^n}|f(x)|\Big[\log^+|f(x)|\Big]^{n-1}dx+1\Big\}. \quad \blacksquare$$

Using (1.1.35), (1.1.36), (1.2.32), (1.2.34), and (1.2.35), we obtain

Lemma 1.2.11. a) *Let $f\in L(\log^+L)^{n-1}(T^n)$. Then for any $p\in\,]0,1[$, we have*

$$\|F_B^*\|_p\le A(p,n,a)\Big\{\int\limits_{T^n}|f(x)|\Big[\log^+|f(x)|\Big]^{n-1}dx+1\Big\}; \tag{1.2.43}$$

b) *Let $B\subset M$ and $f\in L(\log^+L)^{\overline{k}(B)-1}(T^n)$. Then for any $p\in\,]0,1[$ and $B_1\subset B$, it holds*

$$\|t_{B_1}^*\|_p\le A(p,n,a)\Big\{\int\limits_{T^n}|f(x)|\Big[\log^+|f(x)|\Big]^{\overline{k}(B)-1}dx+1\Big\}. \tag{1.2.44}$$

Let now $\overline{f}_B(x,\varepsilon)$ be defined by (1.1.8) with $\varepsilon_i=1-r_i$ for $1\le i\le n$. Then the conclusions of Lemma 1.2.11 enable us to affirm

Lemma 1.2.12. *Let $f\in L(\log^+L)^{n-1}(T^n)$, $B\subset M$ and $B\ne\varnothing$. Then for any $p\in\,]0,1[$, we have*

$$\Big\{\int\limits_{T^n}\Big[\sup_r|\overline{f}(x,r,B)-\overline{f}_B(x,\varepsilon)|\Big]^p dx\Big\}^{1/p}\le$$

$$\le A(p,n,a)\Big\{\int\limits_{T^n}|f(x)|\Big[\log^+|f(x)|\Big]^{n-1}dx+1\Big\}. \tag{1.2.45}$$

Analyzing the proofs of Lemma 1.2.10 and using (1.1.35), (1.1.36), (1.2.32), (1.2.34), and (1.2.35), we make sure that it holds

Lemma 1.2.13. *Let $f \in L(\log^+ L)^n(T^n)$. Then for any $B \subset M$, the following is true.*

$$\text{a)}\quad \|M_{B'}(g_B)\|_1 \le A(n,a)\Big\{ \int_{T^n} |f(x)|\big[\log^+|f(x)|\big]^n dx + 1\Big\}, \tag{1.2.46}$$

$$\text{b)}\quad \|F^*_B\|_1 \le A(n,a)\Big\{ \int_{T^n} |f(x)|\big[\log^+|f(x)|\big]^n dx + 1\Big\}, \tag{1.2.47}$$

$$\text{c)}\quad \|t^*_B\|_1 \le A(n,a)\Big\{ \int_{T^n} |f(x)|\big[\log^+|f(x)|\big]^n dx + 1\Big\}, \tag{1.2.48}$$

$$\text{d)}\quad \int_{T^n} \sup_r |\overline{f}(x,r,B) - \overline{f}_B(x,\varepsilon)|\,dx \le$$

$$\le A(n,a)\Big\{ \int_{T^n} |f(x)|\big[\log^+|f(x)|\big]^n dx + 1\Big\}. \tag{1.2.49}$$

Analogously, basing on (1.2.34) and (1.2.35), we have

Lemma 1.2.14. *Let $\overline{k}(B) \in [1,n[$ and $f \in L(\log^+ L)^{\overline{k}(B)}(T^n)$. Then for any $B_1 \subset B$, the inequality*

$$\|t^*_{B_1}\|_1 \le A(n,a)\Big\{ \int_{T^n} |f(x)|\big[\log^+|f(x)|\big]^{\overline{k}(B)} dx + 1\Big\} \tag{1.2.50}$$

holds.

The relation (1.2.18) implies that if $\tau \in H^p(T^n)$ for some $p \in]0,+\infty[$, then the boundary values of τ are almost everywhere finite in nontangent directions. Using Lemma 1.2.11 (see(1.2.43)), we conclude that it holds

Theorem 1.2.1. *Let $f \in L(\log^+ L)^{n-1}(T^n)$. Then for any $B \subset M$, there almost everywhere exists a limit*

$$\lim_{r\to 1-} \overline{f}(x,r,B).$$

On the basis of Lemma 1.2.12, we have

Theorem 1.2.2. *Let $f \in L(\log^+ L)^{n-1}(T^n)$. Then for any nonempty $B \subset M$, we have*

$$\lim_{r\to 1-} \big[\overline{f}(x,r,B) - \overline{f}_B(x,\varepsilon)\big] = 0$$

almost everywhere.

Using (1.2.44), we make sure that it holds

Theorem 1.2.3. *Let $f \in L(\log^+ L)^{\beta}(T^n)$ for some $\beta \in [0,n-1]$. Then for any $B \subset M$ with $\overline{k}(B) \in [1,\beta+1]$, the function $\overline{f}_B$ exists almost everywhere and*

$$\overline{f}_B(x) = \overline{f}_{i_1,i_1,\dots,i_k}(x) = \cdots = \overline{f}_{i_k,i_{k-1},\dots,i_1}(x) \ \ for\ \overline{k}(B) \in [2,\beta+1] \tag{1.2.51}$$

almost everywhere. Moreover, for any $p \in]0,1[$, *we have*

$$\|\overline{f}_B\|_p \le A(p,n)\Big\{\int\limits_{T^n} |f(x)|\Big[\log^+ |f(x)|\Big]^\beta dx + 1\Big\}. \tag{1.2.52}$$

Note that for $\beta = n-1$, the theorem was proved by Zygmund [11].
It should be noted that if $B_1 \subset B$, $B_1 \ne B$, then

$$\|t^*_{B_1}\|_1 \le A(n,\beta)\Big\{\int\limits_{T^n} |f(x)|\Big[\log^+ |f(x)|\Big]^\beta dx + 1\Big\}.$$

The last relation is implied by (1.2.34) and the inequality

$$\int\limits_{T^n} M_{B\setminus B_1}(x, g_{B_1})\,dx \le A(n,\beta)\Big\{\int\limits_{T^n} |f(x)|\Big[\log^+ |f(x)|\Big]^\beta dx + 1\Big\}$$

which, in its turn, follows from the scheme of the proof of Lemma 1.2.10.

Corollary 1.2.2. *Let* $f \in L(\log^+ L)^{n-1}(T^n)$. *Then for any nonempty* $B \subset M$, *the function* $\overline{f}_B$ *exists almost everywhere and almost everywhere* (1.2.51) *holds.*

In particular, if $f \in L(\log^+ L)^{n-1}(T^n)$, then almost everywhere

$$\overline{f}_M(x) = \overline{f}_{1,2,\dots,n}(x) = \cdots = \overline{f}_{n.n-1,\dots,1}(x).$$

Now we begin to investigate the problem whether Theorem 1.2.3 can be improved. First we prove some auxiliary assertions which are interesting by themselves.

Lemma 1.2.15. *Let* $a = (a_1, a_2) \in \mathbb{R}^2_+$ *and*

$$I^2 = \Big]0, \frac{1}{3}\Big]^2.$$

Put

$$E(\alpha) \equiv E(\alpha, a) =$$
$$= \Big\{x : \Big[x_1x_2|\log x_1|^{a_1}|\log x_2|^{a_2}\Big]^{-1} > \alpha,\ x \in I^2\Big\} \ \text{for}\ \alpha \in]1, +\infty[, \tag{1.2.53}$$

and $\nu = a_1 + a_2$. *Then there is a number* $\alpha_0 \equiv \alpha_0(a) \in]0, +\infty[$ *such that*

$$A(a)\Big[\alpha^{-1}(\log\alpha)^{1-\nu}\Big] \le \mu_2[E(\alpha)] \le$$
$$\le A_1(a)\Big[\alpha^{-1}(\log\alpha)^{1-\nu}\Big] \quad \text{for}\ \nu \in [0,1[,\ \ \alpha \in [\alpha_0, +\infty[; \tag{1.2.54}$$
$$A(a)\Big[\alpha^{-1}\log\log\alpha\Big] \le \mu_2[E(\alpha)] \le$$
$$\le A_1(a)\Big[\alpha^{-1}\log\log\alpha\Big] \quad \text{for}\ \nu = 1,\ \ \alpha \in [\alpha_0, +\infty[; \tag{1.2.55}$$
$$\mu_2[E(\alpha)] \le A(a)\alpha^{-1} \quad \text{for}\ \nu \in]1, \infty[,\ \ \alpha \in [\alpha_0, +\infty[. \tag{1.2.56}$$

Proof. Define

$$E_1(\alpha)=\left\{x:\left[x_1x_2|\log x_1|^{a_1}|\log x_2|^{a_2}\right]^{-1}>\alpha,\ \ x_2\le x_1,\ \ x\in I^2\right\},$$
$$E_2(\alpha)=\left\{x:\left[x_1x_2|\log x_1|^{a_1}|\log x_2|^{a_2}\right]^{-1}>\alpha,\ \ x_2> x_1,\ \ x\in I^2\right\},$$
$$E_3(\alpha)=\left\{x:\left[x_1x_2|\log x_1|^{\nu}\right]^{-1}>\alpha,\ \ x\in I^2\right\},$$
$$E_4(\alpha)=\left\{x:\left[x_1x_2|\log x_2|^{\nu}\right]^{-1}>\alpha,\ \ x\in I^2\right\}.$$

It is easy to see that

$$E(\alpha)=E_1(\alpha)\cup E_2(\alpha)\subset E_3(\alpha)\cup E_4(\alpha). \tag{1.2.57}$$

But

$$E_3(\alpha)=\left\{x:x_2\le\left[\alpha x_1|\log x_1|^{\nu}\right]^{-1},\ \ x\in I^2\right\}. \tag{1.2.58}$$

Let now $y\equiv y(a,\alpha)$ be such that

$$\left[\alpha y|\log y|^{\nu}\right]^{-1}=3^{-1}.$$

We can plainly see that

$$\alpha^{-2}<y<3\alpha^{-1}\quad\text{for}\quad\alpha\in[\alpha_0,+\infty[.$$

Then, by (1.2.58), for $\alpha\in[\alpha_0,+\infty[$ it holds

$$\mu_2[E_3(\alpha)]\le A\alpha^{-1}\int_{1/\alpha^2}^{1/3}\left[x_1|\log x_1|^{\nu}\right]^{-1}dx_1\le$$
$$\le\begin{cases}A(a)\alpha^{-1}(\log\alpha)^{1-\nu}, & \nu\in[0,1[,\\ A(a)\alpha^{-1}\log\log\alpha, & \nu=1,\\ A(a)\alpha^{-1}, & \nu\in]1,+\infty[;\end{cases} \tag{1.2.59}$$

Quite analogously

$$\mu_2[E_4(\alpha)]\le\begin{cases}A(a)\alpha^{-1}(\log\alpha)^{1-\nu}, & \nu\in[0,1[,\\ A(a)\alpha^{-1}\log\log\alpha, & \nu=1,\\ A(a)\alpha^{-1}, & \nu\in[1,+\infty[.\end{cases} \tag{1.2.60}$$

Define $t\equiv t(a,\alpha)$ by

$$t^2|\log t|^{\nu}=\alpha^{-1}.$$

It is easy to see that

$$\alpha^{-1}<t<\alpha^{-1/3}\quad\text{for}\quad\alpha\in[a_0,+\infty[.$$

Then our notation implies

$$\mu_2[E_2(\alpha)] \geq A \int_{3/\alpha}^{\frac{1}{\sqrt[\nu]{\alpha}}} \Big[(\alpha x_1 |\log x_1|^{\nu})^{-1} - x_1\Big]\, dx_1 \geq$$

$$\geq A(a)\Big[\alpha^{-1}(\log\alpha)^{1-\nu}\Big] \quad \text{for } \nu \in [0,1[, \quad \alpha \in [\alpha_0, +\infty[, \tag{1.2.61}$$

and if $\nu = 1$, then

$$\mu_2[E(\alpha)] \geq A\Big[\alpha^{-1}\log\log\alpha\Big] \quad \text{for } \alpha \in [\alpha_0, +\infty[. \tag{1.2.62}$$

Therefore, taking note of (1.2.57) and (1.2.59)–(1.2.63), we make sure that (1.2.54)–(1.2.56) are valid. ■

Remark 1.2.1. Let

$$a \in [0, +\infty[^n, \quad \nu = \sum_{i=1}^{n} a_i, \quad I^n = \Big]0, \frac{1}{3}\Big]^n \quad (n \geq 3)$$

and

$$E(\alpha) \equiv E(\alpha, a, n) = \left\{ x : \left[\prod_{i=1}^{n} x_i |\log x_i|^{a_i}\right]^{-1} > \alpha, \quad x \in I^n \right\} \tag{1.2.63}$$

$$\text{for } \alpha \in [20, +\infty[.$$

In this notation, one can establish the assertions which are analogues (for $n \geq 3$) of the inequalities (1.2.54)–(1.2.56), naturally with appropriate changes. In the first place, the μ_n-measure of Lebesgue should be considered, and in the second place, three cases are possible for ν: a) $\nu \in [0, n-1[$; b) $\nu = n-1$, and c) $\nu \in]n-1, +\infty[$. Each of these cases contains various admissible values for the corresponding a_i, $1 \leq i \leq n$. All these peculiarities have some influence on the estimates of the type (1.2.54)–(1.2.56) in the case $n \geq 3$.

Lemma 1.2.16. *Let $a \in [0, +\infty[$ and*

$$I^n \equiv I^n(\gamma) =]0, \gamma]^n.$$

Put

$$E(\alpha) \equiv E(\alpha, a, n) =$$

$$= \left\{ x : \prod_{i=1}^{n} x_i^{-1} \left[\log\left(\prod_{i=1}^{n} x_i^{-1}\right)\right]^{-a} > \alpha, \; x \in I^n \right\} \quad \textit{for } \alpha \in]10, +\infty[. \tag{1.2.64}$$

Then there is a number $\alpha_0 \equiv \alpha_0(a, n)$ such that

$$A(n,a)\Big[\alpha^{-1}(\log\alpha)^{n-1-a}\Big] \leq \mu_n[E(\alpha)] \leq$$

$$\leq A_1(n,a)\Big[\alpha^{-1}(\log\alpha)^{n-1-a}\Big], \quad a \in [0, n-1[, \quad \alpha \in [\alpha_0, +\infty[, \tag{1.2.65}$$

$$\mu_n[E(\alpha)] \leq A(n,a)\alpha^{-1} \quad a \in]n-1, +\infty[, \quad \alpha \in [\alpha_0, +\infty[. \tag{1.2.66}$$

We are not going to give the proof of Lemma 1.2.16 since its validity is easily seen from the scheme of the proof of the following lemma. We only add that if $a = 0$ and $n = 2$, the estimate (1.2.65) follows from the corresponding inequalities (1.2.54).

Lemma 1.2.17. *Let $n \geq 2$, $a \in]0, +\infty[$ and*

$$I^n \equiv I^n(\gamma) =]0, \gamma]^n. \tag{1.2.67}$$

Put

$$E(\alpha) \equiv E(\alpha, a, n) =$$
$$= \left\{x : \prod_{i=1}^{n} x_i \left[\log\left(\prod_{i=1}^{n} x_i^{-1}\right)^{n-1}\right]^{-1} \left[\log\log\left(\prod_{i=1}^{n} x_i^{-1}\right)\right]^a \geq \alpha, \quad x \in I^n\right\}$$
$$\textit{for} \quad \alpha \in [20, +\infty[.$$

Then there is a number $\alpha_0 \equiv \alpha_0(a, n)$ such that

$$\mu_n[E(\alpha)] \geq A(n, a)\left[\alpha^{-1}(\log\log\alpha)^a\right] \quad \textit{for} \quad \alpha \in [\alpha_0, +\infty[.$$

Proof. Assume

$$q(s) = \left[s(\log s)^{n-1}\right]^{-1}(\log\log s)^a \quad \text{for} \quad s \in [e^{n^2}, +\infty[,$$

and define s_0 from the condition $q(s_0) = \alpha$. It is easy to see that if

$$s_1 = \left[2\alpha(\log\alpha)^{n-1}\right]^{-1}(\log\log\alpha)^a,$$
$$s_2 = 2\left[\alpha(\log\alpha)^{n-1}\right]^{-1}(\log\log\alpha)^a \quad \text{for} \quad \alpha \in [\alpha_0, +\infty[,$$

then $s_1 < s_0 < s_2$. According to our notation, we get

$$\mu_n[E(\alpha)] = \left\{x : q\left(\prod_{i=1}^{n} x_i\right) \geq q(s_0), \quad x \in I^n\right\}. \tag{1.2.68}$$

On the other hand, we can assume that

$$\gamma^{-1}s_1 \in]0, \alpha^{-1}[\quad \text{for} \quad \alpha \in [\alpha_0, +\infty[.$$

Since $s_1 < s_0$, in view of the monotonicity of q and (1.2.68) we find

$$\mu_n[E(\alpha)] \geq \mu_n\left[\left\{x : \prod_{i=1}^{n} x_i \leq s_1 \in I^n\right\}\right] =$$
$$= \mu_n\left[\left\{x : x_n \leq s_1 \prod_{i=1}^{n-1} x_i^{-1}, \quad x \in I^n\right\}\right].$$

Hence, taking note of $\gamma^{-1}s_1 < \alpha^{-1}$, we have

$$\mu_n[E(\alpha)] \geq s_1 \int\limits_{\{x: \prod_{i=1}^{n-1} x_i > \alpha^{-1}\}} \prod_{i=1}^{n-1} x_i^{-1} dx_i \geq$$

$$\geq A(n,\alpha)s_1(\log\alpha)^{n-1}, \quad \alpha\in[\alpha_0,+\infty[. \quad \blacksquare$$

As already noted, in 1961 Stein [1] indicated that the condition $f\in L(\log^+ L)^{1-\varepsilon}(T^2)$ with $\varepsilon\in]0,1]$, in general, does not guarantee the existence of $\overline{f}_M$ ($n=2$, $M=\{1,2\}$) almost everywhere even in the sense of λ-limit ($\lambda=1$). We noted in [8] and [13, Ch. III] that there are some inaccuracies in the scheme of the proof proposed by Stein.

Stein made use of the fact that the function $f:T^2\to\mathbb{R}$ defined as

$$f(x)=f(x_1)f(x_2),$$

where

$$f(x_i)\sim\sum_{p=2}^{\infty}\frac{\cos px_i}{(\log p)^\delta} \quad \text{with } \delta>0 \quad (i=1,2),$$

has the following properties.

a) $f\in L(\log^+ L)^{1-\varepsilon}(T^2)$ for $\varepsilon\in]0,\delta[$;

b)

$$\mu_2\Big[\Big\{x:|\overline{f}_M(x)|\geq\alpha,\ x\in T^2\Big\}\Big]\geq$$
$$\geq A(\delta)\alpha^{-1}(\log\alpha)^{1-\delta} \quad \text{for } \alpha\in\alpha_0,+\infty[, \tag{1.2.69}$$

the latter being a consequence of the inequality

$$|\overline{f}_M(x)|\geq A(\delta)\bigg[|x_1x_2|\Big(\log\frac{2\pi}{|x_1|}\log\frac{2\pi}{|x_2|}\Big)^\delta\bigg]^{-1}$$
$$\text{for } |x_i|\in]0,\delta_0], \quad \delta_0\in]0,\pi] \quad (i=1,2).$$

But, in fact, the conclusion a) is true only for $\delta>1-\varepsilon$, and Lemma 1.2.15 implies that (1.2.69) also fails to be true.

In view of the above said, the argument of Stein does not lead to the required result in the two-dimensional case even for the almost everywhere summability of the series $\sigma_1[f,\{1,2\}]$ by the method of Poisson–Abel.

Remark 1.2.2. The work of Stein is an important contribution to the modern function theory, and the above mentioned inaccuracies do not affect its value. From the further researches we should mention the papers by Garsia [1], Coifman [1], S. Sawyer [1], as well as the work of Nikishin [1]. By now, the main principles concerning the almost everywhere convergence of sequences of certain measurable functions $(f_k)_{k\geq1}$ are known well enough (see Garsia [1], S. Sawyer [1], Stein [1]). These principles claim that in some rather general conditions, the everywhere convergence of $(f_k)_{k\geq1}$ is equivalent to the validity of a certain inequality of the weak type for $\sup_k|f_k(x)|$. In what follows, we will use these principles.

In 1965, on the seminar led by Men'shov and Ul'yanov, Ul'yanov posed the problem: if $f\in L(\log^+ L)^\beta$ ($n\geq2$) for some $\beta\in]0,n-1[$, then which of the conjugate functions $\overline{f}_B$ ($\varnothing\neq B\subset M$) do exist almost everywhere on T^n?

Below the assertions are given which, in particular, solve this problem. Moreover, in the case $n=2$, the scheme of Stein [1] is corrected.

Theorem 1.2.4. a) *Suppose that* $\beta \in [0, n-1]$ *and* $f \in L(\log^+ L)^\beta(T^n)$. *Then all the conjugate functions* $\overline{f}_B$ *with* $\overline{k}(B) \in [1, \beta+1]$ *exist almost everywhere;*

b) *let* $\beta \in [0, n-1[$ *and* $B \subset M$ *with* $\overline{k}(B) \in]\beta+1, n]$. *Suppose that* $\varepsilon_m \in]0, \gamma[$ $(m = 1, 2, \dots)$, $\lim_{m\to\infty} \varepsilon_m = 0$, *and the functions* $\overline{f}_B(x, \varepsilon)$ *are defined by* (1.1.8). *Then there exists a function* $\varphi \in L(\log^+ L)^\beta(T^n)$ *such that*

$$\limsup_{m\to\infty} |\overline{\varphi}_B(x, \varepsilon(\varepsilon_m))| = +\infty$$

almost everywhere.

Proof. The assertion a) for $\beta \in [0, 1[$ (and therefore for $\overline{k}(B) = 1$ as well) is a consequence of the well-known Luzin–Privalov theorem (see, e.g., Bari [2, Ch. VIII]) and Fubini's theorem. If $\beta \in [1, n-1]$, then a) follows from Theorem 1.2.3, the functions $\overline{f}_B$ with $\overline{k}(B) \in [2, \beta+1]$ being considered as $\overline{k}(B)$-fold iterated integrals in the sense of conjugation.

Now we pass to the proof of the assertion b).

Assume that $\eta \in]0, k-\beta-1[$, $a \in]0, +\infty[$, $k \equiv \overline{k}(B)$ and the function $f : T^n \to \mathbb{R}$ is defined by

$$f(x) = \begin{cases} (2\pi)^{2\pi} \prod_{i=1}^n x_i^{-1} \left[\log\left(\prod_{i=1}^n x_i^{-1}\right)\right]^{-n-\beta-\eta} \times \\ \qquad \times \left[\log\log\left(\prod_{i=1}^n x_i^{-1}\right)\right]^a \quad \text{for } x \in I^n(\gamma), \\ 0 \quad \text{for } x \in T^n \setminus I^n(\alpha). \end{cases} \tag{1.2.70}$$

Extend f 2π-periodically in each of the coordinates of the point x. According to the condition, we have $k - \beta - 1 > 0$. Then, by (1.2.70), we get

$$\int_{T^n} f(x)\left[\log^+ f(x)\right]^\beta dx \le$$
$$\le A(n, \beta, \eta, a) \int_{T^n} \prod_{i=1}^n x_i^{-1} \left[\log\left(\prod_{i=1}^n x_i^{-1}\right)\right]^{-n-\eta} \times$$
$$\times \left[\log\log\left(\prod_{i=1}^n x_i^{-1}\right)\right]^a dx < +\infty.$$

Let now $x \in]0, \gamma[^n$. In view of (1.1.8) and (1.2.70), we have

$$(2\pi)^k \sup_m |\overline{f}_B(-x_B, \varepsilon(\varepsilon_m))| \ge \int_{E(x,\gamma,B)} f(s_B + x_{B'}) \prod_{i\in B} \operatorname{ctg} \frac{s_i + x_i}{2}\, ds_B.$$

Hence

$$\sup_m |\overline{f}(-x_B, \varepsilon(\varepsilon_m))| \ge A(n, \beta, \eta) \prod_{i=1}^n x_i^{-1} \left[\log\left(\prod_{i=1}^n x_i^{-1}\right)\right]^{-n+1} \times$$
$$\times \left[\log\log\left(\prod_{i=1}^n x_i^{-1}\right)\right]^a \quad \text{for } x \in I^n(\gamma). \tag{1.2.71}$$

Thus, by virtue of (1.2.71) and Lemma 1.2.17, it holds

$$\mu_n\Big(\Big\{x : \sup_m |\overline{f}_B(x, \varepsilon(\varepsilon_m))| \geq \alpha, \;\; x \in T^n\Big\}\Big) \geq$$
$$\geq A(n, \beta, \eta, a)\alpha^{-1}(\log\log\alpha)^a \;\text{ for }\; \alpha \in [\alpha_0, +\infty]. \tag{1.2.72}$$

If we put

$$\Phi(t) = t\Big[\log(e+t)\Big]^\beta \;\text{ for }\; t \in [0, +\infty[,$$

we can easily see that

$$\int_{T^n} \Phi(\alpha^{-1}Af(x))\,dx \leq \alpha^{-1}A(\Phi, f) \;\text{ for }\; \alpha \in [\alpha_0, +\infty[. \tag{1.2.73}$$

Therefore, using for the sequence

$$\Big(\overline{f}_B(x, \varepsilon(\varepsilon_m))\Big)_{m\geq 1} \tag{1.2.74}$$

the appropriate result of Stein [1] and the relations (1.2.71) and (1.2.73), we make sure that b) is true. ■

In the works of Getsadze [1], [2] and Konyagin [1], [2], the problem of convergence in measure of multiple trigonometric Fourier series is investigated. The same problem arises in connection with the sequence (1.2.74).

From the Theorem 1.2.4, we can deduce

Corollary 1.2.3. a) *Let $\beta \in [0, n-1[$ and $f \in L(\log^+ L)^\beta(T^n)$. Then all the sequences (1.2.74) with $\overline{k}(B) \in [1, \beta+1]$ converge by measure to $\overline{f}_B$ on T^n.*

b) *Let $\beta \in [0, n-1[$ and $B \subset M$ with $\overline{k}(B) \in]\beta+1, n[$. Then there is a function $f \in L(\log^+ L)^\beta(T^n)$ such that the corresponding sequence (1.2.74) does not converge by measure on T^n.*

Note that in the assertions b) of Theorem 1.2.4 and Corollary 1.2.3, the functions are not (in a certain sense) constructed effectively.

Below a scheme used in Getsadze [2] is set forth which makes it possible (in a certain sense) to define the function f from b) of Corollary 1.2.3.

Assume that the function f is defined by (1.2.70). Let a sequence $(\varepsilon_m)_{m\geq 1}$ satisfy the conditions of the assertion b) of Theorem 1.2.4, and let $\varepsilon_m \downarrow 0$. It can be shown that

$$\mu_n\Big(\Big\{x : |\overline{f}_B(x, \varepsilon(\varepsilon_m))| \geq \varepsilon_m^{-1}, \;\; x \in T^n\Big\}\Big) \geq$$
$$\geq A(n, \beta, \eta, a)\varepsilon_m(\log\log\varepsilon_m^{-1})^a \;\text{ for }\; m \geq m_0.$$

Then (see Getsadze [2]) we can define sequences of measurable functions $(\psi_i)_{i\geq 1}$ and positive numbers $(b_i)_{i\geq 1}$ such that the function f_0 given by

$$f_0(x) = \sum_{i=1}^{\infty} b_i\psi_i(x)$$

is just the one the conclusion of b) of Corollary 1.2.3 affirms to exist.

We should add that the function f_0 likewise effectively realizes the corresponding statement of Theorem 1.2.4 on a set $E \subset T^n$ with $\mu_n(E) > 0$.

Let now a function $\Phi : [0, +\infty[\to [0, +\infty[$ be increasing, $\Phi(0) = 0$, and

$$\liminf_{t\to\infty} t^{-1}\Phi(t) > 0.$$

For such Φ put

$$L_\Phi(T^n) = \Big\{ f : \int_{T^n} \Phi(|f(x)|)\,dx < +\infty \Big\}. \tag{1.2.75}$$

Basing on fundamental researches by Saks [1], Busemann and Feller [1], and Jessen, Marcinkiewicz and Zygmund [1], we can ascertain

Theorem 1.2.5. *If for any $f \in L_\Phi(T^n)$ all the functions $\overline{f}_B$ with $\overline{k}(B) \in [1, n]$ exist almost everywhere, then*

$$\Phi(t) \geq At\log^{n-1}(e+t).$$

Remark 1.2.3. In the condition of the theorem, it suffices to require that for any $f \in L_\Phi(T^n)$, the function $\overline{f}_M$ exists almost everywhere, though the existence almost everywhere of the function $\overline{f}_B$ with $\overline{k}(B) \in [3, n]$ does not imply the existence of $\overline{f}_{B_1}$ with $\overline{k}(B_1) \in [2, \overline{k}(B)[$ on a set E with $\mu_n(E) > 0$.

In Theorem 1.2.5, the existence of conjugate functions is meant in the sense of Stolz–Pringsheim. Tkebuchava [3] proved that the conclusion of Theorem 1.2.5 remains true if the existence of conjugate functions is meant in the sense of λ-convergence ($\lambda = 1$). This result improves the corresponding statement of our work [37]. To be more precise, in [37] we have established

Theorem 1.2.6. *Let a function $\Phi : [0, +\infty[\to [0, +\infty[$ satisfy $\Phi \uparrow \infty$ as $t \uparrow \infty$ and*

$$\Phi(t)\left[(\log t)^{n-1}\right]^{-1} \to 0 \quad as \quad t \to \infty.$$

Then there is a function $f \in L(T^n)$ such that

$$\int_{T^n} \Phi(|f(x)|)\,dx < +\infty,$$

yet almost everywhere

$$\limsup_{m\to\infty} |\overline{f}_M(x, \varepsilon(\varepsilon_m))| = +\infty.$$

We have made an essential use of the function f defined by (1.2.70). Below we define and use a function f which in some cases is more convenient.

For any $\eta \in]0, [\beta] + 1 - \beta[$, put

$$f(x) = \begin{cases} \prod_{i=1}^n x_i^{-1}\left[\log\left(\prod_{i=1}^n x_i^{-1}\right)\right]^{-n-\beta-\eta} & \text{for } x \in I^n(\gamma), \\ 0 \quad \text{for} \quad x \in T^n \setminus I^n(\gamma), \end{cases} \tag{1.2.76}$$

where $[\beta]$ denotes the integer part of $\beta \in [0, n-1[$. Extend f 2π-periodically in each of the coordinates of x. We have

$$\int_{T^n} f(x)\left[\log^+ f(x)\right]^\beta dx < +\infty.$$

Further, let

$$x_i \in \left[\frac{\gamma}{2m}, \gamma\right[\quad \text{for} \quad i \in B \quad \text{and} \quad x_j \in]0, \gamma[\quad \text{for} \quad j \in B'.$$

Using (1.1.8) and (1.2.76), we get

$$\left|\overline{f}_B\left(-x_B, \varepsilon\left(\frac{\gamma}{2m}\right)\right)\right| \geq$$
$$\geq A(n, \beta, \eta) \prod_{i=1}^n x_i^{-1} \left[\log\left(\prod_{i=1}^n x_i^{-1}\right)\right]^{-n-\beta-\eta+\overline{k}(B)}. \tag{1.2.77}$$

If $\eta_1 = 1 + [\beta] - \beta - \eta$, then (1.2.76) implies that $\eta_1 > 0$. On the other hand, for $\overline{k}(B) \in]\beta + 1, n]$ we have $\overline{k}(B) \geq [\beta] + 2$. Then (1.2.77) yields

$$\left|\overline{f}_B\left(-x_B, \varepsilon\left(\frac{\gamma}{2m}\right)\right)\right| \geq$$
$$\geq A(n, \beta, \eta) \prod_{i=1}^n x_i^{-1} \left[\log\left(\prod_{i=1}^n x_i^{-1}\right)\right]^{-n+1+\eta_1}. \tag{1.2.78}$$

Using Lemma 1.2.16 (see (1.2.65)), according to (1.2.78) we find

$$\mu_n\left(\left\{x : \left|\overline{f}_B\left(-x_B, \varepsilon\left(\frac{\gamma}{2m}\right)\right)\right| \geq m, \ x \in T^n\right\}\right) \geq$$
$$\geq A(n, \beta, \eta, \eta_1) m^{-1} \log^{\eta_1}(m+2) \quad \text{for} \quad m \geq m_0. \tag{1.2.79}$$

It follows from (1.2.76) that $n + \beta + \eta < n + \beta + 1 + [\beta] - \beta \leq 2n - 1$. If we assume that $n + \beta + \eta = 2n - 1$, i.e.,

$$f(x) = \begin{cases} \prod_{i=1}^n x_i^{-1} \left[\log\left(\prod_{i=1}^n x_i^{-1}\right)\right]^{-2n+1}, & x \in I^n(\gamma), \\ 0, \quad x \in T^n \setminus I^n(\gamma), \end{cases} \tag{1.2.80}$$

and extend f periodically in each of the coordinates of x, then we can easily see that

$$\int_{T^n} f(x)\left[\log f(x)\right]^{n-1} dx = +\infty.$$

In the proof of Theorem 1.2.6, we have made use of a modification of the function f defined by (1.2.80).

Now we should say a few words about the multivariable Hilbert transforms defined by (1.1.12).

It is well-known that if $\mu_n(E) = +\infty$, the Lebesgue classes $L(E)$, $L(\log^+ L)^\beta(E)$ with $\beta \in]0, +\infty[$, $L^p(E)$ with $p \in]1, +\infty[$ are not imbedded in each other as in the case $E = T^n$. Therefore, in studying the existence almost everywhere of the functions $\widetilde{f}_B$, some additional conditions appear.

In 1957, Zygmund [14] noted that if $f \in L(\mathbb{R}^2) \cap L(\log^+ L)(\mathbb{R}^2)$, then $\widetilde{f}_M$ ($M = \{1,2\}$) exists almost everywhere.

In 1972, Fefferman [3] established that Zygmund's assertion is valid for more general two-dimensional singular integrals. Fefferman's method works for any $n \geq 2$. So that the following is true.

If $f \in L(\mathbb{R}^n) \cap L(\log^+ L)^{\beta}(\mathbb{R}^n)$ for some $\beta \in [0, n-1]$, then for any $B \subset M$ with $\overline{k}(B) \in [1, \beta + 1]$, the function $\widetilde{f}_B$ exists almost everywhere.

It should be noted that as the assertion b) of Theorem 1.2.4 shows, this statement cannot be improved in the sense that the function $\widetilde{f}_B$ with $\overline{k}(B) \in]\beta + 1, n]$, in general, does not exist almost everywhere.

Now consider the case where $f \in L^p(\mathbb{R}^n)$ for some $p \in]1, +\infty[$. Although Sokol-Sokolowski [1] did not consider multivariable Hilbert transforms, his method proposed for the conjugate functions of several variables imply the following assertion.

If $f \in L^p(\mathbb{R}^n)$ for $p \in]1, +\infty[$, then all the functions $\widetilde{f}_B$ with $k(B) \in [1, n]$ exist and are finite almost everywhere, and

$$\|\widetilde{f}_B\|_p \leq A(p,n)\|f\|_p. \tag{1.2.81}$$

In this assertion, the function $\widetilde{f}_B$ can almost everywhere be represented as a k-fold iterated Hilbert transform, i.e.,

$$\widetilde{f}_B(x) = \widetilde{f}_{i_1, i_2, \ldots, i_k}(x) = \cdots = \widetilde{f}_{i_k, i_{k-1}, \ldots, i_1}(x)$$

almost everywhere.

By now, more general results are obtained in this direction in the sense that comparatively general n-fold integrals are considered and inequalities of the type (1.2.81) are established with weight functions. In this direction, we mention the works of Kokilashvili [1], and R. Fefferman and Stein [1].

The above mentioned results lead to the following problem due to Ul'yanov: is there a function $f \in L(T^n)$ ($n \geq 2$) such that $\overline{f}_M$ does not exist almost everywhere, but $\overline{f}_M$ exists almost everywhere in the sense of λ-limit ($\lambda = 1$)?

In 1984, Gogoladze [4], [5] solved this problem. In particular, he showed that there is a function $f \in L(T^n)$ for which the limit

$$\limsup_{\varepsilon_\lambda \to 0+} \overline{f}_M(x, \varepsilon)$$

exists almost everywhere for any $\lambda \in [1, +\infty[$, yet

$$\limsup_{\varepsilon \to 0+} |\overline{f}_M(x, \varepsilon)| = +\infty$$

almost everywhere.

1.3. Lebesgue Integrability of Conjugate Functions of Several Variables

Using the appropriate results for the one-dimensional case (see, e.g., Bari [2, Ch. VIII]), we conclude that the function $\overline{f}_B$ need not be integrable in any of the variables regardless whether its index belongs to B or B'. It suffices to define the function $f : T^n \to \mathbb{R}$ as

$$f(x) = g\left(\sum_{i=1}^{n} x_i\right),$$

where $g : T \to \mathbb{R}$ is a Luzin's function (see, e.g., Bari [2, Ch. VIII]) for which $\overline{g}$ is not integrable on any interval $]a, b[$ with $b - a > 0$.

In 1947, Sokol-Sokolowski [1] showed that if $f \in L(\log^+ L)^{n+1}(T^n)$, then $\overline{f}_B \in L(T^n)$, $\overline{k}(B) \in [1, n]$.

In 1949, Zygmund [11] established that if $f \in L(\log^+ L)^n(T^n)$, then $\overline{f}_B \in L(T^n)$, $k(B) \in [1, n]$.

The question of the definitiveness of the result of Zygmund was investigated by Helson [1] in the case $n = 2$, and by the author for any $n \geq 2$. In the works [13, Ch. III, IV], [16], we investigated in detail the integrability of conjugate functions of several variables. In the work [37], some assertions in this direction are likewise pointed out. Below we give the main theorems established in these works. In particular, a problem Ul'yanov posed in 1965 is solved: if $n \geq 2$ and $f \in L(\log^+ L)^\beta(T^n)$ for some $\beta \in [1, n[$, then which conjugate functions $\overline{f}_B$ are summable on T^n?

Theorem 1.3.1. a) *Let $f \in L(\log^+ L)^\beta(T^n)$ for some $\beta \in [1, n]$. Then all the functions $\overline{f}_B$ with $\overline{k}(B) \in [1, \beta]$ are summable on T^n.*

b) *Suppose that $\beta \in [1, n[$ and $B \subset M$ with $\overline{k}(B) \in]\beta, n]$. Then there is a function $f \in L(\log^+ L)^\beta(T^n)$ such that $\overline{f}_B \notin L(T^n)$.*

Proof. a) Using Theorem 1.2.3 (see (1.2.51)) and (1.1.45), we have

$$\int_{T^n} |\overline{f}_B(x)| \left[\log^+ |\overline{f}_B(x)|\right]^{\beta - \overline{k}(B)} dx \leq$$

$$\leq A(n, \beta) \left\{ \int_{T^n} |f_B(x)| \left[\log^+ |f_B(x)|\right]^{\beta} dx + 1 \right\}. \tag{1.3.1}$$

Hence, taking into account that $\beta - \overline{k}(B) \geq 0$, we conclude that a) is true.

b) Define the function $f : T^n \to \mathbb{R}$ by

$$f(x) = \begin{cases} (2\pi)^{2\pi} \prod_{i=1}^{n} x_i^{-1} \left[\log\left(\prod_{i=1}^{n} x_i^{-1}\right)\right]^{-n-\beta-\eta} & \text{for } x \in I^n(\gamma), \\ 0 \quad \text{for} \quad x \in T^n \setminus I^n(\gamma), \quad \eta \in]0, \overline{k}(B) - \beta], \end{cases} \tag{1.3.2}$$

and then extend it 2π-periodically in each of the coordinates of x.

We have

$$\int_{T^n} f(x) \left[\log^+ f(x)\right]^\beta dx \leq$$

$$\leq A(n,\beta) \int\limits_{I^n(\gamma)} \prod_{i=1}^{n} x_i^{-1} \left[\log \left(\prod_{i=1}^{n} x_i^{-1} \right) \right]^{-n-\eta} dx < +\infty,$$

i.e., $f \in L(\log^+ L)^\beta(T^n)$.

Let $x \in]0,\gamma[^n$. Using (1.1.8) and (1.3.2), according to our notation we find

$$|f_B(-x_B,\varepsilon)| \geq A(n) \int\limits_{E(x,\gamma,B)} f(s_B + x_{B'}) \prod_{i\in B} \operatorname{ctg} \frac{s_i + x_i}{2} \, ds_B \geq$$

$$\geq A(n,\beta,\eta) \prod_{i=1}^{n} x_i^{-1} \left[\log \left(\prod_{i=1}^{n} x_i^{-1} \right) \right]^{-n-\beta+\overline{k}(B)-\eta}$$

Hence, taking note of $\overline{k}(B) - \beta - \eta \geq 0$, we obtain

$$|\overline{f}_B(-x_B)| \geq A(n,\beta,\eta) \prod_{i=1}^{n} x_i^{-1} \left[\log \left(\prod_{i=1}^{n} x_i^{-1} \right) \right]^{-n}, \quad x \in]0,\gamma[^n,$$

that is,

$$\int\limits_{T^n} |\overline{f}_B(x)| \, dx \geq$$

$$\geq A(n,\beta,\eta) \int\limits_{T^n(\gamma)} \prod_{i=1}^{n} x_i^{-1} \left[\log \left(\prod_{i=1}^{n} x_i^{-1} \right) \right]^{-n} = +\infty. \quad \blacksquare$$

According to the assertion a) of Theorem 1.2.4, the functions f_B with $\overline{k}(B) \in [1,\beta+1]$ exist almost everywhere on T^n. As to the above example, all the functions $\overline{f}_B$ exist almost everywhere on $[-\gamma,0]^n$.

Note also that if in the definition (1.3.2) of the function f we take $\eta \in]0,[\beta]+1-\beta[$, then for any $B \subset M$ with $\overline{k}(B) \in]\beta,n]$, the function $\overline{f}_B \notin L(T^n)$.

Theorem 1.3.2. *Let the function* $\Phi : [0,+\infty[\to [0,+\infty[$ *satisfy* $\Phi \uparrow +\infty$ *as* $t \uparrow +\infty$ *and*

$$\Phi(t)\left[t(\log t)^n\right]^{-1} \to 0 \quad as \quad t \to +\infty. \tag{1.3.3}$$

Then there is a function $f \in L(T^n)$ *such that*

$$\int\limits_{T^n} \Phi(|f(x)|) \, dx < +\infty,$$

yet $\overline{f}_M \notin L(T^n)$.

Proof. Let

$$\mu(t) \equiv \Phi(t)\left[t(\log t)^n\right]^{-1}.$$

By the conditions of the theorem, $\mu(t) \to 0+$ as $t \uparrow +\infty$. We will assume that $\mu(t) \downarrow 0$ as $t \uparrow +\infty$.

We choose a strictly increasing sequence $(p_i)_{i\geq 1}$ of positive numbers such that

$$p_1 \geq e^{2n}, \quad \sum_{i=1}^{\infty} \mu(p_i \log^{-2n} p_i) < +\infty, \tag{1.3.4}$$

$$\sum_{i=1}^{\infty} \Big(\log\log p_{i+1}\Big)^{-1} < +\infty, \quad \log\log p_{i+1} > 2\log\log p_i \quad \text{for } i \in \mathbb{N}. \tag{1.3.5}$$

Put

$$f(x) = \begin{cases} \Big(\log\log p_{i+1}\Big)^{-1} \prod_{j=1}^{n} x_j^{-1} \Big[\log\Big(\prod_{j=1}^{n} x_j^{-1}\Big)\Big]^{-2n} \\ \qquad \text{for } x \in \Big(\bigcup_{i=1}^{\infty}]p_{i+1}^{-1}, p_i^{-1}[\Big) \times]0,\gamma]^{n-1}, \\ 0 \ - \text{ in the remaining points of } T^n, \end{cases} \tag{1.3.6}$$

and extend the function 2π-periodically in each of the coordinates of x.

We have

$$\int_{T^n} \Phi(|f(x)|)\,dx \leq$$

$$\leq A(n) \sum_{i=1}^{\infty} \frac{\mu(p_i \log^{-2n} p_i)}{\log\log p_{i+1}} \int_{\frac{1}{p_{i+1}}}^{\frac{1}{p_i}} \Bigg[\int_{]0,\gamma]^{n-1}} \prod_{j=1}^{n} x_j^{-1} \times$$

$$\times \Big[\log\Big(\prod_{j=1}^{n} x_j^{-1}\Big)\Big]^{-n} dx_2 \cdots dx_n \Bigg] dx_1 \leq$$

$$\leq A(n) \sum_{i=1}^{\infty} \frac{\mu(p_i \log^{-2n} p_i)}{\log\log p_{i+1}} \int_{\frac{1}{p_{i+1}}}^{\frac{1}{p_i}} \Big[x_1\Big(\log x_1^{-1}\delta^{1-n}\Big)\Big]^{-1} dx_1.$$

Hence, using (1.3.4) and (1.3.5), we get

$$\int_{T^n} \Phi(|f(x)|)\,dx < +\infty.$$

Let now $x_1 \in]0, p_1^{-1}[$ and $x_j \in]0,\gamma]$ for $2 \leq j \leq n$. Then there is $\nu \in \mathbb{N}$ such that

$$p_{\nu+1}^{-1} < x_1 \leq p_\nu^{-1}. \tag{1.3.7}$$

According to our notation, we have

$$|\overline{f}_M(-x)| \geq A(n) \int_{E(x,\gamma,M)} f(s) \prod_{j=1}^{n} \operatorname{ctg} \frac{s_j + x_j}{2}\, ds_B \geq$$

$$\geq A(n) \prod_{j=1}^{n} x_j^{-1} \int_{E(x,\gamma,M)} f(s)\,ds,$$

whence by (1.3.6) and (1.3.7), we obtain

$$|\overline{f}_M(-x)| \geq A(n)\prod_{j=1}^{n} x_j^{-1}\Big\{ \int\limits_{\frac{1}{p_{\nu+1}}}^{x_1} \Big[\int\limits_0^{x_2}\cdots\int\limits_0^{x_n} f(s)\,ds_2\ldots ds_n\Big]ds_1 +$$

$$+\sum_{i=\nu+1}^{\infty}\int\limits_{\frac{1}{p_{i+1}}}^{\frac{1}{p_i}} \Big[\int\limits_0^{x_2}\cdots\int\limits_0^{x_n} f(s)\,ds_2\ldots ds_n\Big]ds_1 \geq$$

$$\geq A_1(n)\prod_{j=1}^{n} x_j^{-1}\Big\{\big(\log\log p_{\nu+1}\big)^{-1}\Big[\ln\Big(\prod_{j=1}^{n} x_j^{-1}\Big)\Big]^{-n} -$$

$$-\Big[\ln\Big(\prod_{j=2}^{n} x_j^{-1}p_{\nu+1}^{-1}\Big)\Big]^{-n} + \sum_{i=\nu+1}^{\infty}\int\limits_{\frac{1}{p_{i+1}}}^{\frac{1}{p_i}} \Big[\int\limits_0^{x_2}\cdots\int\limits_0^{x_n} f(s)\,ds_2\ldots ds_n\Big]ds_1\Big\}.$$

Thus

$$|\overline{f}_M(-x)| \geq A(n)\big(\log\log p_{\nu+1}\big)^{-1}\prod_{j=1}^{n} x_j^{-1}\Big\{\Big[\log\Big(\prod_{j=1}^{n} x_j^{-1}\Big)\Big]^{-n} -$$

$$-\Big[\log\Big(\prod_{j=2}^{n} x_j^{-1}p_{\nu+1}^{-1}\Big)\Big]^{-n}\Big\}. \tag{1.3.8}$$

Then

$$\int\limits_{T^n}|\overline{f}_M(x)| \geq \int\limits_0^{\frac{1}{p_1}}\Big[\int\limits_0^{\gamma}\cdots\int\limits_0^{\gamma}|\overline{f}_M(-x)|\,dx_2\ldots dx_n\Big]dx_1 \geq$$

$$\geq A(n)\sum_{\nu=1}^{\infty}\Big\{\big(\log\log p_{\nu+1}\big)^{-1}\int\limits_{\frac{1}{p_{\nu+1}}}^{\frac{1}{p_\nu}}\int\limits_0^{\gamma}\cdots\int\limits_0^{\gamma}\Big[\prod_{j=1}^{n} x_j^{-1}\Big[\log\Big(\prod_{j=1}^{n} x_j^{-1}\Big)\Big]^{-n} -$$

$$-\prod_{j=1}^{n} x_j^{-1}\Big[\log\Big(\prod_{j=2}^{n} x_j^{-1}p_{\nu+1}^{-1}\Big)\Big]^{-n}\Big]dx_2\ldots dx_n\Big]dx_1\Big\}.$$

Hence, by virtue of (1.3.4) and (1.3.5), we find

$$\int\limits_{T^n}|\overline{f}_M(x)|\,dx = +\infty.$$

Later, statements will be brought which imply that only the class $L(\log^+ L)^n(T^n)$ does guarantee the integrability of the function $\overline{f}_M$ on T^n. However, analyzing the proof of Theorem 1.3.2 and appropriately modifying the function f from (1.3.6), we conclude that it holds

Theorem 1.3.3. *Suppose that* $\Phi : [0,+\infty[\to [0,+\infty[$, $\Phi \uparrow \infty$ *as* $t \uparrow \infty$, *and*

$$\liminf_{t\to\infty} \Phi(t)(t\log^n t)^{-1} = 0. \tag{1.3.9}$$

Then there is a function $f \in L(T^n)$ *such that*

$$\int_{T^n} \Phi(|f(x)|)\,dx < +\infty,$$

yet $\overline{f}_M \notin L(T^n)$.

Finally note that according to Theorem 1.2.3 (see (1.2.52)), if $f \in L(\log^+ L)^\beta(T^n)$ for some $\beta \in [0, n-1]$, then all the functions $\overline{f}_B$ with $\overline{k}(B) \in [1, \beta+1]$ belong to $L^p(T^n)$ for any $p \in]0,1[$.

Consider now the case where $f \in L^p(T^n)$ with $p \in]1,+\infty[$.

As already noted, Sokol-Sokolowski [1] showed that if $f \in L^p(T^n)$ for some $p \in]1,+\infty[$, then all the functions $\overline{f}_B$ exist almost everywhere and

$$\|\overline{f}_B\|_p \le A(p,n)\|f\|_p. \tag{1.3.10}$$

We should add that in obtaining this inequality, he made an essential use of the M. Riesz's inequality (see (1.1.10) from the first part).

Using a result of Ryan [1], we can show that if $L_\Phi(T^n)$ is a reflexive Orlicz space and $f \in L_\Phi(T^n)$, then all the functions $\overline{f}_B$ exist almost everywhere and

$$\|\overline{f}_B\|_\Phi A(\Phi,n)\|f\|_\Phi. \tag{1.3.11}$$

A similar inequality holds for all the multivariable Hilbert transforms as well, naturally, with appriopriate changes in $\|\cdot\|_\Phi$.

1.4. On the Validity of a Theorem of M. Riesz for Functions of Several Variables

As already said, the following theorem is due to M. Riesz [3] (see, e.g., Bari [2, Ch. VIII]): if $f \in L(T)$, $f \ge 0$, and $\overline{f} \in L(T)$, then $\overline{f} \in L\log^+ L(T)$.

In the papers [12], [13, Ch. III], [16], we presented some results connected with the problem of validity of this theorem for functions of several variables. Below these results are set forth in a more general form.

Let $B \subset M$ and $\overline{k}(B) \in [2,n]$. If $f \in L(T^n)$ and $f \ge 0$, then the condition $\overline{f}_B \in L^p(T^n)$ for some $p \in [1,+\infty[$ does not imply that $f \in L(\log^+ L)^\beta(T^n)$ for some $\beta \in]0,+\infty[$. The counterexample is the function defined as

$$f(x) = \sum_{i=1}^n f_i(x_i), \tag{1.4.1}$$

where $f_i \in L(T)$, $f_i \ge 0$ for $1 \le i \le n$, and at least one of them does not belong to $L(\log^+ L)^\beta(T)$ for any $\beta \in]0,+\infty[$.

Further, the conditions $f \in L(T^n)$, $f \ge 0$ and $\overline{f}_B \in L(T^n)$ for all $B \subset M$ such that

$$\overline{k}(B) = 1, \quad \mathrm{Card}\left\{\overline{k}(B) \,:\, B \subset M\right\} < n$$

do not imply the conclusion $f \in L(\log^+ L)^\beta(T^n)$ for some $\beta \in]0, +\infty[$. The counterexample is easy to construct using the functions $f_i : T \to \mathbb{R}$, $1 \le i \le n$, where $f_i \in L\log^+ L(T)$ for $i \ne i_0$, $f_{i_0} \in L(T) \setminus \bigcup_{\beta>0} L(\log^+ L)^\beta(T)$ and $\overline{f}_{i_0} \notin L(T)$. The required function $f : T^n \to \mathbb{R}$ can be defined either as in (1.4.1), or as

$$f(x) = \prod_{i=1}^{n} f_i(x_i).$$

Of course, in these remarks instead of the class $L(\log^+ L)^\beta(T^n)$, the class $L\Phi(L)(T^n)$ can be considered, where $\Phi : [0, +\infty[\to [0, +\infty[$ and $\Phi \uparrow \infty$ as $t \uparrow \infty$.

Thus the problem of the validity of M. Riesz's theorem must be investigated in the terms of conjugate functions $\overline{f}_B$ with $\overline{k}(B) = 1$, i.e., all the functions $\overline{f}_i$, $1 \le i \le n$. To be more precise, it holds

Theorem 1.4.1. *Let $f \in L(T^n)$ and $f \ge 0$. If $\overline{f}_i \in L(T^n)$ for $1 \le i \le n$, then $f \in L\log^+ L(T^n)$.*

Proof. Put

$$\varphi^{(1)}(f, x_2, \dots, x_n) = \frac{1}{2\pi} \int_T f(x)\, dx_1. \tag{1.4.2}$$

It is clear that $\varphi^{(1)} \ge 0$, $\varphi^{(1)} \in L(T^{n-1})$, and almost everywhere on T^{n-1} it holds

$$\overline{\varphi}_2^{(1)}(f, x_2, \dots, x_n) = \frac{1}{2\pi} \int_T \overline{f}_2(x)\, dx_1. \tag{1.4.3}$$

Further, let

$$\begin{aligned} U(f) &\equiv U(f, z_1, x_2, \dots, x_n) = \frac{1}{\pi} \int_T f(s_1, x_2, \dots, x_n) P(r_1, s_1 - x_1)\, ds_1, \\ V(f) &\equiv V(f, z_1, x_2, \dots, x_n) = -\frac{1}{\pi} \int_T f(s_1, x_2, \dots, x_n) Q(r_1, s_1 - x_1)\, ds_1, \\ &\qquad z_1 = r_1 e^{ix_1}, \quad r_1 \in [0, 1[, \end{aligned} \tag{1.4.4}$$

where $p(r_1, t_1)$ and $Q(r_1, t_1)$ are defined by (1.1.29) and (1.1.30), respectively. Without loss of generality, we can assume that $f \ge e$ (otherwise we could deal with the function $g(x) = f(x) + e$). Consider (for almost all fixed $x_2, \dots, x_n$) an analytic in the circle $|z_i| < 1$ function $F(z_1) \equiv F(f, z_1, x_2, \dots, x_n) = U(f, z_1, x_2, \dots, x_n) + iV(f, z_1, x_2, \dots, x_n)$, where $U \ge e$. Using the Cauchy formula, we have

$$\frac{1}{2\pi i} \int_{|z_1| = r_1} \frac{F(z_1) \log F(z_1)}{z_1}\, dz_1 = F(0) \log F(0).$$

Hence, by (1.4.2), we find

$$\frac{1}{2\pi} \int_T \left[U(f) \log \sqrt{U^2(f) + V^2(f)} - \psi V(f) \right] dx_1 =$$

$$= \varphi^{(1)}(f, x_2, \ldots, x_n) \log \varphi^{(1)}(f, x_2, \ldots, x_n),$$

where $\psi = \arg F(z_1)$. But since $|\psi| \leq \frac{\pi}{2}$, the last inequality implies

$$\frac{1}{2\pi} \int_T U(f) \log U(f)\, dx_1 \leq$$

$$\leq \frac{1}{4} \int_T |V(f)|\, dx_1 + \varphi^{(1)}(f, x_2, \ldots, x_n) \log \varphi^{(1)}(f, x_2, \ldots, x_n). \tag{1.4.5}$$

Repeat now the same with respect to $\varphi^{(1)}$ and x_2. We obtain

$$\varphi^{(2)}(\varphi^{(1)}, x_3, \ldots, x_n) = \frac{1}{2\pi} \int_T \varphi^{(1)}(f, x_2, \ldots, x_n)\, dx_2 =$$

$$= \frac{1}{(2\pi)^2} \int_{T^2} f(x)\, dx_1\, dx_2, \tag{1.4.6}$$

$$\overline{\varphi}_3^{(2)}(\varphi^{(1)}, x_3, \ldots, x_n) = \frac{1}{2\pi} \int_T \overline{\varphi}_3^{(1)}(f, x_2, \ldots, x_n)\, dx_2 =$$

$$= \frac{1}{(2\pi)^2} \int_{T^2} \overline{f}_3(x)\, dx_1\, dx_2, \tag{1.4.7}$$

$$\frac{1}{2\pi} \int_T U(\varphi^{(1)}) \log U(\varphi^{(1)})\, dx_2 \leq$$

$$\leq \frac{1}{4} \int_T |V(\varphi^{(1)})|\, dx_2 + \varphi^{(2)}(\varphi^{(1)}, x_3, \ldots, x_n) \log \varphi^{(2)}(\varphi^{(1)}, x_3, \ldots, x_n). \tag{1.4.8}$$

Thus, for any $3 \leq j \leq n-1$, we have

$$\varphi^{(j)}(\varphi^{(j-1)}, x_{j+1}, \ldots, x_n) = \frac{1}{2\pi} \int_T \varphi^{(j-2)}(\varphi^{(j-2)}, x_j, \ldots, x_n)\, dx_j =$$

$$= \frac{1}{(2\pi)^j} \int_{T^j} f(x)\, dx_1 \cdots dx_j, \tag{1.4.9}$$

$$\overline{\varphi}_{j+1}^{(j)}(\varphi^{(j-1)}, x_{j+1}, \ldots, x_n) = \frac{1}{(2\pi)^j} \int_{T^j} \overline{f}_{j+1}(x)\, dx_1 \cdots dx_j, \tag{1.4.10}$$

$$\frac{1}{2\pi} \int_T U(\varphi^{(j-1)}) \log U((\varphi^{(j-1)})\, dx_j \leq \frac{1}{4} \int_T |V(\varphi^{(j-1)})|\, dx_j +$$

$$+ \varphi^{(j)}(\varphi^{(j-1)}, x_{j+1}, \ldots, x_n) \log \varphi^{(j)}(\varphi^{(j-1)}, x_{j+1}, \ldots, x_n). \tag{1.4.11}$$

For $j = n-1$, the relation (1.4.11) writes as

$$\frac{1}{2\pi} \int_T U(\varphi^{(n-2)}) \log U(\varphi^{(n-2)})\, dx_{n-1} \leq \frac{1}{4} \int_T |V(\varphi^{(n-2)})|\, dx_{n-1} +$$

$$+ \varphi^{(n-1)}(\varphi^{(n-2)}, x_n) \log \varphi^{(n-1)}(\varphi^{(n-2)}, x_n). \tag{1.4.12}$$

On the other hand, in view of (1.4.9) and (1.4.10), we get

$$\varphi^{(n-1)}(\varphi^{(n-2)}, x_n) = \frac{1}{(2\pi)^{n-1}} \int\limits_{T^{n-1}} f(x)\, dx_1 \cdots dx_{n-1},$$

$$\overline{\varphi}_n^{(n-1)}(\varphi^{(n-2)}, x_n) = \frac{1}{(2\pi)^{n-1}} \int\limits_{T^{n-1}} \overline{f}_n(x)\, dx_1 \cdots dx_{n-1}.$$

Using the above mentioned theorem of M. Riesz and the conditions of the theorem, we can conclude that the function

$$\varphi^{(n-1)}(\varphi^{(n-2)}) \;\; \log \varphi^{(n-1)}(\varphi^{(n-2)})$$

is summable on T with respect to x_n. Then, using again the conditions of the theorem and (1.4.2)–(1.4.12), we conclude that

$$\int\limits_{T^n} f(x) \log f(x)\, dx < +\infty. \quad \blacksquare$$

Theorem 1.4.2. *Let $f \in L(T^n)$ and $f \geq 0$. Suppose that $\lambda \in [1, +\infty[$ is a natural number. Then if $\overline{f}_i \in L(\log^+ L)^\lambda(T^n)$ for $1 \leq i \leq n$, then $f \in L(\log^+ L)^{\lambda+1}(T^n)$.*

Proof. Using the conclusion of Theorem 1.4.1, we will have $f \in L \log^+ L(T^n)$. Here too we can assume $f \geq e$ and consider the function $F(z_1) \equiv F(f, z_1)$ used in the proof of Theorem 1.4.1. By the Cauchy formula, we have

$$\frac{1}{2\pi_i} \int\limits_{|z_1|=r_1} \frac{F(z_1)[\log F(z_1)]^{\lambda+1} dz_1}{z_1} = F(0)[\log F(0)]^{\lambda+1},$$

whence it follows

$$\begin{aligned} \frac{1}{2\pi} \int\limits_T U(f)\Big[\log U(f)\Big]^{\lambda+1} dx_1 \leq A(\lambda)\Big\{ \int\limits_T |V(f)|\, dx_1 + \\ + \int\limits_T |V(f)| \log^+ |V(f)|\, dx_1 + \cdots + \int\limits_T |V(f)|\Big[\log^+ |V(f)|\Big]^\lambda dx_1 + \\ + \int\limits_T U(f)\Big[\log U(f)\Big]^\lambda dx_1 \Big\} + \\ + \varphi^{(1)}(f, x_2, \ldots, x_n)\Big[\log \varphi^{(1)}(f, x_2, \ldots, x_n)\Big]^{\lambda+1}, \end{aligned} \tag{1.4.13}$$

where $\varphi^{(1)}$ is defined by (1.4.2). If $\lambda = 1$, then, according to our notation, we have

$$\begin{aligned} \frac{1}{2\pi} \int\limits_T U(\varphi^{(1)})\Big[\log U(\varphi^{(1)})\Big]^2 dx_1 \leq A\Big\{ \int\limits_T U(\varphi^{(1)})\, dx_2 + \\ + \int\limits_T |V(\varphi^{(1)})| \log^+ |V(\varphi^{(1)})|\, dx_2 + \int\limits_T U(\varphi^{(1)} \log U(\varphi^{(1)})\, dx_2 \Big\} + \\ + \varphi^{(2)}(\varphi^{(1)}, x_3, \ldots, x_n)\Big[\log \varphi^{(2)}(\varphi^{(1)}, x_3, \ldots, x_n)\Big]^2. \end{aligned}$$

Continuing this process, as in the proof of Theorem 1.4.1 we obtain

$$\varphi^{(1)}\Big[\log\varphi^{(1)}\Big]^2 \in L(T^n).$$

Hence, according to (1.4.13) and the conditions of the theorem, we conclude that $f \in L(\log^+ L)^2$. Moreover, it holds

$$\int\limits_{T^n} f(x)\Big[\log f(x)\Big]^2 dx \le A(n)\Bigg\{\int\limits_{T^n} f(x)\,dx +$$

$$+\int\limits_{T^n} f(x)\log f(x)\,dx + \int\limits_{T^n} |\overline{f}_1(x)\log^+|\overline{f}_1(x)|\,dx\Bigg\} +$$

$$\int\limits_{T^{n-1}} \varphi^{(1)}(f,x_2,\dots,x_n)\Big[\log\varphi^{(1)}(f,x_2,\dots,x_n)\Big]^2 dx_1\cdots dx_n. \qquad (1.4.14)$$

Let now $\lambda \ge 2$. Then along with (4.1.13) the inequality

$$\frac{1}{2\pi}\int\limits_{T} U(f)\Big[\log U(f)\Big]^\lambda dx_1 \le A(\lambda)\Bigg\{\int\limits_{T} |V(f)|\,dx_1 +$$

$$+\int\limits_{T} |V(f)|\log^+|V(f)|\,dx_1 + \cdots + \int\limits_{T} |V(f)|\Big[\log^+|V(f)|\Big]^{\lambda-1} dx +$$

$$+\int\limits_{T} |U(f)|\Big[\log U(f)\Big]^{\lambda-1} dx_1\Bigg\} + \varphi^{(1)}(f,x_2,\dots,x_n)\Big[\log\varphi^{(1)}(f,x_2,\dots,x_n)\Big]^\lambda$$

holds. Hence, applying the above scheme and the conditions of the theorem, we find $\varphi^{(1)} \in L(\log L)^\lambda(T^n)$ and $f \in L(\log L)^\lambda(T^n)$. Then from (1.1.13) we get

$$\int\limits_{T^n} f(x)\Big[\log f(x)\Big]^{\lambda+1} dx < +\infty.$$

To be more precise, an inequality of the type (1.1.14) holds. ∎

The scheme of the proof of Theorem 1.4.2 showes that in this theorem λ need not be natural. To be more precise, it holds

Theorem 1.4.3. *Let* $\beta \in [1,+\infty[$, $f \in L(T^n)$ *and* $f \ge 0$. *If* $\overline{f}_i \in L(\log^+ L)^{\beta-1}(T^n)$ *for* $1 \le i \le n$, *then* $f \in L(\log^+ L)^\beta(T^n)$.

A similar theorem can be proved for some nonreflexive Orlicz spaces.

Finally, note that as the corresponding assertions from the first part show (see Ch.1, Section 1), an analogue of M. Riesz's theorem for the Hilbert transform is of a little interest in the multidimensional case as well.

1.5. On the Validity of Kolmogorov's Theorem for Conjugate Functions and Hilbert Transforms of Functions of Several Variables

In the first part of this book (see Ch.I, (1.1.22) and (1.1.23)), we brought the inequalities of Kolmogorov [3] and Loomis [1] which claim properties of the weak type (1,1) for one-dimensional Hilbert and conjugation operators. Using Theorem 1.2.5 of this chapter, we conclude that in the multi-dimensional case Kolmogorov's and Loomis inequalities, in general, do not hold since $\overline{f}_B$ and $\tilde{f}_B$ may not exist on a set of positive measure.

The question arises: if $f \in L(T^n)$ and all the functions $\overline{f}_B$ with $\varnothing \neq B \subset M$ exist almost everywhere, is the inequality

$$\mu_n\Big[\Big\{x : |\overline{f}_B(x)| \geq y > 0, \quad x \in T^n\Big\}\Big] \leq A(n)y^{-1}\|f\|_1 \tag{1.5.1}$$

true?

A similar problem can be posed for the multivariable Hilbert transform.

In the works [17], [26], and [32], we gave some results answering the posed questions. They are presented below.

Theorem 1.5.1. *There is a function $f \in L(T^n)$ for which all the functions $\overline{f}_B$ exist almost everywhere, yet for any $\overline{f}_B$ with $\overline{k}(B) \in [2, n]$, we have*

$$\mu_n\Big[\Big\{x : |\overline{f}_B(x)| \geq y, \quad x \in T^n\Big\}\Big] \geq Ay^{-1}\log\log y \quad \textit{for} \quad y \in]20, +\infty[.$$

Proof. For some $\eta \in]0, \frac{1}{n}[$, define the function $f : T^n \to \mathbb{R}$ by

$$f(x) = \prod_{i=1}^{n} g(x_i), \tag{1.5.2}$$

where

$$g(x_i) = \begin{cases} x_i^{-1}(\log x_i^{-1})^{-1-\eta} & \text{for } x_i \in]0, \gamma'], \\ 0 \quad \text{for } x_i \in T\backslash]0, \gamma'], \end{cases} \tag{1.5.3}$$

and extend g 2π-periodically, $\gamma' \in]0, \frac{1}{4}[$ being chosen such that $g(x_i) > 1$ for $x_i \in]0, \gamma']$ $(1 \leq i \leq n)$.

It is not difficult to verify that $f \in L(\log^+ L)^{\eta_1}(T^n)$ for any $\eta_1 \in]0, \eta[$. Further, if $x_i \in]0, \frac{\gamma'}{2}]$, then

$$|\overline{g}(-x_i)| \geq Ax_i^{-1}(\log x_i^{-1})^{-\eta}, \quad 1 \leq i \leq n. \tag{1.5.4}$$

Applying a well-known theorem of Luzin–Privalov (see, e.g., Bari [2, Ch. VIII]), in view of (1.5.3) we can conclude that all the functions $\overline{f}_B$ with $B \subset M$ exist almost everywhere. According to (1.1.6), (1.1.7), (1.1.8), and (1.5.2), it follows that almost everywhere

$$\overline{f}_B(x) = \prod_{i\in B} \overline{g}(x_i) \prod_{j\in B'} g(x_j), \tag{1.5.5}$$

where

$$\prod_{j\in B'} g(x_j) \equiv 1 \quad \text{for } B' = \varnothing.$$

Let now $x_i \in]-\frac{\gamma'}{2}, 0[$ for $i \in B$, and $x_j \in]0, \gamma'[$ for $j \in B'$. Then, by (1.5.3)–(1.5.5), we get

$$|\overline{f}_B(x)| \geq A(n) \prod_{i\in B} x_i^{-1}(\log x_i^{-1})^{-\eta}.$$

Put

$$E_B(y) \equiv \left\{x \in T^n \ : \ |\overline{f}_B(x)| \geq y > 0\right\}.$$

Clearly $E_B(y)$ is μ-measurable. Using a corresponding assertion of Lemma 1.2.15 (see also Remark 1.2.1), we see that

$$\mu_n[E_B(y)] \geq A(n,\eta) y^{-1}(\log y)^{1-n\eta}, \quad y \in]20, +\infty].$$

Hence it follows the validity of the theorem. ■

The question arises: in what form is Kolmogorov's theorem valid for conjugate functions of several variables?

The answer is given by

Theorem 1.5.2. a) *Let $\beta \in [0, n-1]$ and $f \in L(\log^+ L)^\beta(T^n)$. Then for all the conjugate functions $\overline{f}_B$ with $\overline{k}(B) \in [1, \beta+1]$, the relations*

$$\mu_n[E_B(y)] \leq A(n) y^{-1} \|f\|_1 \quad \textit{for } \overline{k}(B) = 1, \tag{1.5.6}$$

$$\mu_n[E_B(y)] \leq A(n,\beta)\left\{\int_{T^n} |f(x)|\left[\log^+ |f(x)|\right]^\beta dx + 1\right\} \quad \textit{for } \overline{k}(B) \geq 2, \tag{1.5.7}$$

hold.

b) *The condition $f \in L(\log^+ L)^\beta(T^n)$ with $\beta \in [0, n-1[$ does not guarantee, in general, the validity of* (1.5.7) *for the functions $\overline{f}_B$ with $\overline{k}(B) \in]\beta+1, n]$.*

Proof. The assertion b) of the theorem follows from b) of Theorem 1.2.4 since the functions $\overline{f}_B$ with $\overline{k}(B) \in]\beta+1, n]$, in general, need not exist almost everywhere.

Turn now to the proof of (1.5.6) and (1.5.7). Using Kolmogorov's inequality and Fubini's theorem, we have

$$\mu_n[E_B(y)] \leq A(n) y^{-1} \|f\|_1 \quad \text{for } \overline{k}(B) = 1,$$

that is, (1.5.6) is fulfilled. Further, conditions of a) of the theorem and the conclusion of a) of Theorem 1.2.4 imply that if $\overline{k}(B) \in [1, \beta+1]$, the function $\overline{f}_B$ exists almost everywhere. Moreover, $\overline{f}_B$ can almost everywhere be represented as an n-fold iterated limit in the sense of conjugation. By virtue of the conclusion of the assertion a) of Theorem 1.3.1, all the functions $\overline{f}_B$ with $\overline{k}(B) \in [1, \beta]$ are summable on T^n. Using again Kolmogorov's inequality, (1.1.45) and Fubini's theorem, we get

$$\mu_n[E_B(y)] \leq A(n,\beta) y^{-1}\left\{\int_{T^n} |f(x)|\left[\log^+ |f(x)|\right]^{\overline{k}(B)-1} dx + 1\right\}$$

for $\overline{k}(B) \in [1, \beta+1]$, i.e., (1.5.7) can be considered as proved. ■

Now consider similar questions for the Hilbert transform of functions of several variables. As it is clear from the results stated below, conjugate functions and Hilbert transforms of several variables possess some specific properties which are absent in the one-dimensional case. To be more precise, it holds

Theorem 1.5.3. *For any $p \in [1,+\infty[$, there is a function $f \in L^p(\mathbb{R}^n)$ satisfying*

$$\mu_n\Big[\Big\{x : |\widetilde{f}_B(x)| \geq y,\ x \in \mathbb{R}^n\Big\}\Big] \geq A(n) y^{-1} \log y^{-1} \quad \textit{for } y \in]0, 10^{-2}[,$$

where $\overline{k}(B) \in [2,n]$.

Proof. Define the function $f : \mathbb{R}^n \to \mathbb{R}$ by

$$f(x) = \prod_{i=1}^{n} g(x_i), \tag{1.5.8}$$

where

$$g(x_i) = \begin{cases} (1+x_i^2)^{-1} & \text{for } x_i \in [0,+\infty[, \\ 0 & \text{for } x_i \in]-\infty, 0[, \end{cases} \qquad 1 \leq i \leq n.$$

Clearly $f \in L^p(T^n)$ for all $p \in [1,+\infty[$, that is, all the functions $\widetilde{f}_B$, $B \subset M$, exist almost everywhere on $\mathbb{R}^n$. If $x_i \in]-\infty, 0[$ for $1 \leq i \leq n$, then

$$|\widetilde{g}(x_i)| \geq \frac{1}{\pi} \int_0^{|x_i|} \frac{g(s_i)}{s_i + |x_i|}\, ds_i \geq \frac{A}{|x_i|} \quad \text{for } 1 \leq i \leq n. \tag{1.5.9}$$

Let now $n = 2$ and $\overline{k}(B) = 2$. Without loss of generality, we can assume that $B = \{1,2\}$. Then, according to (1.5.8) and (1.5.9), we find

$$\Big\{x : |\widetilde{f}_B(x)| \geq y > 0,\ x \in \mathbb{R}^2\Big\} \supset \Big\{x : \frac{A}{|x_1 x_2|} \geq y > 0,\ x_i \in [0,+\infty[,\ 1 \leq i \leq 2\Big\}.$$

Therefore

$$\mu_2\Big[\Big\{x : |\widetilde{f}_B(x)| \geq y > 0,\ \ x \in \mathbb{R}^2\Big\}\Big] \geq$$

$$\geq A\Big\{\int_1^{1/y} \Big[(yx_1)^{-1} - 1\Big]\, dx_1\Big\} \geq A_1 y^{-1} \log y^{-1} \quad \text{for } y \in]0, 10^{-2}[.$$

Thus, in the case $n = 2$ and $\overline{k}(B) = 2$, the theorem is proved.

Assume now that $n \geq 3$ and $\overline{k}(B) \in [2,n]$. By (1.5.8) and (1.5.9), we get

$$|\widetilde{f}_B(x)| \geq A(n) \prod_{i \in B} (|x_i|)^{-1} \prod_{j \in B'} g(x_i),$$

where $x_i \in]-\infty, 0[$ for $i \in B$, and $x_j \in]0,+\infty[$ for $j \in B'$.

Thus, using the already proven case and Fubini's theorem, we make sure that the theorem is true. ■

The theorem just proved showes us that the multiple Hilbert operators do not have, in general, the weak type (1,1).

Using (1.2.81), we can conclude that if $f \in L^p(\mathbb{R}^n)$ for some $p \in]1, +\infty[$, then for any nonempty $B \subset M$, the inequality

$$\mu_n\left[\left\{x \in \mathbb{R}^n : |\overline{f}_B(x)| \geq y > 0\right\}\right] \leq A(p,n)y^{-p}\|f\|_p^p$$

holds, i.e., $\overline{f}_B$ has the weak type (p,p).

1.6. On Some Conditions of Existence and Integrability of Conjugate Functions of Several Variables in Terms of Mixed and Partial Integral Moduli of Continuity

In connection with Theorems 1.2.4, 1.3.1, and 1.3.2, the following problems arise.

1) Let $f \in L(T^n)$ and $B \subset M$. Which are the best possible orders of tending to zero of the mixed or partial integral moduli of continuity guaranteeing the almost everywhere existence of one (all) of the functions $\overline{f}_B$?

2) Let $f \in L(T^n)$ and $B \subset M$. Which are the best possible orders of tending to zero of mixed or partial moduli of continuity guaranteeing the integrability on T^n of one (all) of the functions $\overline{f}_B$?

We do not know the complete solution of these problems in terms of the classes $H(M,L)$, $H(n,L)$, and $H(\omega,L)$. A partial solution of the problem 2) under some restrictions on ω_B and ω are indicated in our work [28]. Below we give some assertions connected with the posed problems.

Theorem 1.6.1. a) *Let $f \in L(T^n)$, $B \subset M$, and the series*

$$\sum_{m \in \mathbb{N}_0^n : m = m_B} \omega\left(\left(\frac{1}{m+1}\right)_B, f\right)_1 \prod_{i \in B} \frac{1}{m_i + 1} \tag{1.6.1}$$

be convergent. Then $\overline{f}_B$ exists almost everywhere and $\overline{f}_B \in L(T^n)$.

b) *Let $f \in L(T^n)$ and the series*

$$\sum_{B \subset M}\left\{\sum_{m \in \mathbb{N}_0^n : m = m_B} \omega\left(\left(\frac{1}{m_i+1}\right)_B, f\right)_1 \prod_{i \in B} \frac{1}{m_i + 1}\right\} \tag{1.6.2}$$

be convergent. Then all the functions $\overline{f}_B$, $B \subset M$, exist almost everywhere and all of them are integrable on T^n.

The proof of this theorem is not difficult and is based on the equivalence of (1.6.1) and the condition

$$\int_{[0,\pi]^k} \omega(s_B, f)_1 \prod_{i \in B} s_i^{-1}\, ds_B < +\infty.$$

The condition (1.6.1), in general, does not guarantee the almost everywhere existence of the functions $\overline{f}_{B_1}$ with $B_1 \neq B$, $B_1 \subset M$. The counterexample is easy to construct. Therefore, for the validity of the assertion b) of the theorem, the condition (1.6.1) is in a certain sense unimprovable.

Theorem 1.6.2. *Let $f \in L(T^n)$ and for some $\beta \in [0, n-2]$, the relation*

$$\sum_{p=1}^{\infty} p^{-1}\left[\sum_{i=1}^{n} \omega_i\left(\frac{1}{p}, f\right)_1\right]\left[\log(p+1)\right]^{\beta} < +\infty \tag{1.6.3}$$

be fulfilled. Then all the functions $\overline{f}_B$ with $\overline{k}(B) \in [1, \beta+2]$ exist almost everywhere.

This theorem follows from a) of Theorem 1.2.4 if we take into account that according to results of Panjikidze [1], condition (1.6.3) implies that f belongs to $L(\log^+ L)^{\beta+1}(T^n)$.

Theorem 1.6.3. a) *Let $f \in L(T^n)$ and for any $\beta \in [0, n-1]$, the relation* (1.6.3) *be fullfiled. Then all the functions $\overline{f}_B$ with $\overline{k}(B) \in [1, \beta+1]$ are integrable on T^n.*

b) *Let $\beta \in [0, n-1[$. Then there is a function $f \in L(T^n)$ for which* (1.6.3) *holds with that β, yet the functions $\overline{f}_B$ with $\overline{k}(B) \in]\beta+1, n]$ are not integrable on T^n.*

Proof. a) As above, using results of Panjikidze [1], we conclude that

$$f \in L(\log^+ L)^{\beta+1}(T^n).$$

Then a) follows from a) of Theorem 1.3.1.

b) Define the function $f : T^n \to \mathbb{R}$ roughly likewise to the proof of Theorem 1.3.1 (see (1.3.2)). To be more precise, for $\eta \in [0, \overline{k}(B) - \beta - 1]$ we put

$$f(x) = \begin{cases} (2n)^{2n} \prod_{i=1}^{n} x_i^{-1}\left[\log\left(\prod_{i=1}^{n} x_i^{-1}\right)\right]^{-n-\beta-1-\eta} & \text{if } x \in I^n(\gamma), \\ 0, \quad x \in T^n \setminus I^n(\gamma), \end{cases} \tag{1.6.4}$$

and extend the function f 2π-periodically in each of the coordinates of x. By (1.6.4), we find that $f \in L(T^n)$ and

$$\omega_i(\delta, f)_1 \leq A(n, \beta, \eta)(\log \delta^{-1})^{-\beta-1-\eta} \quad \text{for } \delta \in]0, \gamma], \quad 1 \leq i \leq n.$$

That is, (1.6.3) is fulfilled.

Let $x \in]0, \gamma]^n$ for $1 \leq i \leq n$. Using (1.6.4), quite analogously to the proof of the assertion b) of Theorem 1.3.1, we get

$$|\overline{f}_B(-x_B)| \geq A(n, \beta, \eta) \prod_{i=1}^{n} x_i^{-1}\left[\log\left(\prod_{i=1}^{n} x_i^{-1}\right)\right]^{-n-\beta+\overline{k}(B)-1-\eta}.$$

Then, taking note of b) of Theorem 1.6.1, we obtain

$$|\overline{f}_B(-x_B)| \geq A(n, \beta, \eta) \prod_{i=1}^{n} x_i^{-1}\left[\log\left(\prod_{i=1}^{n} x_i^{-1}\right)\right]^{-n}$$

since $n + \beta + 1 + \eta - \overline{k}(B) \leq n$. That is, $\overline{f}_B \notin L(T^n)$. ■

1.7. Unsolved Problems

1.7.1. Describe all the functions $\omega : T^n \to \mathbb{R}$ for which (1.2.52) holds with the weight ω.

1.7.2. Describe all the weight functions ω, v for which

$$\int\limits_{E_B(y)} \omega(x)\,dx \le A(n)y^{-1}\Big\{\int\limits_{T^n} |f(x)|\Big[\log^+|f(x)|\Big]^{\overline{k}(B)-1} v(x)\,dx + 1\Big\}.$$

1.7.3. Let $f \in H(\omega, B, L)$. Is it possible to "correct" f on a set of arbitrarily small μ_n-measure so that the obtained function and all its conjugate functions of n variables would belong to the class $H(\omega, B, L)$?

1.7.4. Let $f \in H(\omega, L)$. Is it possible to "correct" f on a set of arbitrarily small μ_n-measure so that the obtained function and all its conjugate functions of n variables would belong to the class $H(\omega, L)$?

We should add that Problems 1.7.3 and 1.7.4 are meaningfull for nonreflexive Orlicz spaces as well.

1.7.5. Let $B \subset M$, $f \in L(T^n)$ and $\overline{f}_B \in L(T^n)$. Then is the conclusion

$$\overline{\sigma}_n[f, B] = \sigma_n[\overline{f}_B]$$

true? The problem is due to Ul'yanov.

1.7.6. Give a complete description of pairs of weight functions ω, v of several variables for which inequalities of the types (1.3.10) and (1.2.81) hold.

1.7.7. We have mentioned the result of Kober [1]: $f \in L(\mathbb{R})$ and $\widetilde{f} \in L(\mathbb{R})$ imply

$$\int\limits_{\mathbb{R}} f(x)\,dx = 0.$$

Is an analogue of this assertion true for Hilbert transforms of functions of several variables?

CHAPTER II

Convergence and Summability at a Point or Almost Everywhere of Multiple Trigonometric Fourier Series and Their Conjugates

2.1. Convergence and Summability by Multiple Cesàro Method of Negative Order of the Series $\sigma_n[f]$ and $\overline{\sigma}_n[f,B]$ at Separate Points or Almost Everywhere

In the works [13, Ch. IV], [16], [35, Ch. II], we have set down various questions connected with the convergence and summability at a point or almost everywhere of multiple trigomometric series and their conjugates. Some other results in this direction are presented in our works [15], [24], [37], [39], [41], [42], [45].

These results, in some cases in a more general form, are set forth in this chapter.

Denote by $U_k \equiv U_k(M)$, $k \in [1,n]$, the class of all the subsets of M containing k elements. Put $U_0 = \{-1,1\}$ and $[s \times t] = (s_1 t_1, \dots, s_n t_n)$.

Theorem 2.1.1. *Let $f \in L(T^n)$ and $x \in T^n$ be a fixed point. If*

$$\sum_{B \subset M} \int_{T^n(B)} \prod_{i \in B} (|s_i|)^{-1} |\Delta(f,x,s_B)| \, ds_B < +\infty, \tag{2.1.1}$$

then the series $\sigma_n[f]$ converges to f at x.

Proof. According to (1.1.23), a simple calculation shows that

$$S_m(x,f) - f(x) = \frac{1}{\pi^n} \int_{T^n} [f(x+s) - f(x)] \prod_{i=1}^{n} D_i(s_i) \, ds =$$

$$= \frac{1}{\pi^n} \int_{[0,\pi]^n} \left\{ \sum_{t_1 \in U_0} \cdots \sum_{t_n \in U_0} [f(x + [s \times t]) - 2^n f(x)] \right\} \prod_{i=1}^{n} D_i(s_i) \, ds =$$

$$= \frac{1}{\pi^n} \int_{[0,\pi]^n} \left\{ \sum_{t_1 \in U_0} \cdots \sum_{t_n \in U_0} [\Delta(f,x,[s \times t]_M)] \right\} \prod_{i=1}^{n} D_i(s_i) \, ds +$$

$$+ \frac{2}{\pi^n} \int_{[0,\pi]^n} \left\{ \sum_{B \in U_{n-1}} \sum_{t_{i_1} \in U_0} \cdots \sum_{t_{i_k} \in U_0} \Delta(f,s,[s \times t]_B) \right\} \prod_{i=1}^{n} D_i(s_i) \, ds +$$

$$+\cdots+\frac{2^{n-1}}{\pi^n}\int\limits_{[0,\pi]^n}\Big\{\sum_{B\in U_1}\sum_{t_{i_1}\in U_0}\Delta(f,x,[s\times t]_B)\Big\}\prod_{i=1}^n D_i(s_i)\,ds.$$

Hence, using (2.1.1) and Remark 1.1.1, we get

$$\lim_{m\to\infty} S_m(x,f)=f(x). \quad \blacksquare$$

Quite analogously it can be proved

Theorem 2.1.2. *Let $f\in L(T^n)$ and $x\in T^n$ be a fixed point. If*

$$\sum_{B\subset M}\int\limits_{T^n(B)}\Big(\prod_{i\in B}s_i^{-1}\Big)|\overline{\Delta}(f,x,s_B)|\,ds_B<+\infty, \tag{2.1.2}$$

then for any nonempty $B\subset M$, we have

$$\lim_{m\to\infty}\Big[\overline{S}_m(x,f,B)-\overline{f}_B\Big(x,\frac{\pi}{m}\Big)\Big]=0,$$

where $\overline{S}_m(x,f,B)$ and $\overline{f}_B(x,\varepsilon)$ are defined by (1.1.24) *and* (1.1.8), *respectively.*

The application of Theorem 2.1.1 gives us

Theorem 2.1.3. *Let $f\in L(T^n)$. If*

$$\sum_{B\subset M}\int\limits_{T^n}\Big[\int\limits_{T^n(B)}\Big(\prod_{i\in B}s_i^{-1}\Big)|\overline{\Delta}(f,x,s_B)|\,ds_B\Big]\,dx<+\infty, \tag{2.1.3}$$

then the series $\sigma_n[f]$ converges almost everywhere.

The conclusion of the theorem remains valid also for the series $\overline{\sigma}_n[f,B]$ with $B\subset M$.

Remark 2.1.1. Theorem 2.1.1 is a multidimensional analogue of the well-known Dini test (see, e.g., Zygmund [15, Ch. II]) for the convergence of the series $\sigma[f]$ at a fixed point. In the case $n=2$, Gergen [2] established a test for convergence of the series $\sigma_2[f]$ at a fixed point and compared it with earlier tests. Later, in 1939, Marcinkiewicz [4] proved Theorem 2.1.3 for $n=1$.

Various corollaries can be deduced from Theorems 2.1.1–2.1.3. Below we give some of them.

Note first that according to the scheme of the proof of Theorem 2.1.1, if $f\in C(T^n)$ and (2.1.2) is fulfilled uniformly in T^n, then the series $\sigma_n[f]$ uniformly converges to f.

Corollary 2.1.1. *Let $f\in L(T^n)$ and $\eta\in\,]0,+\infty[$. If*

$$\omega_i(\delta,f)_1\Big(\log\frac{2\pi}{\delta}\Big)^{n+\eta}\le A(f,n)\quad \text{for } \delta\in\,]0,\pi],\ 1\le i\le n, \tag{2.1.4}$$

then the series $\sigma_n[f]$ as well as all the series $\sigma_n[f,B]$ with $B\subset M$ converge almost everywhere.

We should add that (2.1.4) guarantees the almost everywhere existence of all the functions $\overline{f}_B$ and their integrability on T^n.

Corollary 2.1.2. *Let $f \in C(T^n)$ and $\eta \in]0,+\infty]$. If*

$$\omega_i(\delta, f)_C\left(\log\frac{2\pi}{\delta}\right)^{n+\eta} \leq A(f,n) \quad for \ \delta \in]0,\pi], \ 1 \leq i \leq n, \tag{2.1.5}$$

then all the functions $\overline{f}_B$ with $B \subset M$ everywhere exist, belong to $C(T^n)$, and all the series $\sigma_n[f]$ and $\sigma_n[f, B]$ uniformly converge.

In the consequent chapters we will investigate whether the conclusion of Corollary 2.1.2 remains true when in (2.1.5) $\eta = 0$.

As to the analogous question for the conclusion of Corollary 2.1.1 (i.e., when in (2.1.4) $\eta = 0$), it is still an open problem. For $n = 1$, this problem was stated by Zygmund (see [16, Ch. XIII]).

We will need some auxiliary statements.

Lemma 2.1.1. *If $\alpha \in]0,1[$, then for any $m \in \mathbb{N}$ and any $1 \leq \nu \leq m$, we have*

$$\int_T \left|(A_m^{-\alpha})^{-1}\sum_{k=0}^{\nu} A_{m-k}^{-\alpha}(k+1)^{-\alpha}\cos kx\right| dx \leq A(\alpha). \tag{2.1.6}$$

Proof. Let $\nu = m$. Using Abel's transform, we get

$$(A_m^{-\alpha})^{-1}\sum_{k=0}^{n} A_{m-k}^{-\alpha}(k+1)^{-\alpha}\cos kx =$$

$$= (A_m^{-\alpha})^{-1}(m+1)^{-\alpha}\sum_{k=0}^{n} A_{m-k}^{-\alpha}\cos kx +$$

$$+(A_m^{-\alpha})^{-1}\sum_{k=0}^{m-1}\left[(k+1)^{-\alpha}-(k+2)^{-\alpha}\right]\left[\sum_{j=0}^{k} A_{m-j}^{-\alpha}\cos jx\right] \equiv$$

$$\equiv J_1(x,m,\alpha) + J_2(x,m,\alpha). \tag{2.1.7}$$

Basing on (2.2.9), (2.2.11), (2.2.13), and (2.2.15) from the first part of the book, we find

$$\|J_1(\cdot,m,\alpha)\|_1 \leq A(\alpha). \tag{2.1.8}$$

The expression $J_2(x,m,\alpha)$ from (2.1.7) can be written as

$$J_2(x,m,\alpha) = (A_m^{-\alpha})^{-1}\sum_{k=0}^{[\frac{m}{2}]}\left[(k+1)^{-\alpha}-(k+2)^{-\alpha}\right]\left[\sum_{j=0}^{k} A_{m-j}^{-\alpha}\cos jx\right] +$$

$$+(A_m^{-\alpha})^{-1}\sum_{k=[\frac{m}{2}]+1}^{m-1}\left[(k+1)^{-\alpha}-(k+2)^{-\alpha}\right]\left[\sum_{j=0}^{k} A_{m-j}^{-\alpha}\cos jx\right] \equiv$$

$$\equiv J_3(x,m,\alpha) + J_4(x,m,\alpha). \tag{2.1.9}$$

On the other hand, by Abel's transform we have

$$\sum_{j=0}^{k} A_{m-j}^{-\alpha}\cos jx = \sum_{j=0}^{k} A_{m-j-1}^{-\alpha-1}D_i(x) + A_{m-k}^{-\alpha}D_k(x), \tag{2.1.10}$$

where D_i is the Dirichlet kernel. Since in $J_3(x,m,\alpha)$ from (2.1.9) $0 \le k \le m/2$, by (2.1.9) and (2.1.10) we obtain

$$\|J_3(\cdot,m,\alpha)\|_1 \le A(\alpha)\Big\{\frac{1}{m}\sum_{k=0}^{m}(k+1)^{-\alpha}\log(k+2) + \\ + \sum_{k=0}^{\infty}(k+1)^{-1-\alpha}\log(k+2)\Big\}. \tag{2.1.11}$$

Using a well-known scheme (see, e.g., Zygmund [15, Ch. III]), we get

$$\sum_{j=0}^{k-1} A_{m-j-1}^{-\alpha-1} D_j(x) = \Big(2\sin\frac{x}{2}\Big)^{-1} \operatorname{Im}\Big[\sum_{j=0}^{k-1} A_{m-j-1}^{-\alpha-1} e^{i(j+\frac{1}{2})x}\Big] = \\ = \Big(2\sin\frac{x}{2}\Big)^{-1} \operatorname{Im}\Big[e^{i(m-\frac{1}{2})x} \sum_{j=m-k}^{m-1} A_j^{-\alpha-1} e^{-ijx}\Big] = \\ = \Big(2\sin\frac{x}{2}\Big)^{-1} \operatorname{Im}\Big\{e^{i(m-\frac{1}{2})x}\Big[(1-e^{-ix})^{-\alpha} - \\ - \sum_{j=m}^{\infty} A_j^{-\alpha-1} e^{-ijx} - (1-e^{-ix})^{-\alpha} + \sum_{j=m-k}^{\infty} A_j^{-\alpha-1} e^{-ijx}\Big]\Big\} \quad \text{for } x \in]0,\pi[,$$

whence it follows

$$\Big|\sum_{j=0}^{k-1} A_{m-j-1}^{-\alpha-1} D_j(x)\Big| \le A(\alpha)(m-k)^{-\alpha-1}x^{-2} \\ \text{for } 1 \le k \le m-1, \quad x \in]0,\pi[. \tag{2.1.12}$$

On the other hand, we have

$$\Big|\sum_{j=0}^{k-1} A_{m-j-1}^{-\alpha-1} D_j(x)\Big| \le A \sum_{j=0}^{k-1} \big|A_{m-j-1}^{-\alpha-1}\big|(j+1) \text{ for } x \in T. \tag{2.1.13}$$

Therefore, using (2.1.9), (2.1.10), (2.1.12), and (2.1.13), we find

$$\|J_4(\cdot,m,\alpha)\|_1 \le \int_0^{1/m} |J_4(x,m,\alpha)|\,dx + \int_{1/m}^{\pi} |J_4(x,m,\alpha)|\,dx \le \\ \le A(\alpha)\Big\{\frac{1}{m}\Big[\sum_{k=[\frac{m}{2}]+1}^{m-1}\Big(\sum_{j=0}^{k-1}|A_{m-j-1}^{-\alpha-1}| + A_{m-k}^{-\alpha}\Big)\Big] + \\ + \sum_{k=[\frac{m}{2}]+1}^{m-1}\Big[(m-k)^{-\alpha-1}\frac{1}{m}\int_{1/m}^{\pi} x^{-2}dx + \frac{1}{m}A_{m-k}^{-\alpha}\log(k+2)\Big]\Big\} \le \\ \le A_1(\alpha)\Big\{\sum_{k=0}^{\infty}|A_k^{-\alpha-1}| + m^{-\alpha}\log(m+2)\Big\}. \tag{2.1.14}$$

Thus, according to (2.1.7), (2.1.8), (2.1.11), and (2.1.14), we get (2.1.6).

The lemma is proved for $\nu = m$. However, the scheme of the proof shows us that it is true in the case $1 < \nu < m$ as well. ■

The lemma can also be proved using the method of Sunouchi and Yano [1].

Lemma 2.1.2. *Let* $\alpha \in]0,1[$, *m and p be natural, and* $\lambda = \min\{m,p\}$. *Put*

$$u(x,\lambda,\alpha) \equiv u(x,m,p,\alpha) =$$
$$= \left(A_m^{-\alpha}A_p^{-\alpha}\right)^{-1}\sum_{k=0}^{\lambda} A_{m-k}^{-\alpha}A_{p-k}^{-\alpha}(k+1)^{-2\alpha}\cos kx. \tag{2.1.15}$$

Then

$$\|u(\cdot,\lambda,\alpha)\|_1 \le A(\alpha)\left\{\lambda^{\alpha}\int_T\left|\sum_{k=0}^{\lambda} A_{\lambda-k}^{-\alpha}(k+1)^{-\alpha}\cos kx\right|dx\right\}. \tag{2.1.16}$$

Proof. Without loss of generality, we can assume that $\lambda = m$, that is, $p \ge m$. As above, using Abel's transform, we get

$$u(x,\lambda,\alpha) =$$
$$= \left(A_m^{-\alpha}A_p^{-\alpha}\right)^{-1}\sum_{k=0}^{m-1}\Delta\left[(k+1)^{-\alpha}A_{p-k}^{-\alpha}\right]\left[\sum_{j=0}^{k}A_{m-j}^{-\alpha}(j+1)^{-\alpha}\cos jx\right]+$$
$$+\left(A_m^{-\alpha}A_p^{-\alpha}\right)^{-1}(m+1)^{-\alpha}A_{p-m}^{-\alpha}[\sum_{j=0}^{m}A_{m-j}^{-\alpha}(j+1)^{-\alpha}\cos jx \equiv$$
$$\equiv I_1(x,m,p,\alpha)+I_2(x,m,p,\alpha) \quad \text{for} \quad m \ge 2. \tag{2.1.17}$$

Consider the expression

$$(A_p^{-\alpha})^{-1}(m+1)^{-\alpha}A_{p-m}^{-\alpha}.$$

There are two possibilities: either $m \le p \le 2m$ or $p > 2m$. In both cases we have

$$0 < (A_p^{-\alpha})^{-1}(m+1)^{-\alpha}A_{p-m}^{-\alpha} \le A(\alpha).$$

Therefore, using the assertion (2.1.6) of Lemma 2.1.1, by (2.1.17) we obtain

$$\|I_2(\cdot,m,p,\alpha)\|_1 \le A(\alpha)\left\{m^{\alpha}\int_T\left|\sum_{k=0}^{m}A_{m-k}^{-\alpha}(k+1)^{-\alpha}\cos kx\right|dx\right\}. \tag{2.1.18}$$

On the other hand,

$$\Delta\left[(k+1)^{-\alpha}A_{p-k}^{-\alpha}\right] = (k+1)^{-\alpha}\left[A_{p-k}^{-\alpha}-A_{p-k-1}^{-\alpha}\right]+$$
$$+A_{p-k-1}^{-\alpha}\left[(k+1)^{-\alpha}-(k+2)^{-\alpha}\right].$$

Therefore, by virtue of (2.1.17), the expression $I_1(x,m,p,\alpha)$ can be written as

$$I_1(x,m,p,\alpha) =$$
$$= \left(A_m^{-\alpha}A_p^{-\alpha}\right)^{-1}\left[\sum_{k=0}^{[\frac{m}{2}]}+\sum_{k=[\frac{m}{2}]+1}\right]\left\{\left[(k+1)^{-\alpha}\left[A_{p-k}^{-\alpha}-A_{p-k-1}^{-\alpha}\right]+\right.\right.$$

$$+A_{p-k-1}^{-\alpha}\Big[(k+1)^{-\alpha}-(k+2)^{-\alpha}\Big]\Big]\Big[\sum_{j=0}^{k}A_{m-j}^{-\alpha}(j+1)^{-\alpha}\cos js\Big]\Big\}.$$

Hence, using Lemma 2.1.1, we find

$$\|I_1(\cdot,m,p,\alpha)\|_1 \le A(\alpha)m^{\alpha}\int_T\Big|\sum_{k=0}^{m}A_{m-k}^{-\alpha}(k+1)^{-\alpha}\cos kx\Big|\,dx. \tag{2.1.19}$$

Then, on the basis of (2.1.17)–(2.1.19), we get

$$\|u(\cdot,\lambda,\alpha)\|_1 \le A(\alpha)\Big\{\lambda^{\alpha}\int_T\Big|\sum_{k=0}^{\lambda}A_{m-k}^{-\alpha}(k+1)^{-\alpha}\cos kx\Big|\,dx\Big\}. \qquad \blacksquare$$

It is well-known (see, e.g., Zygmund [15, Ch. V]) that

$$\sum_{k=0}^{\infty}\frac{\cos kx}{\log_2(k+2)}$$

is the trigonometric Fourier series of a certain function f, and its partial sums

$$u(x,m)\equiv u(x,m,0)=\sum_{k=0}^{m}\frac{\cos kx}{\log_2(k+2)} \tag{2.1.20}$$

satisfy

$$\|u(\cdot,m)\|_1 \le A(f) \quad \text{for} \quad m\in\mathbb{N}. \tag{2.1.21}$$

Below we will assume that for the trigonometric Fourier series

$$\sum_{k=0}^{\infty}\frac{\cos kx}{(k+1)^{\alpha}} \quad \text{with} \quad \alpha\in]0,1[,$$

the expressions $\ell_m^{-\alpha}(x)$ denote $(C,-\alpha)$-means , i.e.,

$$\ell_m^{-\alpha}(x)=(A_m^{-\alpha})^{-1}\sum_{k=0}^{m}A_{m-k}^{-\alpha}(k+1)^{-\alpha}\cos kx. \tag{2.1.22}$$

We also will assume that

$$\ell_m^{0}(x)=\sum_{k=0}^{m}\frac{\cos kx}{\sqrt{\log_2(k+2)}} \qquad \text{for} \quad m\in\mathbb{N}. \tag{2.1.23}$$

For $p\in\mathbb{N}_0^n$ and $\alpha\in[0,1[^n$, we denote

$$\begin{aligned} v(p,\alpha)&=\prod_{i=1}^{n}\log_2(p_i+2) \ \text{ if } \ \alpha=0,\\ v(p,\alpha)&=\prod_{i=1}^{n}(p_i+1)^{2\alpha_i} \ \text{ if } \ \alpha\in]0,1[^n. \end{aligned} \tag{2.1.24}$$

Theorem 2.1.4. *Let the functions $f \in L^2(T^n)$ and $\rho_p(f)$ be defined by (1.1.16) and let either $\alpha \in]0,1[^n$ or $\alpha = 0$. If*

$$\sum_{p\geq 0}[\rho_p(f)]^2 v(p,\alpha) < +\infty, \tag{2.1.25}$$

then the series $\sigma_n[f]$ is almost everywhere $(C,-\alpha)$-summable.

Proof. First make some notes. For $n = 1$ and $\alpha = 0$, this is the well-known theorem of Kolmogorov and Seliverstov [1], [2] and Plessner [2] (see Bari [2, Ch. V], Zygmund [16, Ch. XIII]). As we have already pointed out in the first part of the book, due to a result of Carleson [1] in this case it may be assumed $v(m,\alpha) \equiv 1$. If $n = 1$ and $\alpha \in]0,1[$, then the theorem follows from a result of Sunouchi and Yano [1]. We have treated this case in the first part of the book (Theorem 2.2.4) using a result of Hunt [2]. The method presented here differs from both of above ones. In 1930, Kaczmarz [1], using another techniques, proved the theorem for $n = 2$ and $\alpha = (0,0)$.

Without loss of generality, we can assume that $a_0^{(M)} = 0$. According to (2.1.25), we can conclude that there are functions $f_i : T^n \to \mathbb{R}$ $(i = 1,2)$ such that

$$a_p^{(B)}(f_1) \equiv a_p^{(B)}(f)\left[\prod_{i=1}^n \log_2(p_i+2)\right]^{1/2},$$

$$a_p^{(B)}(f_2) \equiv a_p^{(B)}(f)\prod_{i=1}^n (p_i+1)^{\alpha_i} \quad \text{for } p \in \mathbb{N}_0^n,\ \ B \subset M.$$

Define the function $g : T^n \to \mathbb{R}$ by

$$g(x) \equiv g(x,f,\alpha) = \begin{cases} f_1(x) & \text{if } \alpha = 0, \\ f_2(x) & \text{if } \alpha \in]0,1[^n. \end{cases}$$

Using (1.1.23), (1.1.25), (2.1.22), and (2.1.23), according to our notation we get

$$\sigma_m^{-\alpha}(x,f) = \frac{1}{\pi^n}\int_{T^n} g(s)\prod_{i=1}^n \ell_{m_i}^{\alpha_i}(s_i - x_i)\,ds. \tag{2.1.26}$$

Let now $m_i : T^n \to [0,+\infty[$ be measurable nonnegative functions with integer values which are bounded from above by a number $\nu \in]0,+\infty[$. We will assume that

$$\lambda_i(x,y) = \min\{m_i(x), m_i(y)\} \quad \text{for } 1 \leq i \leq n,$$
$$m(x) = (m_1(x),\dots,m_n(x)), \quad \lambda(x,y) = (\lambda_1(x,y),\dots,\lambda_n(x,y)).$$

By (2.1.26), we have

$$\left|\int_{T^n} \sigma_{m(x)}^{-\alpha}(x,f)\,dx\right| \leq$$

$$\leq A(n)\|g\|_2 \left\{ \int_{T^n} \left| \frac{1}{\pi^n} \int_{T^n} \prod_{i=1}^{n} \ell_{m_i(x)}^{-\alpha_i}(s_i - x_i)\, dx \right|^2 ds \right\}^{\frac{1}{2}}. \tag{2.1.27}$$

On the other hand,

$$\int_{T^n} \left| \frac{1}{\pi^n} \int_{T^n} \prod_{i=1}^{n} \ell_{m_i(x)}^{-\alpha_i}(s_i - x_i)\, dx \right|^2 ds =$$

$$= \int_{T^n}\int_{T^n} \left\{ \int_{T^n} \left[\prod_{i=1}^{n} \ell_{m_i(x)}^{-\alpha_i}(s_i - x_i) \ell_{m_i(y)}^{-\alpha_i}(s_i - y_i) \right] ds \right\} dx\, dy. \tag{2.1.28}$$

But

$$\frac{1}{\pi} \int_{T} \frac{\cos i(t-\gamma)\cos j(t-\tau)}{(i+1)^{\delta}(j+1)^{\delta}}\, dt = \begin{cases} \dfrac{\cos i(\gamma - \tau)}{(i+1)^{2\delta}} & \text{for } i = j, \\ 0 & \text{for } i \neq j; \end{cases}$$

$$\frac{1}{\pi} \int_{T} \frac{\cos i(t-\gamma)\cos j(t-\tau)}{\sqrt{\log_2(i+2)\log_2(j+2)}}\, dt = \begin{cases} \dfrac{\cos i(\gamma - \tau)}{\log_2(i+2)} & \text{for } i = j, \\ 0 & \text{for } i \neq j. \end{cases}$$

Therefore, according to (2.1.15), (2.1.20), and (2.1.28), we get

$$\int_{T^n} \left| \frac{1}{\pi^n} \int_{T^n} \prod_{i=1}^{n} \ell_{m_i(x)}^{-\alpha_i}(s_i - x_i)\, dx \right|^2 ds \leq$$

$$\leq A(n) \left\{ \int_{T^n}\int_{T^n} \left| \prod_{i=1}^{n} u(x_i - y_i, m_i(x)) \right| dx\, dy + \right.$$

$$\left. + \int_{T^n}\int_{T^n} \left| \prod_{i=1}^{n} u(x_i - y_i, m_i(y)) \right| dx\, dy \right\} \quad \text{for } \alpha = 0, \tag{2.1.29}$$

and

$$\int_{T^n} \left| \frac{1}{\pi^n} \int_{T^n} \prod_{i=1}^{n} \ell_{m_i(x)}^{-\alpha_i}(s_i - x_i)\, dx \right|^2 ds \leq$$

$$\leq A(n) \left\{ \int_{T^n}\int_{T^n} \left| \prod_{i=1}^{n} u(x_i - y_i, m_i(x), \alpha) \right| dx\, dy + \right.$$

$$\left. + \int_{T^n}\int_{T^n} \left| \prod_{i=1}^{n} u(x_i - y_i, m_i(y), \alpha) \right| dx\, dy \right\} \quad \text{for } \alpha \in]0,1[^n. \tag{2.1.30}$$

Using (2.1.21), (2.1.29), and (2.1.30), we obtain

$$\int_{T^n} \left| \frac{1}{\pi^n} \int_{T^n} \prod_{i=1}^{n} \ell_{m_i(x)}^{-\alpha_i}(s_i - x_i)\, dx \right|^2 ds \leq A(n) \quad \text{for } \alpha = 0$$

and

$$\int_{T^n}\left|\frac{1}{\pi^n}\int_{T^n}\prod_{i=1}^{n}\ell_{m_i(x)}^{-\alpha_i}(s_i-x_i)\,dx\right|^2 ds \le A(n) \ \text{ for } \ \alpha\in]0,1[^n.$$

Then, by (2.1.27), we find

$$\left|\int_{T^n}\sigma_{m(x)}^{-\alpha}(x,f)\,dx\right| \le A(n)\|g\|_2. \tag{2.1.31}$$

Let now ν be an arbitrary natural number. Assume

$$F_\nu^{(1)}(x)\equiv F_\nu^{(1)}(x,f,\alpha)=\sup_{0\le m_i\le\nu,\,1\le i\le n}\sigma_m^{-\alpha}(x,f)$$

and

$$F_\nu^{(2)}(x)\equiv F_\nu^{(2)}(x,f,\alpha)=\sup_{0\le m_i\le\nu,\,1\le i\le n}\{-\sigma_m^{-\alpha}(x,f)\}.$$

Clearly $F_\nu^{(i)}\ge 0$ ($\nu=1,2,\dots;\ i=1,2$), and the sequences $(F_\nu^{(1)})_{\nu\ge1}$, $(F_\nu^{(2)})_{\nu\ge1}$ are increasing. Therefore there exist

$$\lim_{\nu\to\infty}F_\nu^{(1)}(x)\equiv F^{(1)}(x),\quad \lim_{\nu\to\infty}F_\nu^{(2)}(x)\equiv F^{(2)}(x).$$

But then, using (2.1.31) and Fatou's theorem, we have

$$\|F^{(i)}\|_2\le A(n,\alpha)\|g\|_2 \quad (i=1,2).$$

The last relations and (1.1.32) yield

$$\|\sigma(\cdot,f,\alpha)\|_2\le A(n,\alpha)\|g\|_2,\quad \|\sup_m|S_m(\cdot,f)|\|_2\le A(n)\|g\|_2, \tag{2.1.32}$$

Let now $\varepsilon\in]0,1[$ be arbitrary and let $\varphi: T^n\to\mathbb{R}$ be defined as an n-fold trigonometric polynomial of the order $m_0\in\mathbb{N}_0^n$ with coefficients $a_p^{(B)}(f)$, the number m_0 being chosen such that

$$\sum_{p\ge0}[\rho_p(f)]^2v(p,\alpha)-\sum_{0\le p\le m_0}[\rho_p(f)]^2v(m,\alpha)<\varepsilon^2. \tag{2.1.33}$$

Put

$$\varphi_1(x)=f(x)-\varphi(x).$$

It is clear that

$$\sigma_m^{-\alpha}(x,f)=\sigma_m^{-\alpha}(x,\varphi)+\sigma_m^{-\alpha}(x,\varphi_1). \tag{2.1.34}$$

Moreover, we can easily see that

$$\lim_{m\to\infty}\sigma_m^{-\alpha}(x,f)=\lim_{m\to\infty}\frac{1}{\pi^n}\int_{T^n}g_1(s)\prod_{i=1}^{n}\ell_{m_i}^{-\alpha_i}(s_i-x_i)\,ds=\varphi(x),$$

where g_1 is defined by (2.1.26) for the function φ.

If

$$g_2(x)=g(x)-g_1(x),$$

then, by (2.1.33), we obtain

$$\|g_2\|_2 < \varepsilon. \tag{2.1.35}$$

Therefore, using (2.1.32) and (2.1.35), we see that the relation

$$\mu\Big[\Big\{x : \sigma(x, \varphi_1, \alpha) > A(n,\alpha)\sqrt{\varepsilon},\ x \in T^n\Big\}\Big] < \sqrt{\varepsilon}$$

holds. But then, basing on (2.1.34), we can conclude that the theorem is true. ∎

Analyzing the proof of Theorem 2.1.4, we make sure that it holds

Theorem 2.1.5. *Let $\alpha \in [0,1[^n$, $B \equiv B(\alpha) = \{i \in M : \alpha_i = 0\}$, and*

$$v(p,\alpha) = \prod_{i\in B} \log_2(p_i + 2) \prod_{j\in B'} (p_j + 1)^{2\alpha_j}.$$

If $f \in L^2(T^n)$ and

$$\sum_{p\geq 0} [\rho_p(f)]^2 v(p,\alpha) < \infty,$$

then the series $\sigma_n[f]$ is almost everywhere $(C, -\alpha)$-summable.

Theorems 2.1.4 and 2.1.5 can also be formulated in the integral form. To this aim, first introduce some notation.

Let $f \in L(T^n)$, $B \subset M$, and $B \neq M$. Put

$$\psi_B(x(B)) = (2\pi)^{\overline{k}(B)} \int\limits_{T^n(B')} f(x)\, dx_{B'}, \tag{2.1.36}$$

$$P(x) = f(x) - \sum_{B\subset M} 2^{\overline{k}(B)} \psi_B(x(B)). \tag{2.1.37}$$

It is not difficult to check up that $a_p^{(B)}(P) = 0$ if at least one coordinate of p is zero.

Theorem 2.1.6. *Let $\alpha \in [0,1[^n$. If $f \in L^2(T^n)$ and*

$$\sum_{B\subset M, B\neq\varnothing} \int\limits_{T^n} \left[\int\limits_{T^n(B)} \frac{|\overline{\Delta}(f, x, s_B)|^2 ds_B}{\prod\limits_{i\in B} |s_i|^{1+2\alpha_i}} \right] dx < +\infty, \tag{2.1.38}$$

then the series $\sigma_n[f]$ is almost everywhere $(C, -\alpha)$-summable.

Proof. First note that using (2.1.38) and (2.1.36), we can easily verify that for each ψ_B, it holds

$$\sum_{B_1\subset B, B_1\neq\varnothing} \int\limits_{T^n(B)} \left[\int\limits_{T^n(B_1)} \frac{|\overline{\Delta}(\psi_B, x(B), s_{B_1})|^2 ds_{B_1}}{\prod\limits_{i\in B_1} |s_i|^{1+2\alpha_i}} \right] dx_B < +\infty. \tag{2.1.39}$$

Using Parceval's equality, we get

$$\int\limits_{T^n} \left[\int\limits_{T^n} \frac{|\overline{\Delta}(f, x, s)|^2 ds}{\prod\limits_{i=1}^{n} |s_i|^{1+2\alpha_i}} \right] dx = A(n) \left\{ \sum_{p\geq 1} \big[\rho_p(f)\big]^2 \int\limits_{T^n} \left[\prod_{i=1}^{n} \frac{\sin^2 p_i s_i}{|s_i|^{1+2\alpha_i}} \right] ds \right\}.$$

Hence, according to (2.1.38), we have

$$\sum_{p\geq 1}[\rho_p(f)]^2[v(p,\alpha)+v(p,\alpha,B)]<+\infty. \tag{2.1.40}$$

Taking note of (2.1.39) and repeating the same argument for the series $\sigma_n[\psi_B]$, according to (2.1.40) we have

$$\sum_{p\geq 0}[\rho_p(f)]^2[v(p,\alpha)+v(p,\alpha,B)]<+\infty.$$

But then, basing on Theorems 2.1.4 and 2.1.5, we can conclude that the theorem is true. ∎

Theorem 2.1.6 is a multi-dimensional analogue for $\alpha=0$ of a theorem of Plessner [2] which he established in 1926.

Theorems 2.1.4–2.1.6 imply various corollaries. We give some of them.

Corollary 2.1.3. *Let $\eta\in]0,+\infty[$. If $f\in L^2(T^n)$ and*

$$\omega_i(\delta,f)_2\Big(\log\frac{2\pi}{\delta}\Big)^{\frac{n}{2}+\eta}\leq A(f,n)\quad for\ \delta\in]0,\pi],\ 1\leq i\leq n,$$

then the series $\sigma_n[f]$ converges almost everywhere.

Remark 2.1.2. In 1971, Fefferman [1] proved that there is a function $f\in C(T^n)$ such that the series $\sigma_n[f]$ diverges almost everywhere. Using this example, in 1973, Bahbuh and Nikishin [1] showed that for $n=2$, in the conditions of the corollary one cannot take $\eta=0$. The analogous problem for $n\geq 3$ still remains open.

Corollary 2.1.4. *Suppose that $\eta\in]0,+\infty[$. If $f\in L^2(T^n)$, $\alpha\in]0,1[$, and*

$$\omega_i(\delta,f)_2\delta^{-\alpha}\Big(\log\frac{2\pi}{\delta}\Big)^{\frac{n}{2}+\eta}\leq A(f,n)\quad for\ \delta\in]0,\pi],\ 1\leq i\leq n,$$

then the series $\sigma_n[f]$ is almost everywhere $(C,-\beta)$-summable, where $\beta=(\frac{\alpha}{n},\dots,\frac{\alpha}{n})$.

Corollary 2.1.5. *Let $\alpha\in]0,1[^n$. If $f\in L^2(T^n)$ and*

$$\left|a_p^{(B)}(f)\right|\prod_{i=1}^{n}(p_i+1)^{\frac{1}{2}+\alpha_i}\leq A(f,n)\quad for\ p\in\mathbb{N}_0^n,\ B\subset M,$$

then the series $\sigma_n[f]$ is almost everywhere $(C,-\beta)$-summable with any β satisfying $\beta_i\in[0,\alpha_i[$ for $1\leq i\leq n$.

Note that the conclusions of Theorems 2.1.4–2.1.6 (hence of Corollaries 2.1.3–2.1.5) remain valid for all the series $\overline{\sigma}_n[f,B]$, $B\subset M$. Moreover, under the conditions of these theorems, we have $\overline{\sigma}_n[f,B]=\sigma_n[\overline{f}_B]$.

As to the problem of summability by Cesàro method of a positive order (for α satisfying $\alpha_i\in]0,+\infty[$ for $1\leq i\leq n$) at a point or almost everywhere, the results obtained in this direction are presented in our work [35].

In 1970, N. Tevzadze [1] proved that if $f\in L^2(T^2)$, then the series $\sigma_2[f]$ converges by squares almost everywhere. According to this theorem of N. Tevzadze, the Fourier series of the functions from $C(T^2)$ constructed in the works of Fefferman [1], Ash and Gluck [1], and Bahbuh and Nikishin [1], converge by squares almost everywhere. In [35], we have pointed out results concerning the divergence by cubes (for $n = 2$ by squares) of the series $\sigma_n[f]$. Moreover, partial integral moduli of continuity of the constructed function f have a definite order of vanishing. Below we present some of those results.

Lemma 2.1.3. *Let*

$$p_k = 2^{2^j} \quad (j = 1, 2, \dots), \quad 1 \le k \le n, \tag{2.1.41}$$

and

$$\begin{aligned} m(i) \equiv m(k, i) \equiv m(k, i, p_k) = \\ = (4p_k + 1)^4 4^i + p_k \; (i = 1, 2, \dots), \quad 1 \le k \le n. \end{aligned} \tag{2.1.42}$$

Then for any $p = (p_1, p_2, \dots, p_n)$, there is a trigonometric polynomial $u \equiv u_j$ of n variables and of order $2m(2p_k)$ with respect to x_k, $1 \le k \le n$, such that:

a) $\quad \|u\|_1 \le 4^n;$ (2.1.43)

b) *for any $\varepsilon \in]0, 1[$, there is a natural number $n_0 \equiv n_0(\varepsilon)$ such that for any p with $p_k > n_0$, $1 \le k \le n$, a set $E_p \subset [0, 2\pi]^n$ with*

$$\mu_n(E_p) > (2\pi)^n - \varepsilon \tag{2.1.44}$$

and a positive number $A(\varepsilon)$ can be found such that for any $x \in E_p$, there is an index $q \equiv q(x, p)$ with

$$|S_q(x, u)| > A(\varepsilon) 2^{jn}. \tag{2.1.45}$$

Proof. First note that the lemma is a multidimensional analogue of Lemma 2.3.3 from Chapter 2 of the first part. In the proof we essentially use the scheme of the proof of Lemma 2.3.3 based on Kolmogorov's method [1].

Put

$$\alpha_\nu \equiv \alpha_\nu(k) \equiv \alpha(\nu, k, p_k) = 4\pi \frac{\nu}{4p_k + 1},$$

$$\beta_\nu \equiv \beta_\nu(k) \equiv \beta(\nu, k, p_k) =]\alpha_\nu + p_k^{-2}, \alpha_{\nu+1} - p_k^{-2}[, \;\; 0 \le \nu \le 2p_k; \;\; 1 \le k \le n,$$

and

$$u(x) = \prod_{i=1}^{n} u\Big(x_k, 2m(2p_k)\Big), \tag{2.1.46}$$

where $u_k \equiv u(x_k)$, $1 \le k \le n$, is defined by (2.3.5) from the first part of the book.

Using (2.3.11) of the first part, we get

$$\|u\|_1 \le 4^n.$$

Let $p' = (p'_1, p'_2, \dots, p'_n)$, where

$$p'_k = 2m(k, i) \quad \text{for} \quad 1 \le k \le n.$$

It is clear that

$$S_{p'}(x,u)=\prod_{k=1}^{n} S_{p'_k}(x_k,u_k).$$

Then, applying b) and c) of Lemma 2.3.3 of the first part to each of $s_{p'_k}(x,u_k)$, $1\le k\le n$, and using Fubini's theorem, we make sure that the assertions a) and b) of the lemma are true. ∎

Lemma 2.1.4. *Suppose that p_k and $m(k,i)$, $i\in\mathbb{N}$, $1\le k\le n$, are defined by* (2.1.41) *and* (2.1.42), *respectively. Put*

$$\ell(p_k)=2(4\cdot 2^{2^j}+1)^4 4^{2^{2^j}}+3\cdot 2^{2^j},\quad j\in\mathbb{N},\ \ 1\le k\le n,$$

and

$$v_1(p_k,x_k)\equiv v_1(p_k(j),x_k)=\frac{1}{2p_k+1}\sum_{i=0}^{2p_k}\lambda_{m(k,i)}(x_k-\alpha_i)\times$$
$$\times\Big[2\cos\ell(p_k)x_k+2\cos 2\ell(p_k)x_k\Big],\quad 1\le k\le n, \tag{2.1.47}$$

where $\lambda_{m(k,i)}$ is defined by (2.3.5) *of Chapter* 2 *of the first part of the book. If $f:[0,2\pi]^n\to\mathbb{R}$ is defined as*

$$f(x)=\sum_{j=1}^{\infty}(2^j)^{-n}\prod_{k=1}^{n}v_1(p_k,x_k), \tag{2.1.48}$$

then f and all its conjugates $\overline{f}_B$ with $B\subset M$ are integrable on T^n. Moreover, $\overline{\sigma}_n[f,B]=\sigma_n[\overline{f}_B]$.

Proof. By (2.1.43) and (2.1.46)–(2.1.48), we have

$$\sum_{j=1}^{\infty}(2^j)^{-n}\int_{T^n}\Big|\prod_{k=1}^{n}v_1(p_k,x_k)\Big|\,dx<+\infty.$$

Hence, using Levi's theorem, we find that the n-fold trigonometric series in the right side of (2.1.48) converges almost everywhere, and its sum $f\in L(T^n)$. Since the different products

$$\prod_{k=1}^{n}v_1(p_k,x_k)$$

do not have similar terms, the series in the right side of (2.1.48) can be represented as an n-fold trigonometric series of the form

$$s(n)\equiv\sum_{p\ge 0}2^{-\lambda(p)}\sum_{B\subset M}\alpha_p^{(B)}\prod_{i\in B}\cos p_ix_i\prod_{j\in B'}\sin p_jx_j. \tag{2.1.49}$$

On the other hand, partial sums of this series with indices $m'=(m'_1,m'_2,\dots,m'_n)$, where $m'_k=\ell(p_k)+2m(p_k)$ for $1\le k\le n$, converge to f in the norm of $L(T^n)$. Therefore (2.1.49) is the Fourier series of f, i.e., $\sigma_n[f]=s(n)$.

Let now $\varnothing \neq B \subset M$. Consider the n-fold trigonometric series

$$\sum_{j=1}^{\infty}(2^j)^{-n}\prod_{i\in B}\overline{v}_1(p_i,x_i)\prod_{j\in B'}v_1(p_j,x_j), \tag{2.1.50}$$

where $\overline{v}_1$ is a conjugate to v_1 with respect to the corresponding variable. Using (2.3.26) and (2.3.27) of the first part, we conclude that

$$\overline{v}_1(p_k,x_k)=\frac{1}{2p_k+1}\sum_{i=0}^{2p_k}\lambda_{m(k,i)}(x_k-\alpha_i)\Big[2\sin\ell(p_k)x_k+2\sin 2\ell(p_k)x_k\Big], \tag{2.1.51}$$

and

$$\frac{1}{2\pi}\int_T|\overline{v}_1(p_k,x_k)|\,dx_k\le 4 \quad \text{for } 1\le k\le n.$$

Then, quite as we have shown that (2.1.49) is $\sigma_n[f]$, we can make sure that $\overline{f}_B\in L(T^n)$ and $\overline{\sigma}_n[f,B]=\sigma_n[\overline{f}_B]$, $B\subset M$. ■

While proving Lemmas 2.1.3. and 2.1.4 (as already noted), we essentialy based on the schemes of the proofs of Lemmas 2.3.3 and 2.3.4 of the first part. Hence the proofs of the lemmas are abbreviated.

Theorem 2.1.7. *For any $n\in\mathbb{N}$ $(n\ge 2)$, there is a function $f\in L(T^n)$ such that*

$$\omega_i(\delta,f)_1\Big(\log\log\frac{2\pi}{\delta}\Big)^n\le A(f,n) \ \ for\ \delta\in]0,\pi],\ 1\le i\le n, \tag{2.1.52}$$

all the $\overline{f}_B$ with $B\subset M$ exist almost everywhere, are integrable and their partial integral moduli of continuity also satisfy (2.1.52), *yet all the series $\sigma_n[f]$, $\sigma_n[\overline{f}_B]$ are almost everywhere λ-divergent (for $\lambda=1$).*

Proof. We make use of the scheme of the proof of the assertion b) of Theorem 2.3.1 of the first part.

Let $h_i\in]0,10^{-2}[$ and let $j_0\equiv j_0(h_i)$ be the largest natural number satisfying

$$\big(\exp_4\exp_2\exp_2 j\big)^{10}\le h_i^{-1},\quad 1\le j\le j_0, \tag{2.1.53}$$

and so

$$\big(\exp_4\exp_2\exp_2(j_0+1)\big)^{10}>h_i^{-1},\quad 1\le i\le n.$$

Then we can conclude that

$$\big(2^{-j_0-1}\log\log h_i^{-1}\big)^n\le A(f,n). \tag{2.1.54}$$

Suppose that the function $f:T^n\to\mathbb{R}$ is defined by (2.1.48). By virtue of Lemma 2.1.4, we have $f\in L(T^n)$. Moreover,

$$\|f(\cdot+h_i)-f(\cdot)\|_1\le$$

$$\le A(n)\Bigg\{\sum_{j=1}^{j_0}(2^j)^{-n}\int_{T^n}\Bigg|v(p_i,x_i+h_i)\prod_{k=1,k\neq i}v(p_k,x_k)-\prod_{1\le k\le n}v(p_k,x_k)\Bigg|\,dx+$$

$$+\sum_{j=j_0+1}^{\infty}(2^j)^{-n}\int_{T^n}\left|\prod_{k=1,\,k\neq i} v(p_i,x_i+h_i)v(p_k,x_k)-\prod_{1\le k\le n} v(p_k,x_k)\right|dx\Big\}\equiv$$

$$\equiv B_1(f,n,h_i)+B_2(f,n,h_i)\quad\text{for } 1\le i\le n. \tag{2.1.55}$$

The expression $B_1(f,n,h_i)$ can be estimated similarly to $B_1(h,f)$ in (2.3.32) of the first part. To be more precise, with respect to x_i Lagrange's theorem and Bernstein's inequality [2] must be used, and with respect to other coordinates x_k – the relation

$$\|v_1(p_k,\cdot)\|\le 4.$$

Therefore

$$B_1(f,n,h_i)\le A(f,n)h_i\big(\exp_4\exp_2\exp_2 j_0\big)^8(2^{-j_0-1})^n.$$

Hence, according (2.1.53) and (2.1.54), we get

$$B_1(f,n,h_i)\big(\log\log h_i^{-1}\big)^n\le A(f,n). \tag{2.1.56}$$

It is clear that

$$B_2(f,n,h_i)\big(\log\log h_i^{-1}\big)^n\le A(f,n). \tag{2.1.57}$$

Thus, by (2.1.55)–(2.1.57), we have

$$\|f(\cdot+h_i)-f(\cdot)\|_1\le A(f,n)\big(\log\log h_i^{-1}\big)^{-n}\quad\text{for}\quad 1\le i\le n,$$

so that the relations

$$\omega_i(\delta,f)_1\left(\log\log\frac{2\pi}{\delta}\right)^n\le A(f,n)\ \text{ for }\ \delta\in]0,\pi],\ \ 1\le i\le n$$

hold.

Quite analogously we can prove the inequalities

$$\omega_i(\delta,\overline{f}_B)_1\left(\log\log\frac{2\pi}{\delta}\right)^n\le A(f,n)\ \text{ for }\ \delta\in]0,\pi],\ \ B\subset M,\ \ 1\le i\le n.$$

Indeed, the expressions $\overline{v}_1(p_k,x_k)$, $1\le k\le n$, possess all the properties we made use of for $v_1(p_k,x_k)$. Therefore we have established that integral moduli of continuity of the functions f and $\overline{f}_B$, $B\subset M$, satisfy (2.1.52).

As to the almost everywhere divergence by cubes (for $n=2$ by squares) of the series $\sigma_n[f]$ and $\sigma_n[\overline{f}_B]$, $B\subset M$, it follows from the assertions a) and b) of Lemma 2.1.3, as well as from the specific character of the constructed function f. ■

Remark 2.1.3. In 1969, Gogoladze [1] established that the well-known theorem of Plessner [4], in general, is not true even for the n-fold series $\sigma_n[f]$. Therefore, while proving Theorem 2.1.7, we cannot use the fact that the divergence of $\sigma_n[f]$ on a set $E\subset T^n$ with $\mu_n(E)>0$ implies that of the series $\sigma_n[\overline{f}_B]$ with $B\subset M$ almost everywhere on T^n. We should add that in the works [2], [6], Gogoladze proved a multidimensional analogue of Plessner's theorem for rectangular partial sums.

2.2. Almost Everywhere Summability of the Series $\overline{\sigma}_n[f,B]$ by Multiple Methods of Poisson–Abel and Cesàro of Positive Order. Integrability of Some Majorants Connected with Multiple Series $\sigma_n[f]$ and $\overline{\sigma}_n[f,B]$

We begin this section by a brief survey of results concerning the summability of conjugate series by multiple methods of Poisson–Abel and Cesàro of positive order. In our works [35, Ch. II] and [46], the reader can find more complete information.

In 1947, Sokol-Sokolowski [1] showed that if $f \in L(\log^+ L)^n(T^n)$, then all the series $\overline{\sigma}_n[f,B]$ are almost everywhere summable by the multiple method $(C,1)$. It follows from the argument of Sokol–Sokolowski that his theorem remains valid for the methods of Poisson–Abel and Cesàro of positive order.

In 1949, Zygmund [11] established that the theorem of Sokol–Sokolowski is also true for functions $f \in L(\log^+ L)^{n-1}(T^n)$.

In the above mentioned work, Sokol–Sokolowski noted without proof that if $f \in L(T^n)$, then

$$\lim_{m_\lambda \to \infty} \left[t^1_m(x,f) - \overline{f}_M\left(x, \varepsilon\left(\frac{\pi}{m}\right)\right)\right] = 0 \tag{2.2.1}$$

almost everywhere, where $t^1_m(x,f)$ is defined (with $\alpha = 1 = (1,\dots,1)$) by (1.1.26).

In 1961, Stein [1] investigated for $n = 2$ the problem of the definitiveness of the above mentioned result of Zygmund, and the problem of Sokol–Sokolowski's hypotheses. However, the scheme of the proof of Stein was based on noncorrect assertions. In the works [2], [16], [35, Ch. II], [37], and [45], we pointed out some results in this direction. They are given below.

In 1935, Jessen, Marcinkiewicz and Zygmund [1] actually proved the following assertions.

a) If $f \in L(\log^+ L)^n(T^n)$, then the functions σ, F, and M_B with $B \subset M$ defined, respectively, by (1.1.32), (1.1.34), and (1.1.38) are summable on T^n.

b) If $f \in L(\log^+ L)^{n-1}(T^n)$, then for any $p \in]0,1[$, the functions σ, F, M_B belong to $L^p(T^n)$. Moreover, it was in fact proved that this assertion is unimprovable.

In 1947, Sokol–Sokolowski [1] established that if $f \in L(\log^+ L)^{n+1}(T^n)$, then the functions t, F^*, and t^* defined, respectively, by (1.1.33), (1.1.35), and (1.1.36) with $B = M$ are integrable on T^n.

In 1949, Zygmund [11] showed that the assertion of Sokol–Sokolowski remains true for the class $L(\log^+ L)^n(T^n)$.

In the works [41], [42], we presented some results concerning the integrability of the majorants connected with the multiple series $\sigma_n[f]$ and $\overline{\sigma}_n[f,B]$. They also are given below with complete proofs.

In 1939, Marcinkiewicz and Zygmund [1] proved the following theorems.

a) Let $f \in L(T^n)$. Then for any $\lambda \in [1,+\infty[$, we have

$$\lim_{r_\lambda \to 1-} f(x,r) = f(x)$$

almost everywhere.

b) Let $f \in L(T^n)$ and $\alpha \in]0,+\infty[^n$. Then for any $\lambda \in [1,+\infty[$, we have

$$\lim_{m_\lambda \to \infty} \sigma_m^\alpha(x,f) = f(x)$$

almost everywhere.

The functions $f(x,r)$ and $\sigma_m^\alpha(x,f)$ are defined by (1.1.27) and (1.1.25), respectively.

In 1947, Zygmund [10] proved the stronger statements.

a) Suppose that the functions $\varphi_i :]0,1] \to]0,1]$ do not decrease on $]0,1]$ and $\varphi_i \to 0$ as $t \to 0$ for $1 \le i \le n$. Let $\lambda \in [1,+\infty[$ be fixed and

$$\lambda^{-1}\varphi_i(t) \le 1 - r_i \le \lambda\varphi_i(t) \ \text{ for } \ 1 \le i \le n. \tag{2.2.2}$$

If $f \in L(T^n)$ and (2.2.2) is fulfilled, then

$$\lim_{r \to 1-} f(x,r) = f(x)$$

almost everywhere, where $r = (r_1, r_2 \ldots, r_n)$.

b) Let the functions $q_i : [0,+\infty[\to \mathbb{N}_0$ do not decrease and let $q_i \to \infty$ as $t \to \infty$ for $1 \le i \le n$. Suppose that $\lambda \in [1,+\infty[$ is fixed and $m = (m_1, m_2, \ldots, m_n)$ satisfies

$$\lambda^{-1}q_i(t) \le m_i \le \lambda q_i(t), \quad 1 \le i \le n. \tag{2.2.3}$$

If $f \in L(T^n)$, then

$$\lim_{m \to \infty} \sigma_m^\alpha(x,f) = f(x)$$

almost everywhere.

Theorem 2.2.1. a) *Let $f \in L(\log^+ L)^{n-1}(T^n)$. Then for any nonempty $B \subset M$,*

$$\lim_{r \to 1-} \overline{f}(x,r,B) = \overline{f}_B(x)$$

almost everywhere.

b) *Let $n \ge 3$ and $\beta \in [1, n-1[$. If $f \in L(\log^+ L)^\beta(T^n)$, then for any $B \subset M$ with $\overline{k}(B) \in [1,\beta]$, it holds*

$$\lim_{r \to 1-} \overline{f}(x,r,B) = \overline{f}_B(x)$$

almost everywhere, where r_i satisfy (2.2.2).

c) *Let $n \ge 2$ and $\beta \in [0, n-1[$. Then for any $B \subset M$ with $\overline{k}(B) \in]\beta, n]$, there is a function $f \in L(\log^+ L)^\beta(T^n)$ such that for $\lambda = 1$,*

$$\lim_{r_\lambda \to 1-} |\overline{f}(x,r,B)| = +\infty$$

almost everywhere.

Proof. The assertion a) follows from Theorems 1.2.1–1.2.3. According to a) of Theorem 1.3.1, all the functions $\overline{f}_B$ with $\overline{k}(B) \in [1,\beta]$ are integrable on T^n. Using the corresponding results of Kolmogorov [3] and Smirnov [1], we can easily conclude that $\overline{\sigma}_n[f,B] = \sigma_n[\overline{f}_B]$ with $\overline{k}(B) \in [1,\beta]$. Applying now the above assertion a) of Zygmund, we make sure that the assertion b) is likewise true.

Prove now the assertion c). Analyzing the scheme of the proof of b) of Theorem 1.2.4, we conclude that it suffices to establish the following statement.

Let $\rho \in [0,1[$ and $r = \varepsilon(\rho)$, i.e., $r_i = \rho$, $1 \le i \le n$. Suppose that $a \in]0,+\infty[$, $\beta \in [0,n-1[$, and $B \subset M$ with $\overline{k}(B) \in]\beta, n]$. There exists a function $f \in L(\log^+ L)^\beta(T^n)$ such that for some $\alpha_0 \in]20,+\infty[$,

$$\mu_n\left[\left\{x : \sup_r |\overline{f}(x,r,B)| \ge \alpha > 0,\ \ x \in T^n\right\}\right] \ge$$
$$\ge A(n,\beta)\alpha^{-1}(\log\log\alpha)^a \quad \text{for } \alpha \in [\alpha_0,+\infty[, \tag{2.2.4}$$

i.e., the analogue of (1.2.72) is valid.

To prove this statement, we consider two cases.

1. Let $B \subset M$ and $k \equiv \overline{k}(B) \in]\beta+1,n]$. Suppose that the function $f : T^n \to \mathbb{R}$ is defined by (1.2.70). According to (1.1.28), we have

$$\overline{f}(x,r,B) = \frac{(-1)^k}{\pi^n} \int_{T^n} f(s) \prod_{j\in B'} P(r_j, s_j - x_j) \prod_{j\in B} Q(r_i, s_i - x_i)\, ds, \tag{2.2.5}$$

for $B = M$ assuming that

$$\prod_{j\in B'} P(r_j, s_j - x_j) \equiv 1.$$

Use the notation introduced on the page 120 with $\gamma_1 = \frac{\gamma}{2}$. Fix $x \in I^n(\gamma_1)$. We will also assume that

$$0 < 1 - \rho < \min_{1\le i\le n} x_i. \tag{2.2.6}$$

Taking into account (1.2.70), (2.2.5) and (2.2.6), we have

$$\pi^n|\overline{f}(-x_B, r, B)| >$$
$$> \int_{E(x,\gamma_1,B)} \left[\int_{E_1(x,\gamma_1,B)} f(s) \prod_{j\in B'} P(r_j, s_j - x_j)\, ds_{B'}\right] \times$$
$$\times \prod_{i\in B} Q(r_i, s_i + x_i)\, ds_B. \tag{2.2.7}$$

Using (1.1.29), (1.1.30), and (2.2.6), we get

$$\int_{x_j}^{\gamma_1} P(r_j, s_j - x_j)\, ds_j > \int_{x_j}^{x_j+1-\rho} P(r_j, s_j - x_j)\, ds_j \ge A,\ \ j \in B', \tag{2.2.8}$$

$$Q(r_i, s_i + x_i) \ge A x_i^{-1}, \quad i \in B. \tag{2.2.9}$$

Note that in the case $B = M$, the relation (2.2.9) would suffice.

Thus, according to (1.2.70) and (2.2.7)–(2.2.9), we find

$$\sup_r |\overline{f}(-x_B, r, B)| \ge A(n) \prod_{i=1}^n x_i^{-1} \int_{E(x,\gamma_1,B)} f(s_B + x_{B'})\, ds_B.$$

Hence it follows that

$$\sup_r |\overline{f}(-x_B, r, B)| \geq A(n,\beta,\eta) \prod_{i=1}^n x_i^{-1} \left[\log\left(\prod_{i=1}^n x_i^{-1}\right)\right]^{-n-\beta-\eta+k} \times$$
$$\times \left[\log\log\left(\prod_{i=1}^n x_i^{-1}\right)\right]^a. \tag{2.2.10}$$

But the conditions of the theorem imply $n+\beta+\eta-k \leq n-1$ (see(1.2.70)). Then by (2.2.10), we get

$$\sup_r |\overline{f}(-x_B, r, B)| \geq A(n,\beta,\eta) \prod_{i=1}^n x_i^{-1} \left[\log\left(\prod_{i=1}^n x_i^{-1}\right)\right]^{-n+1} \times$$
$$\times \left[\log\log\left(\prod_{i=1}^n x_i^{-1}\right)\right]^a \quad \text{for } x \in I^n(\gamma_1). \tag{2.2.11}$$

Therefore, on the basis of Lemma 1.2.17, we conclude that (2.2.4) is true.

2. Suppose now that $B \subset M$ and (without loss of generality) $k \equiv \overline{k}(B) = \beta + 1$. Since $\beta \in [0, n-1[$, we have $\beta + 1 < n$. This means that $B \neq M$, i.e., $B' \neq \varnothing$. In this case, the function f defined by (1.2.70) will be of no use since $\eta = 0$ and $f \notin L(\log^+ L)^\beta(T^n)$. Therefore, for $a \in]0, +\infty[$ and $\eta \in]0, [\beta]+1-\beta[$, we define the function $f : T^n \to \mathbb{R}$ as

$$f(x) = \begin{cases} (2n)^{2n} \prod_{i=1}^n x_i^{-1} \left[\log\left(\prod_{i=1}^n x_i^{-1}\right)\right]^{-n-\beta-\eta} \times \\ \quad \times \left[\log\log\left(\prod_{i=1}^n x_i^{-1}\right)\right]^a \quad \text{for } x \in I^n(\gamma), \\ 0 \quad \text{for } x \in T^n \setminus I^n(\gamma), \end{cases} \tag{2.2.12}$$

and we extend it 2π-periodically in each of coordinates of x. It is not difficult to see that $f \in L(\log^+ L)^\beta(T^n)$.

Let now $l \in B'$. Choose $\rho \in [0, 1[$ such that

$$2^{-1} x_l < 1 - \rho \leq x_l \quad \text{for } x_l \in]0, \gamma_1], \tag{2.2.13}$$

the coordinates x_j with $j \in B' \setminus \{l\}$ being assumed to be arbitrary from $]0, \gamma_1]$. By (1.1.29), we easily conclude that

$$P(r_l, s_l + x_l) \geq A x_l^{-1}. \tag{2.2.14}$$

Taking into account (2.2.13), we can verify that for x_i, $i \in B$, with $x_l \leq x_i$ the relations (2.2.9) are valid. Let $-x_{B,l} \equiv x'$ denote the point of $\mathbb{R}^n$ satisfying $x_i' = -x_i$ for $i \in B \cup \{l\}$ and $x_j' = x_j$ for $j \in B' \setminus \{l\}$. For the function f defined by (2.2.12), write (2.2.5) with $\overline{k}(B) = \beta + 1$. Suppose that $x \in]0, \gamma_1]^n$. Using the introduced notation, (2.2.8) for $j \in B' \setminus \{l\}$, (2.2.9) for $i \in B$ and (2.2.14), we get

$$\sup_r |\overline{f}(-x_{B,l}, r, B)| \geq A(n,\beta,\eta) \prod_{i=1}^n x_i^{-1} \left[\log\left(\prod_{i=1}^n x_i^{-1}\right)\right]^{-n-\beta+k+1-\eta} \times$$

$$\times\left[\log\log\left(\prod_{i=1}^{n}x_i^{-1}\right)\right]^{a}.$$

Hence, taking note of the equality $\overline{k}(B)=\beta+1$, we find

$$\sup_{r}|\overline{f}(-x_{B,l},r,B)|\geq A(n,\beta,\eta)\prod_{i=1}^{n}x_i^{-1}\left[\log\left(\prod_{i=1}^{n}x_i^{-1}\right)\right]^{-n+2-\eta}\times$$
$$\times\left[\log\log\left(\prod_{i=1}^{n}x_i^{-1}\right)\right]^{a}, \tag{2.2.15}$$

where $x_j\in[0,\gamma_1]$ for $j\in B'$ and $x_l\leq x_i$ for $i\in B$. Using again Lemma 1.2.17, we conclude that (2.2.4) is true. ■

Remark 2.2.1. In the proof of the case c) of the theorem, we made use of Lemma 1.2.17 for the expression in the right side of (2.2.15). This is correct since $\eta\in]0,1[$ and $n-2+\eta<n-1$.

Using the assertion a) of Theorem 1.2.4 and c) of Theorem 2.2.1, we arrive at

Theorem 2.2.2. *Let $\beta\in[0,n-1[$ and $B\subset M$ with $\overline{k}(B)=[\beta]+1$. Then there is a function $f\in L(\log^+L)^\beta(T^n)$ such that for $\lambda=1$,*

$$\limsup_{r_\lambda\to 1-}|\overline{f}(x,r,B)-\overline{f}_B(x,\varepsilon)|=+\infty \tag{2.2.16}$$

almost everywhere, where $\overline{f}_B(x,\varepsilon)$ is defined by (1.1.8) with $\varepsilon_i=1-r_i$ for $i\in B$.

Since $\beta\in[0,n-1[$, we have $n-\beta-1>0$. Therefore Theorem 2.2.2 does not involve the case $B=M$. However, in this case it holds

Theorem 2.2.3. *Let $\beta\in[0,n-1[$. Then there is a function $f\in L(\log^+L)^\beta(T^n)$ such that for $\lambda=1$,*

$$\limsup_{r_\lambda\to 1-}|\overline{f}(x,r,M)-\overline{f}_M(x,\varepsilon)|=+\infty$$

almost everywhere.

Proof. By the condition of the theorem, we have $n-\beta-1>0$. Therefore we can assume that the function $f:T^n\to\mathbb{R}$ is defined by (1.2.70) with $\overline{k}(B)=n$ for $\eta\in]0,n-\beta-1]$. By (1.1.28), we obtain

$$\overline{f}(x,r,M)=\left(-\frac{1}{\pi}\right)^n\int_{T^n}f(s)\prod_{i=1}^{n}Q(r_i,s_i-x_i)\,ds. \tag{2.2.17}$$

Let

$$R(r_i,t)=\frac{(1-r_i)^2\operatorname{ctg}\frac{t}{2}}{2(1-2r_i\cos t+r_i^2)}\ \text{for } r_i\in[0,1[,\ 1\leq i\leq n,\ |t|\in]0,\pi[\tag{2.2.18}$$

and let $x\in]0,\gamma_1]^n$. Using (1.1.30), (1.1.31), (1.2.70), (2.2.17), and (2.2.18), we get

$$(-\pi)^n\left[\overline{f}(-x,r,M)-\overline{f}_M(-x,\varepsilon)\right]=$$

$$= \sum_{B\subset M, B\neq M} \frac{(-1)^{\overline{k}(B')}}{2^{\overline{k}(B)}} \int_{T^n} f(s) \prod_{i\in B} \operatorname{ctg}\frac{s_i+x_i}{2} \prod_{j\in B'} R(r_j, s_j+x_j)\, ds, \qquad (2.2.19)$$

for $B=\varnothing$ assuming that

$$\prod_{i\in B} \operatorname{ctg}\frac{s_i+x_i}{2} \equiv 1.$$

Suppose that $\rho\in[0,1[$ and in the relation (2.2.19) we have $r=\varepsilon(\rho)$, i.e., $r_i=\rho$ for $1\le i\le n$. In the integrals on the right side of (2.2.19), the variables s_i, $1\le i\le n$, vary in $[0,\gamma]$. Therefore, by virtue of (1.1.31) and (2.2.18), we can conclude that in establishing the growth order with respect to x of the expression

$$\sup_r |\overline{f}(-x,r,M)-\overline{f}(-x,\varepsilon)|$$

in a vicinity of $0\in T^n$, the main terms are the expressions

$$(-1)^{\overline{k}(B')} \int_{I^n(\gamma)} f(s) \prod_{i\in B} \operatorname{ctg}\frac{s_i+x_i}{2} \prod_{j\in B'} R(r_j, s_j+x_j)\, ds \quad \text{with } \overline{k}(B')=1.$$

Choosing x_l as to satisfy (2.2.13), by (1.2.70) and (2.2.19) we find

$$\sup_r |\overline{f}(-x,r,M)-\overline{f}_{M}(-x,\varepsilon)| > A(n,\beta)\prod_{i=1}^n x_i^{-1}\Big[\log\Big(\prod_{i=1}^n x_i^{-1}\Big)\Big]^{-n+1}\times$$
$$\times\Big[\log\log\Big(\prod_{i=1}^n x_i^{-1}\Big)\Big]^{a} \quad \text{for} \quad x\in I^n(\gamma).$$

According to Lemma 1.2.17, for some $\alpha_0\in]20,+\infty[$ we have

$$\mu_n\Big[\Big\{x:\sup_r|\overline{f}(x,r,M)-\overline{f}_{M}(x,\varepsilon)|\ge\alpha>0,\ \ x\in T^n\Big\}\Big]\ge$$
$$\ge A(n,\beta)\alpha^{-1}(\log\log\alpha)^a, \quad \alpha\in[\alpha_0,+\infty[. \qquad (2.2.20)$$

From this point on, as we have already noted, the proof can be completed following the scheme of the proof of the assertion b) of Theorem 1.2.4. ∎

Now consider the problem of integrability on T^n of the functions F, F^*_B, t^*_B, and M_B with $\varnothing\neq B\subset M$ defined by (1.1.34), (1.1.35), (1.1.36), (1.1.37), and (1.1.38), respectively.

Theorem 2.2.4. a) *Suppose that* $\beta\in[0,n-1]$. *If* $f\in L(\log^+L)^\beta(T^n)$, *then for any* $B\subset M$ *with* $\overline{k}(B)\in[1,\beta+1]$ *and any* $p\in]0,1[$, *the inequalities*

$$\|t^*_B\|\le A(p,n)\Big\{\int_{T^n}|f(x)|\,\big[\log^+|f(x)|\big]^\beta dx+1\Big\}, \qquad (2.2.21)$$

$$\|M_B\|_p\le A(p,n)\Big\{\int_{T^n}|f(x)|\,\big[\log^+|f(x)|\big]^\beta dx+1\Big\} \qquad (2.2.22)$$

hold.

b) *Let* $\beta \in [0, n-1[$ *and* $B \subset M$ *with* $\overline{k}(B) \in]\beta+1, n]$. *Then there are functions* $f : T^n \to \mathbb{R}$ *and* $g : T^n \to \mathbb{R}$ *such that for any* $p \in]0,1[$, *it holds*

$$t^*_B(f) \notin L^p(T^n), \quad M_B(g) \notin L^p(T^n). \tag{2.2.23}$$

Proof. The validity of (2.2.21) follows from (1.2.50) (see Theorem 1.2.3). The inequality (2.2.22) is established in Jessen, Marcinkiewicz and Zygmund [1].

Since

$$t^*_B(x, f) \geq \limsup_{\varepsilon \to 0+} |\overline{f}_B(x, \varepsilon)|,$$

the first relation of (2.2.23) follows from b) of Theorem 1.2.4. The corresponding result of Jessen, Marcinkiewicz and Zygmund [1] enables us to conclude that the second relation of (2.2.23) also holds. ■

Theorem 2.2.5. a) *Let* $f \in L(\log^+ L)^\beta(T^n)$ *for some* $\beta \in [1, n]$. *Then all the functions* t^*_B *and* M_B *with* $\overline{k}(B) \in [1, \beta]$ *are integrable on* T^n.

b) *Let* $\beta \in [1, n[$ *and* $B \subset M$ *with* $\overline{k}(B) \in]\beta, n]$. *Then there are functions* $f \in L(\log^+ L)^\beta(T^n)$ *and* $g \in L(\log^+ L)^\beta(T^n)$ *such that*

$$t^*_B \notin L(T^n), \quad M_B(g) \notin L(T^n). \tag{2.2.24}$$

Proof. The conclusion of the assertion a) for the functions t^*_B follows from Lemma 1.2.14 (see (1.2.50)), and for the functions M_B it is a consequence of the corresponding result of Jessen, Marcinkiewicz and Zygmund [1]. It can be proved by consecutive application of (1.1.46).

Using the assertion b) of Theorem 1.3.1, we make sure that for the functions t^*_B with $\overline{k}(B) \in]\beta, n]$, the assertion b) is true (see the first relation in (2.2.24)).

Suppose now that the function $g : T^n \to \mathbb{R}$ is defined by (1.3.2). Let $x \in]0, \gamma_1]^n$. According to our notation, we have

$$M_B(-x_B, g) > A(n) \prod_{i=1}^{n} x_i^{-1} \int_{E(x,\gamma_1,B)} f(s_B + x_{B'})\, ds_B \geq$$

$$\geq A(n,\beta) \prod_{i=1}^{n} x_i^{-1} \Big[\log\Big(\prod_{i=1}^{n} x_i^{-1}\Big)\Big]^{-n}.$$

Therefore

$$M_B(g) \notin L(T^n). \quad ■$$

Theorem 2.2.6. a) *let* $f \in L(\log^+ L)^{n-1}(T^n)$ *and the functions* F *and* F^*_B *be defined by* (1.1.34) *and* (1.1.35), *respectively. Then for any nonempty* $B \subset M$ *and* $p \in]0,1[$, *there hold*

$$\|F\|_p \leq A(p,n)\Big\{\int_{T^n} |f(x)| \left[\log^+ |f(x)|\right]^{n-1} dx + 1\Big\}, \tag{2.2.25}$$

$$\|F^*_B\|_p \leq A(p,n)\Big\{\int_{T^n} |f(x)| \left[\log^+ |f(x)|\right]^{n-1} dx + 1\Big\}. \tag{2.2.26}$$

b) *Let $f \in L(\log^+ L)^n(T^n)$. Then*

$$\|F\|_1 \le A(n)\Big\{ \int\limits_{T^n} |f(x)| \left[\log^+ |f(x)|\right]^n dx + 1 \Big\}, \tag{2.2.27}$$

and for any nonempty $B \subset M$, we have

$$\|F^*_B\|_1 \le A(n)\Big\{ \int\limits_{T^n} |f(x)| \left[\log^+ |f(x)|\right]^n dx + 1 \Big\}. \tag{2.2.28}$$

c) *Let $\phi : [0, +\infty[\to [0, +\infty[$, $\phi \uparrow +\infty$ as $t \uparrow +\infty$, and*

$$\phi(t)\left[t(\log t)^n\right]^{-1} \to 0 \quad \text{as} \quad t \to +\infty.$$

Then there is a function $f \in L(T^n)$ such that

$$\int\limits_{T^n} \phi(|f(x)|)\, dx < +\infty,$$

yet

$$F(f) \notin L(T^n), \quad F^*_M(f) \notin L(T^n), \quad M_M(f) \notin L(T^n). \tag{2.2.29}$$

Proof. The validity of (2.2.25) follows from the corresponding result of Jessen, Marcinkiewicz and Zygmund [1]. As to (2.2.26), it is a consequence of the assertion a) (see (1.2.43)) of Lemma 1.2.11.

The inequality (2.2.27) can likewise be deduced from the corresponding result of Jessen, Marcinkiewicz and Zygmund [1], and (2.2.28) can be obtained with the help of Lemma 1.2.13 (see (1.2.47)).

Now we turn to the proof of c).

Consider the function $f : T^n \to \mathbb{R}$ constructed while proving Theorem 1.3.2 (see (1.3.6)). Let $x \in]0, \gamma_1]^n$ and choose r_i such that

$$2^{-1}(1 - r_i) < x_i \le 1 - r_i \quad \text{for} \quad 1 \le i \le n. \tag{2.2.30}$$

Using (1.1.29), (1.1.30), and (2.2.30), we get

$$P(r_i, s_i + x_i) > A x_i^{-1}, \tag{2.2.31}$$

$$Q(r_i, s_i + x_i) > A x_i^{-1} \quad \text{for} \quad s_i \in [0, x_i], \quad 1 \le i \le n. \tag{2.2.32}$$

Hence, basing on (1.1.27), (1.1.28), (1.1.34), (1.1.35), (1.3.6), (2.2.31), and (2.2.32), we find

$$F(-x, f) > A(n) \prod_{i=1}^{n} x_i^{-1} \int\limits_{E(x,\gamma_1,M)} f(s)\, ds, \tag{2.2.33}$$

$$F^*_M(-x, f) > A(n) \prod_{i=1}^{n} x_i^{-1} \int\limits_{E(x,\gamma_1,M)} f(s)\, ds. \tag{2.2.34}$$

By (1.1.38) and (1.3.6), we have

$$M_M(-x,f) > A(n)\prod_{i=1}^{n} x_i^{-1} \int\limits_{E(x,\gamma_1,M)} f(s)\,ds. \tag{2.2.35}$$

Acting as while proving Theorem 1.3.2 (see (1.3.8)), we obtain

$$\int\limits_{E(x,\gamma_1,M)} f(s)\,ds >$$

$$> A(n)\Big(\log\log p_{\nu+1}\Big)^{-1}\bigg\{\Big[\log\Big(\prod_{i=1}^{n} x_i^{-1}\Big)\Big]^{-n} - \Big[\log\Big(\prod_{i=1}^{n} x_i^{-1}p_{\nu+1}^{-1}\Big)\Big]^{-n}\bigg\},$$

where $x_1 \in]p_{\nu+1}^{-1}, p_\nu^{-1}[$ and $x_j \in]0,\gamma_1]$ for $2 \le j \le n$. Therefore, according to (2.2.33)–(2.2.35) and Theorem 1.3.2, we conclude that (2.2.29) is true. ■

Consider now Cesàro means of positive order of the series $\sigma_n[f]$ and $\overline{\sigma}_n[f,B]$ (see (1.1.25), (1.1.26)). The main theorems of this section concerning Poisson–Abel means of the series $\sigma_n[f]$ and $\overline{\sigma}_n[f,B]$, $B \subset M$, remain true (naturally, with appropriate modification) for the means $\sigma_m^\alpha(x,f)$, $t_n^\alpha(x,f,B)$ with $\alpha \in]0,+\infty[^n$. In the proofs, we have frequently used the fact that the Poisson–Abel kernel is nonnegative. The Cesàro–Fejér kernel, however, possesses such a property only for some α. Therefore we have to use some other argument. Not willing to repeat the above reasoning while proving the theorems stated below, we will only outline the principal moments.

Theorem 2.2.7. a) *Let $f \in L(\log^+ L)^{n-1}(T^n)$. Then for any nonempty $B \subset M$ and $\alpha \in]0,+\infty[^n$*

$$\lim_{m\to\infty} t_m^\alpha(x,f,B) = \overline{f}_B(x)$$

almost everyhere.

b) *Let $n \ge 3$ and $\beta \in [1, n-1[$. If $f \in L(\log^+ L)^\beta(T^n)$, then for any $B \subset M$ with $\overline{k}(B) \in [1,\beta]$ and any $\alpha \in]0,+\infty[^n$, the relation*

$$\lim_{m\to\infty} t_m^\alpha(x,f,B) = \overline{f}_B(x)$$

holds almost everywhere, where m_i satisfy (2.2.3).

c) *Let $n \ge 2$ and $\beta \in [1, n-1[$. Then for any $B \subset M$ with $\overline{k}(B) \in]\beta,n]$ and any $\alpha \in]0,+\infty[^n$, there is a function $f \in L(\log^+ L)^\beta(T^n)$ such that for $\lambda = 1$,*

$$\limsup_{m_\lambda\to\infty} |t_m^\alpha(x,f,B)| = +\infty$$

almost everywhere.

Proof. Check the validity of a). According to (1.1.26) and our notation, we have

$$t_m^\alpha(x,f,B) = \frac{(-1)^k}{\pi^n}\Bigg[\int\limits_{E(\frac{1}{m},\pi,M)} + \sum_{B_1\subset M}\ \int\limits_{E(\frac{1}{m},\pi,B_1)}\ \int\limits_{E_1(\frac{1}{m},\pi,B_1')} +$$

$$+ \int\limits_{E_1(\frac{1}{m},\pi,M)} \Bigg] f(x+s) \prod_{j\in B'} K_{m_j}^{\alpha_j}(s_j) \prod_{i\in B} \tau_{m_i}^{\alpha_i}(s_i)\, ds \equiv$$

$$\equiv \sum_{i=1}^{3} J_i(x,f,\alpha,B,m). \tag{2.2.36}$$

Using (2.2.13) and (2.2.14) from the first part, by (2.2.36) we find

$$|J_1(x,f,\alpha,B,m)| \le A(n,\alpha) M_1 \cdots M_n(x,f). \tag{2.2.37}$$

Basing on (2.2.9)–(2.2.12), (2.2.15), and (2.2.16) from the first part of the book, in view of (2.2.36), (1.2.30), and (1.2.31) we get

$$|J_3(x,f,\alpha,B,m)| \le A(n,\alpha)\Big\{ \sum_{B_1\subset B} M_{B\setminus B_1}(x,g_{B_1})\Big\}. \tag{2.2.38}$$

Consider now the expression $J_2(x,f,\alpha,B,m)$ in (2.2.36). To estimate its addends, use the appropriate inequality from (2.2.13)–(2.2.16) of the first part (depending on the set of integration). We will have

$$|J_2(x,f,\alpha,B,m)| \le A(n,\alpha)\Big\{ \sum_{B_1\subset B} M_{B\setminus B_1}(x,g_{B_1})\Big\}. \tag{2.2.39}$$

Therefore, by virtue of (2.2.36)–(2.2.39), we obtain

$$|t_m^{\alpha}(x,f,B)| \le A(n,\alpha)\Big\{ \sum_{B_1\subset B} M_{B\setminus B_1}(x,g_{B_1})\Big\}.$$

Hence, by (1.1.33), it follows that

$$t_B(x) \equiv t(x,f,\alpha,B)| \le A(n,\alpha)\Big\{ \sum_{B_1\subset B} M_{B\setminus B_1}(x,g_{B_1})\Big\}.$$

Using Lemma 1.2.10, we conclude that for any $B \subset M$, the inequality

$$\|t_B\|_p \le A(p,n,\alpha)\Big\{ \int\limits_{T^n} |f(x)| \left[\log^+ |f(x)|\right]^{n-1} dx + 1\Big\} \tag{2.2.40}$$

holds. Therefore the assertion a) of Theorem 1.2.3 implies that the assertion a) of the theorem is true.

The assertion b) can be verified analogously to the corresponding part of Theorem 2.2.1. To be more precise, by a) of theorem 1.3.1, all the functions $\overline{f}_B$ with $\overline{k}(B) \in [1,\beta]$ are integrable on T^n, so that $\overline{\sigma}_n[f,B] = \sigma_n[\overline{f}_B]$. Using (2.2.3), we make sure that b) is valid.

Now we turn to the proof of c). As while proving the assertion c) of Theorem 2.2.1, it suffices to prove the following statement.

Let $m = (m_1, m_2, \dots, m_n)$ and $m_i = m \in \mathbb{N}_0$ for $1 \le i \le n$. Suppose that $a \in]0, +\infty[$, $\beta \in [0, n-1[$, $\alpha \in]0, +\infty[^n$ and $B \subset M$ with $\overline{k}(B) \in]\beta, n]$. Then there is a function f such that for some $y_0 \in]20, +\infty[$, it holds

$$\mu_n\Big[\Big\{x \in T^n : \sup_m |t_m^\alpha(x, f, B)| \ge y > 0\Big\}\Big] \ge$$
$$\ge A(n, \alpha, \beta) y^{-1} (\log\log y)^a \quad \text{for} \quad y \in [y_0, +\infty]. \tag{2.2.41}$$

First we note that since the coordinates of α can vary independently, some of α_i, $1 \le i \le n$, can be less then 1, some of them can equal it, and the remaining ones will be greater than 1. As we have several times noted in the first part of the book, in the case $\alpha_i \in [1, +\infty[$ the kernels $K_{m_i}^{\alpha_i}$ and $\tau_{m_i}^{\alpha_i}$ behave in a certain sense better than in the case $\alpha_i \in]0, 1[$. For instance, if $\alpha_i = 1$, then

$$K_{m_i}^{\alpha_i}(s_i) = \frac{1}{2(m_i+1)} \left[\frac{\sin(m_i+1)\frac{s_i}{2}}{\sin\frac{s_i}{2}}\right]^2,$$
$$\tau_{m_i}^{\alpha_i}(s_i) = \frac{1}{2}\operatorname{ctg}\frac{s_i}{2} - \frac{1}{m_i+1}\frac{\sin(m_i+1)s_i}{4\sin^2\frac{s_i}{2}} \quad \text{for} \quad |s_i| \in]0, \pi],$$

similar simple formulas being valid for $\alpha_i = 2, 3, \dots$ as well. We should add that for $\alpha_i \in [1, +\infty[$, we have $K_{m_i}^{\alpha_i} \ge 0$, and in this case, the means $\tau_{m_i}^{\alpha_i}$ also behave "better". In the proof of c), we will assume that α_i, $1 \le i \le n$, do not take natural values. The scheme of the proof can be more easily carried out in the remaining cases.

I. Let $B \subset M$ and $\overline{k}(B) \in]\beta + 1, n]$. Suppose that the function $f : T^n \to \mathbb{R}$ is defined by (1.2.70). In view of (1.1.26), we have

$$t_m^\alpha(x, f, B) =$$
$$= \frac{(-1)^k}{\pi^n} \int_{I^n(\gamma)} f(s) \prod_{j \in B'} K_{m_j}^{\alpha_j}(s_j - x_j) \prod_{i \in B} \tau_{m_i}^{\alpha_i}(s_i - x_i)\, ds, \tag{2.2.42}$$

for $B = M$ it being assumed that

$$\prod_{j \in B'} K_{m_j}^{\alpha_j}(s_j - x_j) \equiv 1.$$

Using (2.2.8), (2.2.10), (2.2.12), and (2.2.16) from the first part (see Zygmund [15]), we can conclude that

$$\tau_{m_i}^{\alpha_i}(s_i) = \frac{1}{2}\operatorname{ctg}\frac{s_i}{2} + H_{m_i}^{\alpha_i}(s_i) \quad \text{for} \quad |s_i| \in]0, \pi], \quad 1 \le i \le n, \tag{2.2.43}$$

where

$$|H_{m_i}^{\alpha_i}(s_i)| \le A(\alpha_i) m_i^{-\alpha_i} s_i^{-1-\alpha_i} \quad \text{for} \quad s_i \in [m_i^{-1}, \pi], \quad 1 \le i \le n. \tag{2.2.44}$$

According to (3.2.7), (2.2.9), (2.2.11), and (2.2.13) of the first part, we have

$$K_{m_i}^{\alpha_i}(s_i) \ge A m_i \quad \text{for} \quad |s_i| \in [0, 2^{-1} m_i^{-1}], \tag{2.2.45}$$
$$K_{m_i}^{\alpha_i}(s_i) \le A_1(\alpha_i) m_i^{-\alpha_i} s_i^{-1-\alpha_i} \quad \text{for} \quad |s_i| \in [2^{-1} m_i^{-1}, \pi], \ m_i \in \mathbb{N}, \ 1 \le i \le n. \tag{2.2.46}$$

Let now $A(\alpha_i)$ and $A_1(\alpha_i)$, $1 \le i \le n$, be constants (depending only on the indicated parameters) appearing in (2.2.44) and (2.2.46), respectively. Put

$$a(\alpha) = \min_{1\le i\le n} \alpha_i, \quad A_2(\alpha) = \max_{1\le i\le n} A(\alpha_i), \quad A_3(\alpha) = \max_{1\le i\le n} A_1(\alpha_i), \tag{2.2.47}$$

$$A(n,\alpha) = \max\Big\{\big[4^{7n+1}A_2(\alpha)A_3(\alpha)\big]^{\frac{1}{a(\alpha)}}, [4^{10n+1}]^{\frac{1}{a(\alpha)}}\Big\}. \tag{2.2.48}$$

Using (2.2.45), we have

$$\int\limits_{x_j-2^{-1}m_j^{-1}}^{x_j+2^{-1}m_j^{-1}} K_{m_j}^{\alpha_j}(s_j - x_j)\,ds_j \ge A \text{ for } 1 \le j \le n. \tag{2.2.49}$$

Now consider the possible cases:

1) Let $B = M$. As already noted, we can assume that $m = (m_1, m_2, \ldots, m_n)$ and $m_i = m \in \mathbb{N}$ for $1 \le i \le n$. As above, we assume that $x \in I^n(\gamma)$ and $\gamma_1 = 2^{-1}\gamma$. Choose m satisfying

$$A(n,\alpha)m^{-1} < \min_{1\le i\le n} x_i, \tag{2.2.50}$$

and

$$0 < \frac{x_i}{2} < x_i - \frac{1}{m} < \gamma_1. \tag{2.2.51}$$

Then

$$0 < s_i + x_i \le \gamma \quad \text{for} \quad s_i \in [0,\gamma_1], \quad 1 \le i \le n. \tag{2.2.52}$$

By (2.2.42), we find

$$t_m^\alpha(-x,f,M) = \Big(-\frac{1}{\pi}\Big)^n \int\limits_{I^n(\gamma)} f(s) \prod_{i=1}^n \tau_{m_i}^{\alpha_i}(s_i + x_i)\,ds.$$

Hence, using (2.2.43), we get

$$\begin{aligned} t_m^\alpha(-x,f,M) &= \Big(-\frac{1}{2\pi}\Big)^n \int\limits_{I^n(\gamma)} f(s) \prod_{i=1}^n \operatorname{ctg}\frac{s_i + x_i}{2}\,ds + \\ &+ \sum_{B_1\subset M} \frac{2^{-\bar{k}(B_1)}}{(-\pi)^n} \int\limits_{I^n(\gamma)} f(s) \prod_{i\in B_1} \operatorname{ctg}\frac{s_i + x_i}{2} \prod_{j\in B_1'} H_{m_j}^{\alpha_j}(s_j + x_j)\,ds \equiv \\ &\equiv L_1(x,f,\alpha,M,m) + L_2(s,f,\alpha,M,m). \end{aligned} \tag{2.2.53}$$

According to our notation, we get

$$\begin{aligned} I^n(\gamma) = E(x,\gamma,M) \bigcup E_1(x,\gamma,M) \bigcup \\ \bigcup \Big\{ \bigcup_{B_2\subset M} E(x,\gamma,B_2) \times E_1(x,\gamma,B_2)\Big\}. \end{aligned} \tag{2.2.54}$$

Then, by (2.2.53), the expressions $L_i(x,f,\alpha,M,m)$ $(i=1,2)$ can be represented as

$$L_1(x,f,\alpha,M,m)=\Big(-\frac{1}{2\pi}\Big)^n\Big[\int\limits_{E(x,\gamma,M)}+\int\limits_{E_1(x,\gamma,M)}+$$

$$+\sum_{B_2\subset M}\int\limits_{E(x,\gamma,B_2)\times E_1(x,\gamma,B_2')}\Big]f(s)\prod_{i=1}^n\operatorname{ctg}\frac{s_i+x_i}{2}\,ds, \tag{2.2.55}$$

$$L_2(x,f,\alpha,M,m)=\sum_{B_1\subset M}\frac{2^{-\bar{k}(B_1)}}{(-\pi)^n}\Big\{\Big[\int\limits_{E(x,\gamma,M)}+\int\limits_{E_1(x,\gamma,M)}+$$

$$+\sum_{B_2\subset M}\int\limits_{E(x,\gamma,B_2)\times E_1(x,\gamma,B_2')}\Big]f(s)\prod_{i\in B_1}\operatorname{ctg}\frac{s_i+x_i}{2}\times$$

$$\times\prod_{j\in B_1'}H_{m_j}^{\alpha_j}(s_j+x_j)\,ds\Big\}. \tag{2.2.56}$$

Taking note of (2.2.50)–(2.2.53), (2.2.55), and (2.2.56), we make sure that

$$\sup_m|t_m^\alpha(-x,f,M)|\geq A(n,\alpha)\int\limits_{E(x,\gamma_1,M)}f(s)\prod_{i=1}^n\operatorname{ctg}\frac{s_i+x_i}{2}\,ds\geq$$

$$\geq A(n,\alpha,\beta,\eta)\prod_{i=1}^n x_i^{-1}\Big[\ln\Big(\prod_{i=1}^n x_i^{-1}\Big)\Big]^{-\beta-\eta}\times$$

$$\times\Big[\ln\ln\Big(\prod_{i=1}^n x_i^{-1}\Big)\Big]^a\quad\text{for } x\in I^n(\gamma_1). \tag{2.2.57}$$

2) Suppose that $B\neq M$ and $K_{m_j}^{\alpha_j}\geq 0$ for $j\in B'$. According to (2.2.42), we have

$$t_m^\alpha(-x_B,f,B)=$$

$$=\frac{(-1)^{\bar{k}(B)}}{\pi^n}\int\limits_{I^n(\gamma)}f(s)\prod_{j\in B'}K_{m_j}^{\alpha_j}(s_j-x_j)\prod_{i\in B}\tau_{m_i}^{\alpha_i}(s_i+x_i)\,ds.$$

Hence, using (2.2.43), we get

$$\pi^n(-1)^{\bar{k}(B)}t_m^\alpha(-x_B,f,B)=$$

$$=2^{-\bar{k}(B)}\int\limits_{T^n(\gamma)}f(s)\prod_{j\in B'}K_{m_j}^{\alpha_j}(s_j-x_j)\prod_{i\in B}\operatorname{ctg}\frac{s_i+x_i}{2}\,ds+$$

$$+\sum_{B_1\subset B}2^{-\bar{k}(B_1)}\int\limits_{T^n(\gamma)}f(s)\prod_{j\in B'}K_{m_j}^{\alpha_j}(s_j-x_j)\prod_{i\in B_1}\operatorname{ctg}\frac{s_i+x_i}{2}\times$$

$$\times\prod_{l\in B\setminus B_1}H_{m_l}^{\alpha_l}(s_l+x_l)\,ds. \tag{2.2.58}$$

Then, basing on (2.2.49)-(2.2.52) and (2.2.54), we can conclude that

$$\sup_m |t_m^\alpha(-x_B, f, B)| > A(n,\alpha,\beta,\eta)\Big(\prod_{i=1}^n x_i^{-1}\Big)\Big[\log\Big(\prod_{i=1}^n x_i^{-1}\Big)\Big]^{-n+1} \times$$
$$\times\Big[\log\log\Big(\prod_{i=1}^n x_i^{-1}\Big)\Big]^a \quad \text{for } x \in I^n(\gamma_1). \tag{2.2.59}$$

3) Let now $B \neq M$, and let $K_{m_j}^{\alpha_j}$ be with alterneting signs for some or all $j \in B'$. In this case too, if $m \in \mathbb{N}$ is appropriately chosen, a relation of the type (2.2.59) is valid. We can ascertain this by repeating the above argument. We can also use the appropriate results from Chelidze [1, Ch. VIII] and Gogoladze [3] and reduce this case to an already considered one.

II. Suppose that $B \subset M$ and (without loss of geneality) $\overline{k}(B) = \beta + 1$. By the conditions of the theorem, $\beta \in [0, n-1[$, i.e., $\beta + 1 < n$. Therefore $B \neq M$ and $B' \neq \varnothing$. Let the function $f : T^n \to \mathbb{R}$ be defined by (2.2.12). As it has been already noted, $f \in L(\log^+ L)^\beta(T^n)$. For some coordinate x_l, $l \in B'$, of the point x choose $m \in \mathbb{N}$ such that

$$x_l < \pi(8m)^{-1} < 2x_l. \tag{2.2.60}$$

By virtue of (2.2.7) of the first part of the book and (2.2.60), we find

$$K_{m_l}^{\alpha_l}(s_l + x_l) \geq A(\alpha_l)x_l^{-1}, \quad x_l \in]0, \gamma_1]. \tag{2.2.61}$$

In view of (2.2.8) of the first part, we have

$$\tau_{m_i}^{\alpha_i}(t) = \sum_{\nu=0}^{m_i} A_{m_i-\nu}^{\alpha_i-1} \frac{\cos\frac{t}{2} - \cos(\nu+\frac{1}{2})t}{2\sin\frac{t}{2}} =$$
$$= (A_{m_i}^{\alpha_i})^{-1}\sum_{\nu=0}^{m_i} A_{m_i-\nu}^{\alpha_i-1}\frac{1-\cos\nu t}{2\operatorname{tg}\frac{t}{2}} + (A_{m_i}^{\alpha_i})^{-1}\sum_{\nu=0}^{m_i} A_{m_i-\nu}^{\alpha_i-1}\sin\nu t. \tag{2.2.62}$$

But

$$\Big|(A_{m_i}^{\alpha_i})^{-1}\sum_{\nu=0}^{m_i} A_{m_i-\nu}^{\alpha_i-1}\sin\nu t\Big| \leq 1.$$

Therefore, in estimates of the types (2.2.57) and (2.2.59), the nonnegative part of the expression $\tau_{m_i}^{\alpha_i}(s_i + x_i)$ is the essential one (see (2.2.62)) since by (2.2.52), $(s_i + x_i) \in]0, \gamma]$ for $i \in B$. Taking note of (2.2.60) and (2.2.61), quite as while proving the case 2) of c) of Theorem 2.2.1, we make sure that in the case 2) too, the inequality (2.2.41) holds (in general, for another f). Therefore c) is proved. ■

According to a) of Theorem 1.2.4 and b) of Theorem 2.2.7, it holds

Theorem 2.2.8. *Let $\beta \in [0, n-1[$, $B \subset M$ with $\overline{k}(B) = [\beta]+1$ and $\alpha \in]0, +\infty[^n$. Then there is a function $f \in L(\log^+ L)^\beta(T^n)$ such that for $\lambda = 1$,*

$$\limsup_{m_\lambda\to\infty}\Big|t_m^\alpha(x, f, B) - \overline{f}_B\Big(x, \varepsilon\Big(\frac{\pi}{m}\Big)\Big)\Big| = +\infty$$

almost everywhere, where $\overline{f}_B(x, \varepsilon)$ is defined by (1.1.8).

Analyzing the schemes of the proofs of Theorems 2.2.3 and 2.2.7, we arrive at

Theorem 2.2.9. *Let* $\beta \in [0, n-[$ *and* $\alpha \in]0, +\infty[^n$. *Then there is a function* $f \in L(\log^+ L)^\beta(T^n)$ *such that for* $\lambda = 1$,

$$\limsup_{m_\lambda \to \infty} |t_m^\alpha(x, f, M) - \overline{f}_M(x, \varepsilon(\varepsilon_m))| = +\infty$$

almost everywhere.

Using (2.2.40), we conclude that it holds

Theorem 2.2.10. *Let* $f \in L(\log^+ L)^{n-1}(T^n)$. *Then for any* $B \subset M$, $\alpha \in]0, +\infty[^n$ *and* $p \in]0, 1[$, *the relation*

$$\|t_B\|_p \leq A(p, n, \alpha) \left\{ \int_{T^n} |f(x)| \left[\log^+ |f(x)|\right]^{n-1} dx + 1 \right\} \tag{2.2.63}$$

holds.

Remark 2.2.2. Let $\beta \in [0, n-1[$. Then, in the class $L(\log^+ L)^\beta(T^n)$, the relations (2.2.25), (2.2.26), and (2.2.63), in general, are not valid for $B \subset M$ with $\overline{k}(B) \in]\beta + 1, n]$ and $\alpha \in]0, +\infty[^n$. This is confirmed by a corresponding result of Jessen, Marcinkiewicz and Zygmund [1], as well as by the arguments presented in this chapter.

Theorem 2.2.11. a) *Let* $f \in L(\log^+ L)^n(T^n)$. *Then for any nonempty* $B \subset M$ $(B \neq \varnothing)$ *and* $\alpha \in]0, +\infty[^n$, *the inequality*

$$\|t_B\|_1 \leq A(n, \alpha) \left\{ \int_{T^n} |f(x)| \left[\log^+ |f(x)|\right]^n dx + 1 \right\} \tag{2.2.64}$$

holds.

b) *Let* $\phi : [0, +\infty[\to [0, +\infty[$, $\phi \uparrow +\infty$ *as* $t \uparrow +\infty$, *and*

$$\phi(t) \left[t (\log t)^n\right]^{-1} \to 0 \quad \text{as} \quad t \to +\infty.$$

Then there is a function $f \in L(T^n)$ *such that*

$$\int_{T^n} \phi(|f(x)|)\, dx < +\infty,$$

yet for any $B \subset M$ *and* $\alpha \in]0, +\infty[^n$, *we have*

$$t_B \notin L(T^n). \tag{2.2.65}$$

Proof. The assertion a) follows from the proof of the inequality (2.2.40) and Lemma 1.2.13.

Let us begin the proof of b).

1) If $B \subset M$, then b) follows from Theorem 1.3.2 since the function $\overline{f}_M$ exists almost everywhere and

$$t_M(x) \geq |\overline{f}_M(x)|.$$

2) Let now $B = \varnothing$, i.e.,

$$t_m^\alpha(x, f, B) \equiv \sigma_m^\alpha(x, f).$$

Let the function $f : T^n \to \mathbb{R}$ is defined by (1.3.6). Therefore

$$\int_{T^n} \phi(|f(x)|)\, dx < +\infty.$$

Moreover, let $m = (m_1, m_2, \dots, m_n)$ and $x \in]0, \gamma_1]^n$. Choose the coordinates of m such that

$$(m_i + 1)^{-1} < 4x_i < m_i^{-1} \quad \text{for} \quad 1 \le i \le n. \tag{2.2.66}$$

By (1.1.25), we have

$$\begin{aligned}\sigma_m^\alpha(-x, f) &= \frac{1}{\pi^n} \int_{T^n(\gamma)} f(s) \prod_{i=1}^n K_{m_i}^{\alpha_i}(s_i + x_i)\, ds = \\ &= \frac{(-1)^k}{\pi^n} \Bigg[\int_{E(x,\gamma,M)} + \int_{E_1(x,\gamma,M)} + \\ &+ \sum_{B:B\subset M} \int_{E(x,\gamma,B)\times E_1(x,\gamma,B')} \Bigg] f(s) \prod_{i=1}^n K_{m_i}^{\alpha_i}(s_i + x_i)\, ds.\end{aligned} \tag{2.2.67}$$

Basing on (2.2.66), we find

$$\begin{aligned}&\int_{E(x,\gamma,M)} f(s) \prod_{i=1}^n K_{m_i}^{\alpha_i}(s_i + x_i)\, ds \ge \\ &\ge A(n, \alpha) \prod_{i=1}^n x_i^{-1} \int_{E(x,\gamma,M)} f(s)\, ds,\end{aligned} \tag{2.2.68}$$

and using (2.2.46), we get

$$\begin{aligned}&\left| \int_{E_1(x,\gamma,M)} f(s) \prod_{i=1}^n K_{m_i}^{\alpha_i}(s_i + x_i)\, ds \right| \le \\ &\le A(n, \alpha) \int_{E_1(x,\gamma,M)} \frac{f(s)\, ds}{\prod_{i=1}^n m_i^{\alpha_i}(s_i + x_i)^{1+\alpha_i}}.\end{aligned}$$

Hence

$$\begin{aligned}&\left| \int_{E_1(x,\gamma,M)} f(s) \prod_{i=1}^n K_{m_i}^{\alpha_i}(s_i + x_i)\, ds \right| \le \\ &\le A(n, \alpha) \Big(\prod_{i=1}^n x_i^{-1} \Big) \Big[\ln \Big(\prod_{i=1}^n x_i^{-1} \Big) \Big]^{-2n} \quad \text{for} \quad x \in I^n(\gamma_1).\end{aligned} \tag{2.2.69}$$

Now consider the expression

$$L_3(x,f,\alpha,M,m)\equiv$$

$$\equiv \frac{1}{\pi^n}\sum_{\varnothing\neq B_1\subset M}\int\limits_{E(x,\gamma,B_1)\times E_1(x,\gamma,B_1')} f(s)\prod_{i=1}^{n} K_{m_i}^{\alpha_i}(s_i+x_i)\,ds. \tag{2.2.70}$$

In each of the addends of (2.2.70), with respect to at least one variable s_j the set of integration is $[x_j,\gamma]$. Therefore

$$\left|\int\limits_{E(x,\gamma,B_1)\times E_1(x,\gamma,B_1')} f(s)\prod_{i=1}^{n} K_{m_i}^{\alpha_i}(s_i+x_i)\,ds\right|\leq$$

$$\leq A(n,\alpha)\Big(\prod_{i=1}^{n} x_i^{-1}\Big)\Big[\ln\Big(\prod_{i=1}^{n} x_i^{-1}\Big)\Big]^{-n-1} \quad\text{for}\quad x\in I^n(\gamma_1). \tag{2.2.71}$$

Thus, by (2.2.67)-(2.2.71), we obtain

$$t_B(x)\equiv\sigma(x)>A(n,\alpha)\prod_{i=1}^{n} x_i^{-1}\int\limits_{E(x,\gamma,M)} f(s)\,ds-$$

$$-A_1(n,\alpha)\Big(\prod_{i=1}^{n} x_i^{-1}\Big)\Big[\ln\Big(\prod_{i=1}^{n} x_i^{-1}\Big)\Big]^{-n-1} \quad\text{for}\quad x\in I^n(\gamma_1). \tag{2.2.72}$$

Hence, as noted while proving Theorem 2.2.6 (see (2.2.35)), it follows (2.2.65).

3) Suppose that $B\subset M$ and $B\neq M$, $B\neq\varnothing$. Then according to (1.1.26), we have

$$\pi^n(-1)^k t_m^{\alpha}(-x,f,B)=$$

$$=\int\limits_{I^n(\gamma)} f(s)\prod_{j\in B'} K_{m_j}^{\alpha_j}(s_j+x_j)\prod_{i\in B}\tau_{m_i}^{\alpha_i}(s_i+x_i)\,ds=\Bigg[\int\limits_{E(x,\gamma,M)}+$$

$$+\int\limits_{E_1(x,\gamma,M)}+\sum_{B_1\subset M}\int\limits_{E(x,\gamma,B_1)\times E_1(x,\gamma,B_1')}\Bigg] f(s)\prod_{j\in B'} K_{m_j}^{\alpha_j}(s_i+x_i)\times$$

$$\times\prod_{i\in B}\tau_{m_i}^{\alpha_i}(s_i+x_i)\,ds.$$

By (2.2.66), we obtain

$$\tau_{m_i}^{\alpha_i}(s_i+x_i)\geq A(\alpha_i)x_i^{-1} \quad\text{for}\quad i\in B.$$

Taking note of (2.2.44) and (2.2.43), and using the same method as while proving (2.2.72), we get

$$t_B(x)>A(n,\alpha)\Big(\prod_{i=1}^{n} x_i^{-1}\Big)\int\limits_{E(x,\gamma,M)} f(s)\,ds-$$

$$-A_1(n,\alpha)\Big(\prod_{i=1}^{n} x_i^{-1}\Big)\Big[\ln\Big(\prod_{i=1}^{n} x_i^{-1}\Big)\Big]^{-n-1} \quad\text{for}\quad x\in I^n(\gamma_1). \qquad\blacksquare$$

Remark 2.2.3. The scheme of the proof of the assertion b) of Theorem 2.2.11 shows that the conclusion of c) of Theorem 2.2.6 can be improved in the sense that in (2.2.29), the relations

$$F_B^* \notin L(T^n), \quad B \subset M, \quad \overline{k}(B) \in [1, n[,$$

can be added.

Analyzing the schemes of the proofs of Theorems 2.2.6 and 2.2.11 (as we have done for Theorem 1.3.3), we conclude that for the functions t_B (see(1.1.32)) and F_B^* (see (1.1.35)), it holds

Theorem 2.2.12. *Let the function ϕ satisfy the conditions of Theorem* 1.3.3. *Then there is a function $f \in L(T^n)$ such that*

$$\int\limits_{T^n} \phi(|f(x)|)\, dx < +\infty,$$

yet

$$t_B \notin L(T^n), \quad F_B^* \notin L(T^n), \quad \overline{k}(B) \in [1, n].$$

Remark 2.2.4. According to (1.1.33), t_B depends on α as well. However, in some above assertions, α does not appear. This is caused by the fact that the assertion holds uniformly with respect to α.

Analyzing the results presented in Sections 1.2 and 2.2, we arrive to

Theorem 2.2.13. a) *Let the class $L_\phi(T^n)$ be defined by* (1.2.75). *If for any function $f \in L_\phi(T^n)$, the series $\overline{\sigma}_n[f, M]$ is summable almost everywhere by the n-fold method of Poisson–Abel, then*

$$\phi(t) \geq At(\log t)^{n-1} \quad \textit{for} \quad t \in [t_0, +\infty[. \tag{2.2.73}$$

b) *Let for any function $f \in L_\phi(T^n)$ and some $\alpha \in]0, +\infty[^n$, the series $\overline{\sigma}_n[f, M]$ be almost everywhere summable by n-fold method of Cesàro. Then the relation* (2.2.73) *is true.*

In the assertions a) and b) of Theorem 2.2.13, the convergence at a point of the n-fold means of Poisson–Abel and Cesàro is meant in the sense of Stolz–Pringsheim.

Tkebuchava [3] showed that Theorem 2.2.13 remains true for λ-convergence ($\lambda = 1$). His results improve the corresponding assertions pointed out by us in [37].

Use the notation (1.1.8), (1.1.19), (1.1.20), (1.1.21), (1.1.22), and put for $\varepsilon = (\varepsilon_1, \varepsilon_2, \dots, \varepsilon_n)$, $m = (m_1, m_2, \dots, m_n)$ and $r = (r_1, r_2, \dots, r_n)$ with $\varepsilon_1 = \varepsilon_2 = \cdots = \varepsilon_n$, $m_1 = m_2 = \cdots = m_n$ and $r_1 = r_2 = \cdots = r_n$,

$$\begin{aligned}
F_1(x) &\equiv F_1(x, f, B) = \sup_{\varepsilon} |\overline{f}_B(x, \varepsilon)|, \\
F_2(x) &\equiv F_2(x, f, \alpha) = \sup_{m} |\sigma_m^\alpha(x, f)|, \\
F_3(x) &\equiv F_3(x, f, B, \alpha) = \sup_{m} |t_m^\alpha(x, f, B)|, \\
F_4(x) &\equiv F_4(x, f) = \sup_{r} |f(x, r)|, \\
F_5(x) &\equiv F_5(x, f, B) = \sup_{r} |\overline{f}(x, r, B)|.
\end{aligned} \tag{2.2.74}$$

On the basis of the works of Marcinkiewicz and Zygmund [1], Stein [2], and Guzmán and Welland [1], we can conclude that if $f \in L\log^+ L(T^n)$, then $F_i \in L(T^n)$ $(i = 2, 4)$.

In 1972, at the seminar of Men'shov and Ul'yanov, the latter posed the following question.

Let $n \geq 2$ and $\beta \in [1, n[$. If $f \in L(\log^+ L)^\beta(T^n)$, then for which nonempty subsets $B \subset M$ are the functions F_1, F_3, and F_5 integrable on T^n?

Some statements given below answer the question of Ul'yanov.

Theorem 2.2.14. a) *Suppose that* $n \geq 2$ *and* $\beta \in [1, n[$. *If* $f \in L(\log^+ L)^\beta(T^n)$, *then for any* $B \subset M$ *with* $\overline{k}(B) \in [1, \beta]$, *the function* t^*_B *defined by* (1.1.36) *is integrable on* T^n.

b) *Let* n *and* β *satisfy the conditions of* a). *Then there is a function* $f \in L(\log^+ L)^\beta(T^n)$ *such that for any* $B \subset M$ *with* $\overline{k}(B) \in]\beta, n]$, *we have* $F_1 \notin L(T^n)$.

Proof. a) follows from Lemma 1.2.14.

Now turn to the proof of b). Let the function f be defined by (1.2.76). As already pointed out, it belongs to $L(\log^+ L)^\beta(T^n)$. If $x \in]0, \gamma[^n$, then it is not difficult to check up that

$$F_1(-x_B) \geq A(n, \beta, \eta)\Big(\prod_{i=1}^n x_i^{-1}\Big)\Big[\log\Big(\prod_{i=1}^n x_i^{-1}\Big)\Big]^{-n-\beta-\eta+\overline{k}(B)}. \tag{2.2.75}$$

But, by the conditons of the theorem, $\overline{k}(B) > \beta$, so that $\overline{k}(B) \geq [\beta] + 1$ and $n + \beta + \eta - \overline{k}(B) < n$. Then, by (2.2.75), we get

$$F_1(-x_B) \geq A(n, \beta, \eta)\Big(\prod_{i=1}^n x_i^{-1}\Big)\Big[\log\Big(\prod_{i=1}^n x_i^{-1}\Big)\Big]^n \quad \text{for } x \in I^n(\gamma).$$

Therefore

$$\int_{T^n} F_1(x)\,dx = +\infty.$$

Thus b) is proved. ■

Theorem 2.2.15. a) *Let* $n \geq 3$ *and* $\beta \in [2, n[$. *If* $f \in L(\log^+ L)^\beta(T^n)$, *then for any* $B \subset M$ *with* $\overline{k}(B) \in [1, \beta - 1]$, *the functions* F_3 *and* F_5 *in* (2.2.74) *are integrable on* T^n.

b) *Let* $n \geq 2$ *and* $\beta \in [1, n[$. *Then there is a function* $f \in L(\log^+ L)^\beta(T^n)$ *such that for any* $B \subset M$ *with* $\overline{k}(B) \in]\beta - 1, n]$, *the functions* F_3 *and* F_5 *are not integrable on* T^n.

Proof. We only outline a brief scheme of the proof. Otherwise we should have to repeat the arguments of Theorems 2.2.1 and 2.2.7.

a) Let $n \geq 3$, $\beta \in [2, n[$ and $B \subset M$ with $\overline{k}(B) \in [1, \beta]$. Then according to Theorem 1.3.1, $\overline{f}_B \in L(T^n)$. If $\overline{k}(B) \in [1, \beta - 1]$, then by (1.3.1), we have

$$\int_{T^n} |\overline{f}_B(x)| \log^+ |\overline{f}_B(x)| dx \leq$$

$$\leq A(n,\beta)\left\{\int\limits_{T^n} |f(x)|\left(\log^+|f(x)|\right)^\beta dx + 1\right\}. \tag{2.2.76}$$

Using the results of Marcinkiewicz and Zygmund [1], Guzmán and Welland [1] (see also Guzmán [1, Ch. II]), and (2.2.76), we conclude that a) is true.

b) Let $n \geq 2$, $\beta \in [1,n[$, and $B \subset M$ with $\overline{k}(B) \in]\beta-1,n]$. Let the function f be defined by (1.2.76). Following (in the main) the schemes of the proofs of Theorems 2.2.1 and 2.2.7 for $\overline{k}(B) \geq \beta$, we get convinced that b) is valid. ■

Now we give a result connected with the behavior of the n-fold Schwarz integral. Using Theorem 2.2.1, we arrive to

Theorem 2.2.16. *Let $z_j = r_j e^{ix_j}$, $\tau_j = e^{is_j}$, $r_j \in [0,1[$ for $1 \leq i \leq n$, and suppose that $n \geq 2$ and $\beta \in [0, n-1[$. Then there is a function $f : U^n \to \mathbb{C}$ which can be represented by the n-fold Schwarz integral*

$$f(z) = \left(\frac{1}{2\pi}\right)^n \int\limits_{T^n} \varphi(s) \prod_{j=1}^{n} \frac{\tau_j + z_j}{\tau_j - z_j}\, ds$$

with $\varphi \in L(\log^+ L)^\beta(T^n)$, yet for $\lambda = 1$,

$$\limsup_{r_\lambda \to 1-} |f(z)| = +\infty$$

almost everywhere.

Finally, note that many of theorems related to the summability of the series $\sigma_n[f]$ and $\overline{\sigma}_n[f,B]$ remain true also for some more general n-fold linear means of these series. We did not go into this since it was more important to us to establish new regularities connected with the dimension.

2.3. Unsolved Problems

2.3.1. Let $f \in L(T^n)$ and

$$\omega_i(\delta, f)_1 \left(\log \frac{2\pi}{\delta}\right)^n \leq A(f,n) \text{ for } \delta \in]0,\pi],\ 1 \leq i \leq n.$$

Is the series $\sigma_n[f]$ almost everywhere convergent? As already noted, in the case $n = 1$, this is Zygmund's problem.

2.3.2. Let $f \in L^2(T^n)$ and $n \geq 3$. If

$$\omega_i(\delta, f)_2 \left(\log \frac{2\pi}{\delta}\right)^{\frac{n}{2}} \leq A(f,n) \text{ for } \delta \in]0,\pi],\ 1 \leq i \leq n,$$

then is the series $\sigma_n[f]$ almost everywhere convergent?

We have discussed the case $n = 2$ of this problem in this chapter (see Corollary 2.1.3 and Remark 2.1.2).

2.3.3. Let $n \geq 2$, $p \in]1,n[$, and $f \in L^p(T^n)$. If

$$\omega_i(\delta, f)_p \left(\log \frac{2\pi}{\delta}\right)^{n/p} \leq A(f,n) \text{ for } \delta \in]0,\pi],\ 1 \leq i \leq n,$$

then is the series $\sigma_n[f]$ almost everywhere convergent?

We should add that the restriction $p \in]1, n[$ is essential.

2.3.4. Suppose that $f \in L^p(T^n)$ for some $p \in]1, 2[$, and

$$\sum_{B \subset M} \int_{T^n} \left[\int_{T^n(B)} \left(\prod_{i \in B} |s_i|^{-1} \right)^{-1} |\overline{\Delta}(f, x, s_B)|^p ds_B \right] dx < +\infty.$$

Is the series $\sigma_n[f]$ almost everywhere convergent?

This is a generalization of the well-known theorem of Marcinkiewicz [4] (see, e.g., Bari [2, Ch. V]). The theorem was established by Marcinkiewicz in 1939, although, due to results of Carleson [1] and Hunt [2], it has lost its significance. However, in the multidimensional case, its analogue in the form of problem 2.3.4 seems to us to be very important. Note that in 1977, Bloshanskiĭ [1] established a kind of analogue of Marcinkiewicz's theorem for $n = 2$.

2.3.5. Let $f \in L(T^n)$. What is the minimal rate of vanishing of $\omega_i(\delta, f)_1$, $1 \le i \le n$, which guarantees the almost everywhere summability of $\sigma_n[f]$ by the n-fold methods of Poisson–Abel and Cesàro of positive order? A similar problem can be posed for the series $\overline{\sigma}_n[f, B]$ with $B \subset M$.

Note that if $f \in L(T^n)$ and there is $\eta \in]0, +\infty[$ such that

$$\omega_i(\delta, f)_1 \left(\log \frac{2\pi}{\delta} \right)^{n+\eta} \le A(f, n, \eta) \text{ for } \delta \in]0, \pi], \quad 1 \le i \le n,$$

then all the series $\sigma_n[f]$ and $\overline{\sigma}_n[f, B]$ with $B \subset M$ are almost everywhere summable by the n-fold methods of Poisson–Abel and Cesàro of positive order.

2.3.6. Suppose that $f \in L(T^n)$.

a) Does it hold

$$S_m(x, f) \left[\prod_{i=1}^{n} \log(m_i + 2) \right]^{-1} \to 0 \quad \text{as } m \to \infty$$

almost everywhere?

b) Investigate the problem of establishing the best possible growth orders of rectangular partial sums of $\sigma_n[f]$ and $\overline{\sigma}[f, B]$ on a set of full measure.

In connection with this problem, we should note the following.

If $f \in L(\log^+ L)^{n-1}(T^n)$, then it is easy to show that

$$S_m(x, f) \left[\prod_{i=1}^{n} \log(m_i + 2) \right]^{-1} \to 0 \quad \text{as } m \to \infty \tag{2.3.1}$$

almost everywhere.

If, however, $f \in L(\log^+ L)^n(T^n)$, then, in addition,

$$|S_m(x, f)| \left[\prod_{i=1}^{n} \log(m_i + 2) \right]^{-1} \le A(f, n) M_M(x, f) \tag{2.3.2}$$

almost everywhere.

The problem whether the relations (2.3.1) and (2.3.2) can be improved still remains open.

It likewise remains open the problem on analogues of (2.3.1) and (2.3.2) in the case where $f \in L^p(T^n)$ for some $p \in]1, +\infty[$. In particular, an analogue of the well-known result of Littlewood and Paley [1]–[3] is not yet established. Note that for $p = 2$, one can easily obtain from Theorem 2.1.4 an estimate of the type of Kolmogorov and Seliverstov [1], [2] and Plessner [2].

If $f \in L^p(T^n)$ for some $p \in]1, +\infty[$ and $m = (m_1, m_2, \dots, m_n)$ with $m_i = p \in \mathbb{N}$ for $1 \leq i \leq n$, then for any sequence $(\lambda_m)_{m\geq 1}$ of positive real numbers tending to $+\infty$ as $m \to \infty$, the relation

$$S_m(x, f)\lambda_m^{-1} \to 0 \quad \text{as} \quad m \to \infty \tag{2.3.3}$$

holds almost everywhere.

The last statemant for $p \in [2, +\infty]$ follows from a result of N. Tevzadze [1], and for $p \in]1, 2[$, from an assertion of Fefferman [2]. Moreover, the relation (2.3.3) likewise remains true if $f \in L(\log^+ L)^n \log^+ \log^+ L(T^n)$. This follows from a result of Sjolin [2].

2.3.7. Let $f \in L(\log^+ L)^\beta(T^n)$ for some $\beta \in [0, n-1[$ and $\alpha \in]-1, +\infty[^n$. Investigate (depending on β and α) the problem of establishing the best possible growth orders of the expressions $\sigma_m^\alpha(x, f)$ and $t_m^\alpha(x, f, B)$ on a set of full measure.

CHAPTER III

Some Approximating Properties of n-Fold Cesàro Means of the Series $\sigma_n[f]$ and $\overline{\sigma}_n[f, B]$

3.1. Approximating Properties of n-Fold Cesàro Means of the Series $\sigma_n[f]$

In Chapter 4 of the first part, we have presented some results connected with approximating properties of Cesàro means of the series $\sigma[f]$ and $\overline{\sigma}[f]$. In this chapter, their multidimensional analogues are given. We have treated these items in [4], [5], [13, Ch. IV], [35], [46].

In what follows, we will need the notation

$$H(B) \equiv H(B,m) = \mathop{\times}_{i\in B}\left[0, \frac{\pi}{m_i}\right] \equiv \left[0, \frac{\pi}{m_{i_1}}\right] \times \left[0, \frac{\pi}{m_{i_2}}\right] \times \cdots \times \left[0, \frac{\pi}{m_{i_k}}\right],$$

$$H_1(B) \equiv H_1(B,m) = \mathop{\times}_{i\in B}\left[\frac{\pi}{m_i}, \pi\right] \equiv \left[\frac{\pi}{m_{i_1}}, \pi\right] \times \left[\frac{\pi}{m_{i_2}}, \pi\right] \times \cdots \times \left[\frac{\pi}{m_{i_k}}, \pi\right],$$

$$H_2(B) \equiv H_2(B,m) = \mathop{\times}_{i\in B}\left[\pi - \frac{\pi}{m_i}, \pi\right] \equiv$$
$$\equiv \left[\pi - \frac{\pi}{m_{i_1}}, \pi\right] \times \left[\pi - \frac{\pi}{m_{i_2}}, \pi\right] \times \cdots \times \left[\pi - \frac{\pi}{m_{i_k}}, \pi\right],$$

$$H_3(B) \equiv H_3(B,m) = \mathop{\times}_{i\in B}\left[\frac{\pi}{m_i}, \pi - \frac{\pi}{m_i}\right] \equiv$$
$$\equiv \left[\frac{\pi}{m_{i_1}}, \pi - \frac{\pi}{m_{i_1}}\right] \times \left[\frac{\pi}{m_{i_2}}, \pi - \frac{\pi}{m_{i_2}}\right] \times \cdots \times \left[\frac{\pi}{m_{i_k}}, \pi - \frac{\pi}{m_{i_k}}\right],$$

$$H_4(B) \equiv H_4(B,m) = \mathop{\times}_{i\in B}\left[0, \pi - \frac{\pi}{m_i}\right] \equiv$$
$$\equiv \left[0, \pi - \frac{\pi}{m_{i_1}}\right] \times \left[0, \pi - \frac{\pi}{m_{i_2}}\right] \times \cdots \times \left[0, \pi - \frac{\pi}{m_{i_k}}\right],$$

$$H_5(B) \equiv H_5(B,m) = [H_4(B) \setminus H_3(B)] \times [H_1(B') \setminus H_3(B')],$$

where $\varnothing \neq B \subset M$, $m \in \mathbb{N}$, and $m_i \geq 2$ for $1 \leq i \leq n$.

By $U_i(B)$ we will denote the class of all subsets of a set $B \subset M$ ($B \neq M$, $B \neq \varnothing$) containing i elements, $1 \leq i \leq \overline{k}(B)$.

The following notation will also be used.

$$\Phi_n(x,s,f,M,m)=\Delta(f,x,s)+(-1)^n\Delta\Big(f,x,s+\frac{\pi}{m}\Big)+$$
$$+(-1)^{n-1}\sum_{B\in U_{n-1}}\Delta\Big(f,x,s+\Big(\frac{\pi}{m}\Big)_B\Big)+\cdots+$$
$$+(-1)^{n-k}\sum_{B\in U_{n-k}}\Delta\Big(f,x,s+\Big(\frac{\pi}{m}\Big)_B\Big)+\cdots+$$
$$+(-1)\sum_{B\in U_1}\Delta\Big(f,x,s+\Big(\frac{\pi}{m}\Big)_B\Big), \qquad (3.1.1)$$

$$\Phi_k^{(1)}(x,s,f,B,M,m)=\Delta(f,x,s)+(-1)^k\Delta\Big(f,x,s+\Big(\frac{\pi}{m}\Big)_B\Big)+$$
$$+(-1)^{k-1}\sum_{B\in U_{k-1}(B)}\Delta\Big(f,x,s+\Big(\frac{\pi}{m}\Big)_B\Big)+\cdots+$$
$$+(-1)^{k-i}\sum_{B\in U_{k-i}(B)}\Delta\Big(f,x,s+\Big(\frac{\pi}{m}\Big)_B\Big)+\cdots+$$
$$+(-1)\sum_{B\in U_1(B)}\Delta\Big(f,x,s+\Big(\frac{\pi}{m}\Big)_B\Big), \qquad (3.1.2)$$

$$\Phi_k^{(2)}(x,s,f,B_1,B,m)=\Delta(f,x,s_B)+(-1)^k\Delta\Big(f,x,s_B+\Big(\frac{\pi}{m}\Big)_B\Big)+$$
$$+(-1)^{k-1}\sum_{B_1\in U_{k-1}(B)}\Delta\Big(f,x,s_B+\Big(\frac{\pi}{m}\Big)_{B_1}\Big)+\cdots+$$
$$+(-1)^{k-i}\sum_{B_1\in U_{k-i}(B)}\Delta\Big(f,x,s_B+\Big(\frac{\pi}{m}\Big)_{B_1}\Big)+\cdots+$$
$$+(-1)\sum_{B_1\in U_1(B)}\Delta\Big(f,x,s_B+\Big(\frac{\pi}{m}\Big)_{B_1}\Big), \qquad (3.1.3)$$
$$B_1\subset B\subset M \quad (B_1\neq\varnothing,\quad B\neq M).$$

Everywhere below if either B or B_i, $1\le i\le 4$, is a subset of M, it is always assumed that $B\neq\varnothing$, $B_i\neq\varnothing$, $1\le i\le 4$.

In order that all the definitions (3.1.1)–(3.1.3) be correct, it is assumed that if $n\ge 4$, then $k\equiv\overline{k}(B)\in[2,n-2]$ and $i\in[2,k-2]$; if $n=2$ or $n=3$, then (3.1.1)–(3.1.3) consist only of the terms containing $(-1)^{n-k}$ with $0\le n-k\le 1$, and $(-1)^{k-i}$ with $0\le k-i\le 1$.

In what follows, we will frequently use assertions and relations from the first part of the book. In order that not to overload the text by the phrase "of the first part", we will use the notation (1; 2.2.7),…,(1; 2.2.16), and so on, which means that we refer to relations (2.2.7),…,(2.2.16) of the first part.

First prove some auxiliary statements.

Lemma 3.1.1. *Let $f\in L^p(T^n)$ for some $p\in[1,+\infty]$, and let $\alpha\in]-1,+\infty[^n$.*

Then for any $B \subset M$,

$$\sum_{t_{i_1}\in U_0}\cdots\sum_{t_{i_k}\in U_0}\left\| \int_{H(B)} \Delta(f,\cdot,[s\times t]_B)\prod_{i\in B} K_{m_i}^{\alpha_i}(s_i)\,ds_B\right\|_p \le$$

$$\le A(p,n,\alpha)\omega_B\left(\frac{1}{m},f\right)_p, \quad m\in\mathbb{N}^n, \tag{3.1.4}$$

$$\sum_{t_{i_1}\in U_0}\cdots\sum_{t_{i_k}\in U_0}\left\| \int_{H_1(B)} \Delta(f,\cdot,[s\times t]_B)\prod_{i\in B} r_{m_i}^{\alpha_i}(s_i)\,ds_B\right\|_p \le$$

$$\le A(p,n,\alpha)\int_{H_1(B)} \omega_B(s,f)_p\prod_{i\in B}(m_i s_i^2)^{-1}ds_B. \tag{3.1.5}$$

Proof. Using (1; 2.2.13) and the generalized Minkowski's inequality (for $p \in [1,+\infty[$), we find

$$\sum_{t_{i_1}\in U_0}\cdots\sum_{t_{i_k}\in U_0}\left\| \int_{H(B)} \Delta(f,\cdot,[s\times t]_B)\prod_{i\in B} K_{m_i}^{\alpha_i}(s_i)\,ds_B\right\|_p \le$$

$$\le A(p,n,\alpha)\prod_{i\in B} m_i\left\|\int_{H(B)} |\Delta(f,\cdot,[s\times t]_B)|\,ds_B\right\|_p.$$

Hence, by (1.1.1) and (1.1.2), it follows (3.1.4).

The inequality (3.1.5) can be established analogously. ■

The following two lemmas can be proved quite similarly.

Lemma 3.1.2. *Let $f \in L^p(T^n)$ for some $p \in [1,+\infty]$ and let $\alpha \in]-1,+\infty[^n$. Then for any $B \subset M$,*

$$\sum_{t_{i_1}\in U_0}\cdots\sum_{t_{i_k}\in U_0}\left\| \int_{H(B_1)\times H_1(B\setminus B_1)} \Delta(f,\cdot,[s\times t]_B)\prod_{i\in B'} K_{m_i}^{\alpha_i}(s_i)\prod_{j\in B\setminus B_1} r_{m_j}^{\alpha_j}(s_j)\,ds_B\right\|_p \le$$

$$\le A(p,n,\alpha)\omega_{B_1}\left(\frac{1}{m},f\right)_p. \tag{3.1.6}$$

Lemma 3.1.3. *Let $f \in L^p(T^n)$ for some $p \in [1,+\infty]$ and let $\alpha \in]-1,+\infty]^n$. Then for any $B \subset M$,*

$$\sum_{t_{i_1}\in U_0}\cdots\sum_{t_{i_k}\in U_0}\left\| \int_{H(B_1)\times H_2(B\setminus B_1)} \Delta(f,\cdot,[s\times t]_B)\prod_{i\in B_1'} K_{m_i}^{\alpha_i}(s_i)\prod_{j\in B\setminus B_1} r_{m_j}^{\alpha_j}(s_j)\,ds_B\right\|_p \le$$

$$\le A(p,n,\alpha)\omega_{B_1}\left(\frac{1}{m},f\right)_p, \tag{3.1.7}$$

$$\sum_{t_{i_1}\in U_0}\cdots\sum_{t_{i_k}\in U_0}\left\| \int_{H_2(B_1)\times H_1(B\setminus B_1)} \Delta(f,\cdot,[s\times t]_B)\prod_{i\in B_1} K_{m_i}^{\alpha_i}(s_i)\prod_{j\in B\setminus B_1} r_{m_j}^{\alpha_j}(s_j)\,ds_B\right\|_p \le$$

$$\leq A(p,n,\alpha) \int\limits_{H_1(B\setminus B_1)} \prod_{j\in B_1'} (m_j s_j^2)^{-1} \omega_{B_1}(s,f)_p ds_{B_1'}. \tag{3.1.8}$$

Denote for $m_i \in \mathbb{N}$, $1 \leq i \leq n$,

$$\lambda(m_i,\alpha_i) = \begin{cases} m_i^{-\alpha_i} & \text{for } \alpha_i \in]-1,0[, \\ \log(m_i+2) & \text{for } \alpha_i = 0, \\ 1 & \text{for } \alpha_i \in]0,+\infty[. \end{cases} \tag{3.1.9}$$

Lemma 3.1.4. *Let $f \in L^p(T^n)$ for some $p \in [1,+\infty]$ and let $\alpha \in]-1,+\infty[^n$. Then for any $B \subset M$,*

$$\sum_{t_{i_1}\in U_0} \cdots \sum_{t_{i_k}\in U_0} \left\| \int\limits_{H_1(B)} \Delta(f,\cdot,[s\times t]_B) \prod_{i\in B} \varphi_{m_i}^{\alpha_i}(s_i)\, ds_B \right\|_p \leq$$

$$\leq A(p,n,\alpha)\left[\sum_{B_1\subset B} \prod_{i\in B_1} \lambda(m_i,\alpha_i)\omega_{B_1}\left(\frac{1}{m},f\right)_p\right]. \tag{3.1.10}$$

Proof. We will assume that $B = M$ and $t_i = 1$ for $1 \leq i \leq n$. The scheme of the proof shows that it can all the more be carried out in the case where $B \subset M$, $B \neq M$, $t_i \in U_0$, $1 \leq i \leq n$, and not all the t_i are equal to 1. As shown in the first part of the book (see Chapters 2–4), we may assume that $\alpha \in]-1,1[^n$. The scheme of the proof remains valid and gets simpler in the case where $\alpha_i \in]-1,1[$ for $i \in B$ and $\alpha_j \in [1,+\infty[$ for $j \in B'$, $B' \neq \varnothing$.

Using (1; 2.2.11) and (1; 2.2.21), we conclude that it suffices to show that

$$\prod_{i=1}^n m_i^{-\alpha_i} \left\| \int\limits_{H_1(M)} \Delta(f,\cdot,s) \prod_{i=1}^n \omega(\alpha_i,s_i) \cos m_i s_i\, ds \right\|_p \leq$$

$$\leq A(p,n,\alpha)\left[\sum_{B\subset M} \prod_{i\in B} \lambda(m_i,\alpha_i)\omega_B\left(\frac{1}{m},f\right)_p\right]. \tag{3.1.11}$$

Put

$$J(x,f,m,\alpha) =$$

$$= \prod_{i=1}^n m_i^{-\alpha_i} \int\limits_{H_1(M)} \Delta(f,x,s) \prod_{i=1}^n \omega(\alpha_i,s_i) \cos m_i s_i\, ds. \tag{3.1.12}$$

If we substitute s_i by $s_i + \frac{\pi}{m_i}$ for $1 \leq i \leq n$, then $J(x,f,m,\alpha)$ can be written as

$$J(x,f,m,\alpha) = (-1)^n \prod_{i=1}^n m_i^{-\alpha_i} \times$$

$$\times \int\limits_{H_4(M)} \Delta\left(f,x,s+\frac{\pi}{m}\right) \prod_{i=1}^n \omega\left(\alpha_i, s_i + \frac{\pi}{m_i}\right) \cos m_i s_i\, ds. \tag{3.1.13}$$

More generally, given $B \in U_{n-k}$ with $k \in [1, n-1]$, the substitution of s_i by $s_i + \frac{\pi}{m_i}$ for $i \in B$ leads us to

$$J(x,f,m,\alpha) = (-1)^{n-k} \prod_{i=1}^{n} m_i^{-\alpha_i} \int\limits_{H_4(B)\times H_1(B')} \Delta\Big(f, x, s + \Big(\frac{\pi}{m}\Big)_B\Big) \times$$
$$\times \prod_{i\in B} \omega\Big(\alpha_i, s_i + \frac{\pi}{m_i}\Big) \cos m_i s_i \prod_{j \in B'} \omega(\alpha_j, s_j) \cos m_j s_j \, ds. \tag{3.1.14}$$

Therefore, by (3.1.12)–(3.1.14), we get

$$J(x,f,m,\alpha) = 2^{-n} \prod_{i=1}^{n} m_i^{-\alpha_i} \Big\{ \int\limits_{H_1(M)} \Delta(f,x,s) \prod_{i=1}^{n} \omega(\alpha_i, s_i) \cos m_i s_i \, ds +$$
$$+(-1)^n \int\limits_{H_4(M)} \Delta\Big(f, x, s + \frac{\pi}{m}\Big) \prod_{i=1}^{n} \omega\Big(\alpha_i, s_i + \frac{\pi}{m_i}\Big) \cos m_i s_i \, ds +$$
$$+(-1)^{n-1} \sum_{B \in U_{n-1}} \int\limits_{H_4(B)\times H_1(B')} \Delta\Big(f, x, s + \Big(\frac{\pi}{m}\Big)_B\Big) \times$$
$$\times \prod_{i\in B} \omega\Big(\alpha_i, s_i + \frac{\pi}{m_i}\Big) \cos m_i s_i \prod_{j \in B'} \omega(\alpha_j, s_j) \cos m_j s_j \, ds + \cdots +$$
$$+(-1)^{n-k} \sum_{B \in U_{n-k}} \int\limits_{H_4(B)\times H_1(B')} \Delta\Big(f, x, s + \Big(\frac{\pi}{m}\Big)_B\Big) \times$$
$$\times \prod_{i\in B} \omega\Big(\alpha_i, s_i + \frac{\pi}{m_i}\Big) \cos m_i s_i \prod_{j \in B'} \omega(\alpha_j, s_j) \cos m_j s_j \, ds + \cdots +$$
$$+(-1) \sum_{B \in U_1} \int\limits_{H_4(B)\times H_1(B')} \Delta\Big(f, x, s + \Big(\frac{\pi}{m}\Big)_B\Big) \times$$
$$\times \prod_{i\in B} \omega\Big(\alpha_i, s_i + \frac{\pi}{m_i}\Big) \cos m_i s_i \prod_{j \in B'} \omega(\alpha_j, s_j) \cos m_j s_j \, ds \Big\},$$

whence, according to (3.1.1), we find

$$J(x,f,m,\alpha) =$$
$$= 2^{-n} \prod_{i=1}^{n} m_i^{-\alpha_i} \Big\{ \int\limits_{H_3(M)} \Phi_n(x,s,f,M,m) \prod_{i=1}^{n} \omega(\alpha_i, s_i) \cos m_i s_i \, ds +$$
$$+ \int\limits_{H_1(M)\setminus H_3(M)} \Delta(f,x,s) \prod_{i=1}^{n} \omega(\alpha_i, s_i) \cos m_i s_i \, ds +$$
$$+(-1)^n \int\limits_{H_4(M)\setminus H_3(M)} \Delta\Big(f, x, s + \frac{\pi}{m}\Big) \prod_{i=1}^{n} \omega\Big(\alpha_i, s_i + \frac{\pi}{m_i}\Big) \cos m_i s_i \, ds +$$

$$+(-1)^{n-1}\sum_{B\in U_{n-1}}\int_{H_5(B)}\Delta\Big(f,x,s+\Big(\frac{\pi}{m}\Big)_B\Big)\times$$

$$\times\prod_{i\in B}\omega\Big(\alpha_i,s_i+\frac{\pi}{m_i}\Big)\cos m_is_i\prod_{j\in B'}\omega(\alpha_j,s_j)\cos m_js_j\,ds+\cdots+$$

$$+(-1)^{n-k}\sum_{B\in U_{n-k}}\int_{H_5(B)}\Delta\Big(f,x,s+\Big(\frac{\pi}{m}\Big)_B\Big)\times$$

$$\times\prod_{i\in B}\omega\Big(\alpha_i,s_i+\frac{\pi}{m_i}\Big)\cos m_is_i\prod_{j\in B'}\omega(\alpha_j,s_j)\cos m_js_j\,ds+\cdots+$$

$$+(-1)\sum_{B\in U_1}\int_{H_5(B)}\Delta\Big(f,x,s+\Big(\frac{\pi}{m}\Big)_B\Big)\times$$

$$\times\prod_{i\in B}\omega\Big(\alpha_i,s_i+\frac{\pi}{m_i}\Big)\cos m_is_i\prod_{j\in B'}\omega(\alpha_j,s_j)\cos m_js_j\,ds+$$

$$+(-1)^n\int_{H_3(M)}\Delta\Big(f,x,s+\frac{\pi}{m}\Big)\Big[\prod_{i=1}^n\omega\Big(\alpha_i,s_i+\frac{\pi}{m_i}\Big)-$$

$$-\prod_{i=1}^n\omega(\alpha_i,s_i)\Big]\prod_{i=1}^n\cos m_is_i\,ds+$$

$$+(-1)^{n-1}\sum_{B\in U_{n-1}}\int_{H_3(M)}\Delta\Big(f,x,s+\Big(\frac{\pi}{m}\Big)_B\Big)\Big[\prod_{i\in B}\omega\Big(\alpha_i,s_i+\frac{\pi}{m_i}\Big)-$$

$$-\prod_{i\in B}\omega(\alpha_i,s_i)\Big]\prod_{j\in B'}\omega(\alpha_j,s_j)\prod_{i=1}^n\cos m_is_i\,ds+\cdots+$$

$$+(-1)^{n-k}\sum_{B\in U_{n-k}}\int_{H_3(M)}\Delta\Big(f,x,s+\Big(\frac{\pi}{m}\Big)_B\Big)\Big[\prod_{i\in B}\omega\Big(\alpha_i,s_i+\frac{\pi}{m_i}\Big)-$$

$$-\prod_{i\in B}\omega(\alpha_i,s_i)\Big]\prod_{j\in B'}\omega(\alpha_j,s_j)\prod_{i=1}^n\cos m_is_i\,ds+\cdots+$$

$$+(-1)\sum_{B\in U_1}\int_{H_3(M)}\Delta\Big(f,x,s+\Big(\frac{\pi}{m}\Big)_B\Big)\Big[\prod_{i\in B}\omega\Big(\alpha_i,s_i+\frac{\pi}{m_i}\Big)-$$

$$-\prod_{i\in B}\omega(\alpha_i,s_i)\Big]\prod_{j\in B'}\omega(\alpha_j,s_j)\prod_{i=1}^n\cos m_is_i\,ds\Big\}. \tag{3.1.15}$$

Using (1; 2.2.21) and (3.1.9), according to (3.1.15) we have

$$\prod_{i=1}^n m_i^{-\alpha_i}\Big\|\int_{H_3(M)}\Phi(\cdot,s,f,M,m)\prod_{i=1}^n\omega(\alpha_i,s_i)\cos m_is_i\,ds\Big\|_p\le$$

$$\le A(p,n,\alpha)\prod_{i=1}^n\lambda(m_i,\alpha_i)\omega_M\Big(\frac{1}{m},f\Big)_p, \tag{3.1.16}$$

i.e., this kind of estimate is valid for the first integral expression in the right side of (3.1.15). As to the other integral expressions, they have at least one of the following properties:

a) with respect to at least one variable s_i, the integration is performed over $[0, \frac{\pi}{m_i}]$;

b) with respect to at least one variable s_j, the integration is performed over $[\pi - \frac{\pi}{m_j}, \pi]$;

c) with respect to at least one variable s_l, the integration is performed either over $[\frac{\pi}{m_l}, \pi]$ or over $[\frac{\pi}{m_l}, \pi - \frac{\pi}{m_l}]$, and the integrand is reduced to the expression

$$\omega\left(\alpha_l, s_l + \frac{\pi}{m_l}\right) - \omega(\alpha_l, s_l)$$

satisfying (1; 2.2.22).

Thus, if in every integral term (except the first one) in the right side of (3.1.15) we substitute the coordinates s_i not having one of the properties a)–c) by $s_i + \frac{\pi}{m_i}$, and if we repeat this process, we conclude that for the completion of the proof it suffices to establish estimates for integral expressions of the types

$$J_1(x,f,m,\alpha) = \prod_{i=1}^{n} m_i^{-\alpha_i} \int\limits_{H(M)} \Delta(f,x,s) \prod_{i=1}^{n} \omega\left(\alpha_i, s_i + \frac{\pi}{m_i}\right) \cos m_i s_i \, ds,$$

$$J_2(x,f,m,\alpha) = \prod_{i=1}^{n} m_i^{-\alpha_i} \int\limits_{H_2(M)} \Delta(f,x,s) \prod_{i=1}^{n} \omega(\alpha_i, s_i) \cos m_i s_i \, ds,$$

$$J_3(x,f,m,\alpha) = \prod_{i=1}^{n} m_i^{-\alpha_i} \int\limits_{H(B)\times H_2(B')} \Delta(f,x,s) \prod_{i\in B} \omega\Big(\alpha_i, s_i + \\ + \frac{\pi}{m_i}\Big) \prod_{j\in B'} \omega(\alpha_j, s_j) \prod_{i=1}^{n} \cos m_i s_i \, ds,$$

$$J_4(x,f,m,\alpha) = \prod_{i=1}^{n} m_i^{-\alpha_i} \int\limits_{H(B)\times H_3(B')} \Phi^{(1)}_{n-k}(x,s,f,B',M,m) \prod_{i\in B} \omega\Big(\alpha_i, s_i + \\ + \frac{\pi}{m_i}\Big) \prod_{j\in B'} \omega(\alpha_j, s_j) \prod_{i=1}^{n} \cos m_i s_i \, ds,$$

$$J_5(x,f,m,\alpha) = \\ = \prod_{i=1}^{n} m_i^{-\alpha_i} \int\limits_{H_2(B)\times H_3(B')} \Phi^{(1)}_{n-k}(x,s,f,B',M,m) \prod_{i\in B} \omega(\alpha_i, s_i) \cos m_i s_i \, ds,$$

$$J_6(x,f,m,\alpha) = \\ = \prod_{i=1}^{n} m_i^{-\alpha_i} \int\limits_{H(B)\times H_3(B_2)\times H_2(B_3)} \Phi^{(1)}_{n-k_1-k_3}(x,s,f,B_2,M,m) \prod_{i\in B_1} \omega\Big(\alpha_i, s_i + \\ + \frac{\pi}{m_i}\Big) \prod_{j\in B'} \omega(\alpha_j, s_j) \prod_{i=1}^{n} \cos m_i s_i \, ds,$$

$$J_7(x,f,m,\alpha)=$$

$$=\prod_{i=1}^{n} m_i^{-\alpha_i} \int\limits_{H_3(B)\times H_3(B')} \Phi_{n-k}^{(1)}(x,s,f,B,M,m)\prod_{i\in B}\omega(\alpha_i,s_i)\times$$

$$\times\prod_{j\in B'}\left[\omega(\alpha_j,s_j)-\omega\left(\alpha_j,s_j+\frac{\pi}{m_j}\right)\right]\prod_{i=1}^{n}\cos m_is_i\,ds,$$

$$J_8(x,f,m,\alpha)=$$

$$=\prod_{i=1}^{n} m_i^{-\alpha_i} \int\limits_{H(B_1)\times H_2(B_2)\times H_3(B_3)\times H_3(B_4)} \Phi_{n-k_1-k_2-k_4}^{(1)}(x,s,f,B_3,M,m)\times$$

$$\times\prod_{i\in B_1}\omega\left(\alpha_i,s_i+\frac{\pi}{m_i}\right)\prod_{j\in B_4}\left[\omega(\alpha_j,s_j)-\omega\left(\alpha_j,s_j+\right.\right.$$

$$\left.\left.+\frac{\pi}{m_j}\right)\right]\prod_{l\in B_2\cup B_3}\omega(\alpha_l,s_l)\prod_{i=1}^{n}\cos m_is_i\,ds,$$

$$J_9(x,f,m,\alpha)=$$

$$=\prod_{i=1}^{n} m_i^{-\alpha_i} \int\limits_{H(B_1)\times H_3(B_2)\times H_3(B_3)} \Phi_{n-k_1-k_3}^{(1)}(x,s,f,B_2,M,m)\times$$

$$\times\prod_{i\in B_1}\omega\left(\alpha_i,s_i+\frac{\pi}{m_i}\right)\prod_{j\in B_2}\left[\omega(\alpha_j,s_j)-\omega\left(\alpha_j,s_j+\right.\right.$$

$$\left.\left.+\frac{\pi}{m_j}\right)\right]\prod_{l\in B_3}\omega(\alpha_l,s_l)\prod_{i=1}^{n}\cos m_is_i\,ds,$$

$$J_{10}(x,f,m,\alpha)=$$

$$=\prod_{i=1}^{n} m_i^{-\alpha_i} \int\limits_{H_2(B_1)\times H_3(B_2)\times H_3(B_3)} \Phi_{n-k_1-k_3}^{(1)}(x,s,f,B_2,M,m)\prod_{i\in B_3}\left[\omega(\alpha_i,s_i)-\right.$$

$$\left.-\omega\left(\alpha_i,s_i+\frac{\pi}{m_i}\right)\right]\prod_{j\in B_1\cup B_2}\omega(\alpha_j,s_j)\prod_{i=1}^{n}\cos m_is_i\,ds,$$

$$J_{11}(x,f,m,\alpha)=$$

$$=\prod_{i=1}^{n} m_i^{-\alpha_i} \int\limits_{H(B_1)\times H_3(B_2)\times H_2(B_3)} \Delta(f,x,s)\prod_{i\in B_1}\omega\left(\alpha_i,s_i+\frac{\pi}{m_i}\right)\times$$

$$\times\prod_{j\in B_2}\left[\omega(\alpha_j,s_j)-\omega\left(\alpha_j,s_j+\frac{\pi}{m_j}\right)\right]\prod_{l\in B_3}\omega(\alpha_l,s_l)\prod_{i=1}^{n}\cos m_is_i\,ds,$$

$$J_{12}(x,f,m,\alpha)=$$

$$= \prod_{i=1}^{n} m_i^{-\alpha_i} \int\limits_{H(B)\times H_3(B')} \Delta(f,x,s) \prod_{i\in B} \omega\left(\alpha_i, s_i + \frac{\pi}{m_i}\right) \times$$

$$\times \prod_{j\in B'} \left[\omega(\alpha_j, s_j) - \omega\left(\alpha_j, s_j + \frac{\pi}{m_j}\right)\right] \prod_{i=1}^{n} \cos m_i s_i \, ds,$$

$$J_{13}(x,f,m,\alpha) =$$

$$= \prod_{i=1}^{n} m_i^{-\alpha_i} \int\limits_{H_2(B)\times H_3(B')} \Delta(f,x,s) \prod_{i\in B} \omega(\alpha_i, s_i) \times$$

$$\times \prod_{j\in B'} \left[\omega(\alpha_j, s_j) - \omega\left(\alpha_j, s_j + \frac{\pi}{m_j}\right)\right] \prod_{i=1}^{n} \cos m_i s_i \, ds.$$

Using (1; 2.2.21), (1; 2.2.22), (1.1.1), (1.1.2), and (3.1.9), according to our notation we find

$$\|J_1(\cdot,f,m,\alpha)\|_p \le A(p,n,\alpha)\omega_M\left(\frac{1}{m},f\right)_p,$$

$$\|J_2(\cdot,f,m,\alpha)\|_p \le A(p,n,\alpha)\left[\sum_{i=1}^{n}\left(m_i^{-\alpha_i}+1\right)\omega\left(\frac{1}{m},f\right)_p\right],$$

$$\|J_3(\cdot,f,m,\alpha)\|_p \le A(p,n,\alpha)\omega_B\left(\frac{1}{m},f\right)_p,$$

$$\|J_4(\cdot,f,m,\alpha)\|_p \le A(p,n,\alpha)\prod_{i\in B'}\lambda(m_i,\alpha_i)\omega_{B'}\left(\frac{1}{m},f\right)_p,$$

$$\|J_5(\cdot,f,m,\alpha)\|_p \le A(p,n,\alpha)\prod_{i\in B'}\lambda(m_i,\alpha_i)\omega_{B'}\left(\frac{1}{m},f\right)_p,$$

$$\|J_6(\cdot,f,m,\alpha)\|_p \le A(p,n,\alpha)\prod_{i\in B_2}\lambda(m_i,\alpha_i)\omega_{B_2}\left(\frac{1}{m},f\right)_p,$$

$$\|J_7(\cdot,f,m,\alpha)\|_p \le A(p,n,\alpha)\prod_{i\in B}\lambda(m_i,\alpha_i)\omega_{B}\left(\frac{1}{m},f\right)_p,$$

$$\|J_8(\cdot,f,m,\alpha)\|_p \le A(p,n,\alpha)\prod_{i\in B_3}\lambda(m_i,\alpha_i)\omega_{B_3}\left(\frac{1}{m},f\right)_p,$$

$$\|J_9(\cdot,f,m,\alpha)\|_p \le A(p,n,\alpha)\prod_{i\in B_2}\lambda(m_i,\alpha_i)\omega_{B_2}\left(\frac{1}{m},f\right)_p,$$

$$\|J_{10}(\cdot,f,m,\alpha)\|_p \le A(p,n,\alpha)\prod_{i\in B_2}\lambda(m_i,\alpha_i)\omega_{B_2}\left(\frac{1}{m},f\right)_p,$$

$$\|J_{11}(\cdot,f,m,\alpha)\|_p \le A(p,n,\alpha)\omega_{B_1}\left(\frac{1}{m},f\right)_p,$$

$$\|J_{12}(\cdot,f,m,\alpha)\|_p \le A(p,n,\alpha)\omega_{B}\left(\frac{1}{m},f\right)_p,$$

$$\|J_{13}(\cdot,f,m,\alpha)\|_p \le A(p,n,\alpha)\Big[\sum_{i=1}^{n}\big(m_i^{-\alpha_i}+1\big)\omega_i\Big(\frac{1}{m},f\Big)_p\Big].$$

Thus, taking note of (3.1.12), (3.1.15), (3.1.16), and the estimates for $J_j(x,f,m,\alpha)$ with $1 \le j \le 13$, we make sure that (3.1.11) is true. This means that for $B = M$ and $t_i = 1$ for $1 \le i \le n$, the relation (3.1.10) is established. Clearly the method of the proof also works in the cases $B = M$, $t_i \in U_0$ for $1 \le i \le n$, or $B \subset M$, $B \ne M$ and $t_i \in U_0$, $i \in B$. We should add that for $B \subset M$, $B \ne M$, the notation (3.1.3) should also be used, the proof getting simpler since one will deal with a $k \equiv k(B)$-fold integral ($k \in [1, n-1]$). ■

Analyzing the scheme of the proof of Lemma 3.1.4, we can conclude that it likewise holds

Lemma 3.1.5. *Let $f \in L^p(T^n)$ for some $p \in [1,+\infty]$ and $\alpha \in]-1,+\infty[^n$. Then for any $B \subset M$,*

$$\sum_{t_{i_1}\in U_0}\cdots\sum_{t_{i_k}\in U_0}\Big\|\int_{H(B)\times H_1(B')}\Delta(f,\cdot,[s\times t]_B)\prod_{i\in B}K_{m_i}^{\alpha_i}(s_i)\prod_{j\in B'}\varphi_{m_j}^{\alpha_j}(s_j)\,ds\Big\|_p \le$$
$$\le A(p,n,\alpha)\Big[\sum_{B_1\subset B'}\prod_{j\in B_1}\lambda(m_j,\alpha_j)\omega_{B_1}\Big(\frac{1}{m},f\Big)_p\Big].$$

Theorem 3.1.1. *Suppose that $f \in L^p(T^n)$ for some $p \in [1,+\infty]$. Let, moreover, $\alpha \in]-1,+\infty[^n$ and $\lambda(m_i,s_i)$ for $1 \le i \le n$ be defined by (3.1.9). Then*

$$\|\sigma_m^\alpha(f)-f\|_p \le A(p,n,\alpha)\Big[\sum_{B\subset M}\prod_{i\in B}\lambda(m_i,\alpha_i)\omega_B\Big(\frac{1}{m},f\Big)_p +$$
$$+\sum_{i=1}^{m} m_i^{-1}\int_{\frac{1}{m_i}}^{\pi} s_i^{-2}\omega_i(s_i,f)_p ds_i\Big]. \tag{3.1.17}$$

Proof. By (1.1.25), we have

$$\pi^n[\sigma_m^\alpha(x,f)-f(x)] =$$
$$=\int_{[0,\pi]^n}\Big\{\sum_{t_2\in U_0}\cdots\sum_{t_n\in U_0}\Delta(f,x,[s\times t])\Big\}\prod_{i=1}^{n}K_{m_i}^{\alpha_i}(s_i)\,ds +$$
$$+2\int_{[0,\pi]^n}\Big\{\sum_{t_{i_1}\in U_0}\cdots\sum_{t_{i_k}\in U_0}\Big[\sum_{B\in U_{n-1}}\Delta(f,x,[s\times t]_B)\Big]\Big\}\prod_{i=1}^{n}K_{m_i}^{\alpha_i}(s_i)\,ds+\cdots+$$
$$+2^{n-1}\int_{[0,\pi]^n}\Big\{\sum_{t_1\in U_0}\Big[\sum_{B\in U_1}\Delta(f,x,[s\times t]_B)\Big]\Big\}\prod_{i=1}^{n}K_{m_i}^{\alpha_i}(s_i)\,ds. \tag{3.1.18}$$

Put

$$J_{14}(x,f,m,\alpha)=\int\limits_{[0,\pi]^n}\Delta(f,x,[s\times t])\prod_{i=1}^{n}K_{m_i}^{\alpha_i}(s_i)\,ds. \tag{3.1.19}$$

According to our notation, we get

$$J_{14}(x,f,m,\alpha)=$$

$$=\Bigg[\int\limits_{H(M)}+\int\limits_{H_1(M)}+\sum_{B\subset M}\int\limits_{H(B)\times H_1(B')}\Bigg]\Delta(f,x,[s\times t])\prod_{i=1}^{n}K_{m_i}^{\alpha_i}(s_i)\,ds\equiv$$

$$\equiv J_{15}(x,f,m,\alpha)+J_{16}(x,f,m,\alpha)+J_{17}(x,f,m,\alpha). \tag{3.1.20}$$

Hence, by virtue of Lemma 3.1.1, we find

$$\|J_{15}(\cdot,f,m,\alpha)\|_p\le A(p,n,\alpha)\omega_M\Big(\frac{1}{m},f\Big)_p. \tag{3.1.21}$$

Using (1; 2.2.9) and (3.1.20), we obtain

$$J_{16}(x,f,m,\alpha)=$$

$$=\sum_{B\subset M}\int\limits_{H_1(M)}\Delta(f,x,[s\times t])\prod_{i\in B}\varphi_{m_i}^{\alpha_i}(s_i)\prod_{j\in B'}r_{m_j}^{\alpha_j}(s_j)\,ds+$$

$$+\int\limits_{H_1(M)}\Delta(f,x,[s\times t])\prod_{i=1}^{n}r_{m_i}^{\alpha_i}(s_i)\,ds,\quad \prod_{j\in\varnothing}r_{m_j}^{\alpha_j}(s_j)\equiv 1,$$

whence, on the basis of Lemmas 3.1.4 and 3.1.1 (see (3.1.2)), it follows

$$\|J_{16}(\cdot,f,m,\alpha)\|_p\le A(p,n,\alpha)\Bigg[\sum_{B\subset M}\prod_{i\in B}\lambda(m_i,\alpha_i)\omega_B\Big(\frac{1}{m},f\Big)_p+$$

$$+\sum_{i=1}^{m}m_i^{-1}\int\limits_{\frac{1}{m_i}}^{\pi}s_i^{-2}\omega_i(s_i,f)_p ds_i\Bigg]. \tag{3.1.22}$$

Analogously, basing on (1; 2.2.9) and (3.1.20), we get

$$J_{17}(x,f,m,\alpha)=$$

$$=\sum_{B\subset M}\sum_{B_1\subset B'}\int\limits_{H(B)\times H_1(B')}\Delta(f,x,[s\times t])\prod_{i\in B}K_{m_i}^{\alpha_i}(s_i)\prod_{j\in B_1}\varphi_{m_j}^{\alpha_j}(s_j)\prod_{l\in B\setminus B_1}r_{m_l}^{\alpha_l}(s_l)\,ds+$$

$$+\sum_{B\subset M}\int\limits_{H(B)\times H_1(B')}\Delta(f,x,[s\times t])\prod_{i\in B}K_{m_i}^{\alpha_i}(s_i)\prod_{j\in B'}r_{m_j}^{\alpha_j}(s_j)\,ds,$$

whence, by Lemmas 3.1.5 and 3.1.2, we find

$$\|J_{17}(\cdot,f,m,\alpha)\|_p\le A(p,n,\alpha)\Bigg[\sum_{B\subset M}\prod_{i\in B}\lambda(m_i,\alpha_i)\omega_B\Big(\frac{1}{m},f\Big)_p+$$

$$+\sum_{i=1}^{m} m_i^{-1} \int_{\frac{1}{m_i}}^{\pi} s_i^{-2}\omega_i(s_i,f)_p ds_i\Bigg]. \tag{3.1.23}$$

Thus, in view of (3.1.19)–(3.1.23), it holds

$$\|J_{14}(\cdot,f,m,\alpha)\|_p \le A(p,n,\alpha)\Bigg[\sum_{B\subset M}\prod_{i\in B}\lambda(m_i,\alpha_i)\omega_B\left(\frac{1}{m},f\right)_p +$$

$$+\sum_{i=1}^{n} m_i^{-1} \int_{\frac{1}{m_i}}^{\pi} s_i^{-2}\omega_i(s_i,f)_p ds_i\Bigg]. \tag{3.1.24}$$

We can easily see that the last inequality remains valid for all the integral expressions appearing on the right side of (3.1.18). Therefore (3.1.17) is fulfilled. ∎

Remark 3.1.1. We have noted that an estimate of the type (3.1.24) is true for each of the expressions in the sums of the right side of (3.1.18). But, in fact, sharper estimates are valid in some cases. To be more precise, let us consider the expression

$$J_{18}(x,f,m,\alpha) =$$

$$= \int_{[0,\pi]^n} \Delta(f,x,[s\times t]_B)\prod_{i=1}^{n} K_{m_i}^{\alpha_i}(s_i)\,ds \text{ with } B\in U_{n-k},\ \ k\in[1,n-1].$$

For each $B\in U_{n-k}$, the expression $J_{18}(x,f,m,\alpha)$ is an $(n-k)$-fold integral since $\Delta(f,s,[s\times t]_B)$ depends (along with f and x) only on the coordinates s_i with $i\in B$. Therefore

$$\|J_{18}(\cdot,f,m,\alpha)\|_p \le A(p,n,\alpha)\Bigg[\sum_{B_1\subset B}\prod_{i\in B_1}\lambda(m_i,\alpha_i)\omega_{B_1}\left(\frac{1}{m},f\right)_p +$$

$$+\sum_{i\in B} m_i^{-1} \int_{\frac{1}{m_i}}^{\pi} s_i^{-2}\omega_i(s_i,f)_p ds_i\Bigg].$$

Remark 3.1.2. All the terms in the right side of (3.17) are essential. Using appropriate results of Chapter IV of the first part of the book, we can make sure that the estimates (3.1.17) are best possible in the classes $H(M,C)$ and $H(\omega,C)$. The corresponding examples are easy to construct with the help of n functions of one variable. A similar conclusion can be made in the case $p=1$ (imposing the same restriction on ω_B or ω as for $n=1$, and taking

$$\omega_B(\delta(B)) = \prod_{i\in B}\omega(\delta_i)\Big).$$

If $p\in\,]1,+\infty[$, then the conclusions of Theorem 3.1.1 can be made more precise. Consider some facts confirming this.

1) Let $f \in L^p(T^n)$, $p \in]1,+\infty[$. Then, using the inequality (1.3.10) of Sokol–Sokolowski [1], we easily find that

$$\|S_m(f) - f\|_p \leq A(p,n)E_m(f)_p \quad \text{for} \quad m \in \mathbb{N}^n, \tag{3.1.25}$$

where $E_m(f)_p$ is a best approximation of f in $L^p(T^n)$ by n-fold trigonometric polynomials of order $\leq m_i$ with respect to x_i for $1 \leq i \leq n$.

We should add that, with appropriate changes, (3.1.25) is true for reflexive Orlicz spaces as well. To be more precise, if $f \in L_\Phi(T^n)$, then

$$\|S_m(f) - f\|_\Phi \leq A(\Phi,n)E_m(f)_\Phi \quad \text{for} \quad m \in \mathbb{N}^n. \tag{3.1.26}$$

2) Suppose that $\alpha_i \in [0,+\infty[$ for $1 \leq i \leq n$. Using (3.1.25) for $\|\sigma_m^\alpha(f) - f\|_p$, we can obtain an upper estimate of the Stechkin [5] type by means of $E_k(f)_p$. On the other hand, the inequality

$$\|\sigma_m^\alpha(f) - f\|_p \leq A(p,n,\alpha)\left[\sum_{i=1}^{n} \omega_i\left(\frac{1}{m_i}, f\right)_p\right] \quad \text{for} \quad m \in \mathbb{N}^n \tag{3.1.27}$$

also holds. We proved this inequality in [13, Ch. IV] in the case $n = 2$ and $\alpha_i \in [1,+\infty[$ $(i = 1,2)$. The method of the proof also works for any $n \geq 3$ and $\alpha_i \in [1,+\infty[$, $1 \leq i \leq n$. But, in fact, (3.1.27) is true for any $n \geq 2$ and $\alpha_i \in]0,+\infty]$, $1 \leq i \leq n$.

3) Let $\alpha_i \in]-1,0[$ for $1 \leq i \leq n$. In this case, one can obtain no essential improvement of (3.1.17) for all the class $L^p(T^n)$ since the factors of the type $\prod_{i\in B} m_i^{-\alpha_i}$ with $B \subset M$ are essential. Of course, this does not mean that sharper estimates cannot be obtained; the question is of the essentiality of the expressions $\prod_{i\in B} m_i^{-\alpha_i}$.

4) Let $\alpha_i \in]-1,0[$ for $i \in B_1$, $\alpha_j = 0$, $j \in B_2$, $\alpha_l \in]0,+\infty[$ for $l \in B_3$ and $B_1 \cup B_2 \cup B_3 = M$, at most one of the B_i $(i = 1,2,3)$ being empty. In this case too, more precise estimates can be obtained, and in some cases this can be done using the same arguments as those leading to (3.1.26) and (3.1.27).

Various corollaries can be obtained from Theorem 3.1.1. We should not dwell on this since in Chapter IV of this part, we are going to turn back to this circle of problems.

Theorem 3.1.2. *Let* $\alpha \in]0,1]$, $\alpha p \in]n,+\infty[$ $(n \geq 2)$, *and* $f \in C(T^n) \cap Lip(\alpha,p)$.

a) *If* $\beta \in [0,p^{-1}[^n$, *then*

$$\|\sigma_m^{-\beta}(f) - f\|_C \leq A(p,f,n,\alpha)\left[\sum_{i=1}^{n} m_i^{\frac{n}{p}-\alpha}\right] \quad \textit{for} \quad m \in \mathbb{N}^n; \tag{3.1.28}$$

b) *If* $\beta \in [0,1[^n$ *with* $\sum_{i=1}^n \beta_i \equiv a(\beta) \equiv a < \alpha$, *then*

$$\|\sigma_m^{-\beta}(f) - f\|_C \leq A(p,f,n,\alpha)\left[\sum_{i=1}^{n} m_i^{a(\beta)-\alpha}\right] \quad \textit{for} \quad m \in \mathbb{N}^n. \tag{3.1.29}$$

Proof. According to well-known imbedding theorems of Nikol'skiĭ [4] (see also Nikol'skiĭ [5]), we can see that under the conditions of the theorem, the following assertions are true:

a) $f \in Lip(\alpha - \frac{n}{p})$;

b) for any $B \subset M$,

$$\left\| \left[\int\limits_{T^n(B)} |f(x+h_B) - f(x)|^p dx_B \right]^{1/p} \right\|_{C(T^n(B'))} \le$$
$$\le A(p,f,n,\alpha) \|h_B\|^{\alpha - \frac{n-k}{p}}. \tag{3.1.30}$$

Thus, by (3.1.30), for any $B \subset M$ we have

$$\left\| \left[\int\limits_{T^n(B)} |\Delta(f,x,h_B)|^p dx_B \right]^{1/p} \right\|_{C(T^n(B'))} \le$$
$$\le A(p,f,n,\alpha) \prod_{i\in B} |h_i|^{\frac{\alpha}{k} - \frac{n-k}{p}}, \quad |h_i| \in [0,\pi] \text{ for } i \in B, \quad k \equiv \overline{k}(B). \tag{3.1.31}$$

Since $f \in Lip\,(\alpha - \frac{n}{p})$,

$$m_i^{-1} \int\limits_{\frac{1}{m_i}}^{\pi} s_i^{-2} \omega_i(s_i,f)_C ds_i \le A(p,f,n,\alpha) m_i^{\frac{n}{p}-\alpha}, \tag{3.1.32}$$

$$m_i^{\beta_i - 1} \int\limits_{\frac{1}{m_i}}^{\pi} s_i^{\beta_i - 2} \omega_i(s_i,f)_C ds_i \le A(p,f,n,\alpha) m_i^{\frac{n}{p}-\alpha} \text{ for } 1 \le i \le n. \tag{3.1.33}$$

Therefore, analyzing the scheme of the proof of Theorem 3.1.1, we conclude that it suffices to establish the appropriate estimates for the expressions

$$J_{19}(x,f,B,m,\beta) =$$
$$= \prod_{i\in B} m_i^{-\beta_i} \int\limits_{H_1(B)} \Delta\left(f, x+s, \left(\frac{\pi}{m}\right)_B\right) \prod_{i\in B} \omega(\beta_i,s_i)\, ds_B. \tag{3.1.34}$$

Using Hölder's inequality, we get

$$\|J_{19}(\cdot,f,B,m,\beta)\|_C \le$$
$$\le \prod_{i\in B} m_i^{\beta_i} \left\| \left[\int\limits_{H_1(B)} \left| \Delta\left(f, x+s, \left(\frac{\pi}{m}\right)_B\right)\right|^p ds_B \right]^{1/p} \right\|_{C(T^n(B'))} \times$$
$$\times \left[\int\limits_{H_1(B)} \prod_{i\in B} s_i^{-(1-\beta_i)\frac{p}{p-1}} ds_B \right]^{\frac{p-1}{p}}, \tag{3.1.35}$$

whence, by (3.1.31), we find

$$\|J_{19}(\cdot,f,B,m,\beta)\|_C \le A(p,f,n,\alpha) \left[\sum_{i\in B} m_i^{\frac{n}{p}-\alpha} \right] \tag{3.1.36}$$
$$\text{for} \quad \beta_i \in [0,p^{-1}[, \ \ i \in B,$$

$$\|J_{19}(\cdot, f, B, m, \beta)\|_C \le A(p, f, n, \alpha)\left[\sum_{i\in B} m_i^{a(\beta)-\alpha}\right] \tag{3.1.37}$$
$$\text{for} \quad a(\beta) \in [0, \alpha[.$$

Therefore, using (3.1.32), (3.1.33), (3.1.34), (3.1.36), and (3.1.37), we make sure that (3.1.28) and (3.1.29) are true. ■

Let all the conditions of Theorem 3.1.2. be fulfilled, and let $\beta_i = p^{-1}$ for $1 \le i \le n$. Then, by (3.1.35), we obtain

$$\|J_{19}(\cdot, f, B, m, \beta)\|_C \le A(p, f, n, \alpha)\left[\sum_{B_1\subset B}\prod_{i\in B_1} \frac{[\log(m_i+2)]^{1-1/p}}{m_i^{\frac{\alpha p-1}{kp}}}\right] \tag{3.1.38}$$
$$\text{for} \quad B \subset M, \quad \text{where } k \equiv \overline{k}(B).$$

As we have already noted, in the case $p = +\infty$ we assume that $Lip\,(\alpha, p) \equiv Lip\,\alpha$, $\alpha \in]0, 1]$. It is not difficult to see that if $p = +\infty$, then (3.1.38) remains true. To be more precise,

$$\|J_{19}(\cdot, f, B, m, \beta)\|_C \le A(f, n, \alpha)\left[\sum_{B_1\subset B}\prod_{i\in B_1} \frac{\log(m_i+2)}{m_i^{\frac{\alpha}{k}}}\right]. \tag{3.1.39}$$

The inequalities (3.1.32) and (3.1.33) are obviously true for $p = +\infty$ as well. Thus, basing on (3.1.32), (3.1.33), (3.1.38), and (3.1.39), we arrive at

Theorem 3.1.3. *Let $\alpha \in]0, 1]$ and $\alpha p \in]n, +\infty[$. If $f \in C(T^n) \cap Lip\,(\alpha, p)$, then for any $\beta = (\beta_1, \beta_2, \ldots, \beta_n)$ with $\beta_i = p^{-1}$, $1 \le i \le n$, the inequality*

$$\|\sigma_m^{-\beta}(f) - f\|_C \le$$
$$\le A(p, f, n, \alpha)\left[\sum_{B\subset M}\prod_{i\in B} \frac{[\log(m_i+2)]^{1-1/p}}{m_i^{\frac{\alpha}{k}-\frac{n}{kp}}}\right] \quad \text{for } m \in \mathbb{N}^n \tag{3.1.40}$$

holds.

We should add that in the conditions of the theorem, the restriction $f \in C(T^n)$ is unnecessary in the case $p = +\infty$, and the constant $A(p, f, n, \alpha)$ in (3.1.40) can be substituted by $A(f, n, \alpha)$. But we think that to involve all the cases under consideration, the above given form of the theorem is justified. Note also that for $p = +\infty$, the estimate (3.1.40) is a multidimensional analogue of a well-known result of Lebesgue [4].

From Theorems 3.1.2 and 3.1.3, one can easily deduce the corresponding estimates in the case where $\beta = (\beta_1, \beta_2, \ldots, \beta_n)$ with $\beta_i \in [0, p^{-1}[$ for $i \in B_1$, $\beta_j = p^{-1}$ for $j \in B_2$, $\beta_l \in]p^{-1}, \alpha[$ for $l \in B_3$, $B_1 \cup B_2 \cup B_3 = M$, $a \in]0, \alpha[$, and at most one of the sets B_i $(i = 1, 2, 3)$ can be empty.

Analyzing the schemes of the proofs of Theorems 3.1.2 and 3.1.3, we conclude that there holds

Theorem 3.1.4. a) *Let $f \in Lip\,\alpha$ for some $\alpha \in]0,1[$. Then for any $\beta \in]0,+\infty[^n$, we have*

$$\|\sigma_m^\beta(f) - f\|_C \le A(f,n,\alpha)\left[\sum_{i=1}^n m_i^{-\alpha}\right] \quad \text{for } m \in \mathbb{N}^n. \tag{3.1.41}$$

b) *Let $f \in Lip\,1$. Then for any $\beta \in]0,+\infty[^n$, we have*

$$\|\sigma_m^\beta(f) - f\|_C \le A(f,n,\alpha)\left[\sum_{i=1}^n m_i^{-1}\log(m_i+2)\right] \quad \text{for } m \in \mathbb{N}^n. \tag{3.1.42}$$

Theorem 3.1.4 remains true for the space $L(T^n)$. Note also that the inequalities (3.1.41) and (3.1.42) are best possible (in the sense of order), and they do not depend on β. The same is true for the space $L(T^n)$. Confirming examples can be constructed by means of appropriate results of Ul'yanov [3].

Various corollaries can be deduced from Theorems 3.1.2–3.1.4. We give one of them which is a consequence of Theorems 3.1.2 and 3.1.4.

Corollary 3.1.1. *Let $\alpha \in]0,1[$ and $\alpha p \in]n,+\infty[$. Suppose that $f \in C(T^n) \cap Lip\,(\alpha,p)$. Then for any $\beta \in]0,+\infty[^n$,*

$$\|\sigma_m^\beta(f) - f\|_C \le A(p,f,n,\alpha)\left[\sum_{i=1}^n m_i^{\frac{n}{p}-\alpha}\right] \quad \text{for } m \in \mathbb{N}^n. \tag{3.1.43}$$

Note that, as above, the right-hand side does not depend on β, and the estimate is exact in the sense of order for any $\beta \in \mathbb{R}_+^n$. This can be deduced from the exactness of the corresponding imbedding theorems of Nikol'skiĭ [5, Ch. VII].

For $\beta = 0$, Corollary 3.1.1 implies a multidimensional analogue of a result of Izumi [2] (see (1; 4.1.10)). To be more precise, if $\alpha \in]0,1[$, $\alpha p \in]n,+\infty]$ and $f \in C(T^n) \cap Lip\,(\alpha,p)$, then

$$\|S_m(f) - f\|_C \le A(p,f,n,\alpha)\left[\sum_{i=1}^n m_i^{\frac{n}{p}-\alpha}\right] \quad \text{for } m \in \mathbb{N}^n. \tag{3.1.44}$$

The scheme of the proof of Theorem 3.1.1 leads us to

Theorem 3.1.5. *Let $\alpha \in]0,1[$ and $f \in Lip\,(\alpha,1)$. Then*

a) *for any $\beta \in]0,\alpha[^n$ and $a \equiv a(\beta) \in]0,\alpha[$, we have*

$$\|\sigma_m^{-\beta}(f) - f\|_1 \le A(f,n,\alpha)\left[\sum_{i=1}^n m_i^{a-\alpha}\right], \quad m \in \mathbb{N}^n;$$

b) *if $\beta = 0$, then*

$$\|S_m(f) - f\|_1 \le A(f,n,\alpha)\left[\sum_{B\subset M}\prod_{i\in B}\frac{\log(m_i+2)}{m_i^{\frac{\alpha}{k}}}\right] \quad \text{for } m \in \mathbb{N}^n.$$

Let now $\alpha \in]0,1[$ and $p \in [1,+\infty]$. Suppose that $\varepsilon(h) > 0$ for $|h| \in]0,\pi[$ and $\varepsilon(h) \to 0$ as $h \to 0$. A function $f \in L^p(T^n)$ is said to be of the class $Lip^*(\alpha,p)$, if

$$\|f(\cdot + h) - f(\cdot)\|_p \le \varepsilon(h)\|h\|^\alpha,$$

it being assumed that $Lip^*(\alpha,+\infty) \equiv Lip\alpha$.

For $n = 1$, the class $Lip^*(\alpha,p)$ was introduced by Hardy and Littlewood [6] in 1928.

The schemes of the proofs of Theorems 3.1.1 and 3.1.2 enable us to conclude that the following theorems are true.

Theorem 3.1.6. *Let $\alpha \in]0,1[$ and either $\beta \in [0,\frac{\alpha}{n}]^n$ or $\beta \in]0,\alpha[^n$ with $a(\beta) = \alpha$.*

a) *If $f \in Lip(\alpha,p)$ and $\alpha p \in]n,+\infty[$, then*

$$\|\sigma_m^{-\beta}(f) - f\|_C \le A(p,f,n,\alpha) \quad for \ m \in \mathbb{N}_0^n. \tag{3.1.45}$$

b) *If $f \in Lip^*(\alpha,p)$ and $\alpha p \in]n,+\infty[$, then*

$$\|\sigma_m^{-\beta}(f) - f\|_C \to 0 \quad as \quad m \to \infty. \tag{3.1.46}$$

Theorem 3.1.7. *Let $\alpha \in]0,1[$, $p \in [1,+\infty[$, and let β satisfy the conditions of Theorem* 3.1.6.

a) *If $f \in Lip(\alpha,p)$, then*

$$\|\sigma_m^{-\beta}(f) - f\|_p \le A(p,f,n,\alpha) \quad for \quad m \in \mathbb{N}_0^n. \tag{3.1.47}$$

b) *If $f \in Lip^*(\alpha,p)$, then*

$$\|\sigma_m^{-\beta}(f) - f\|_p \to 0 \quad as \quad m \to \infty. \tag{3.1.48}$$

We should stress that the restriction $\alpha p \in]n,+\infty[$ appearing in Theorem 3.1.6 is unnecessary for Theorem 3.1.7.

Remark 3.1.3. Unlike our work [13, Ch. IV], where some of results deal with the case $n = 2$, in the second part of this book (except Chapter V), all the results are valid for any finite dimension. The methods of the proofs of the theorems presented in this chapter give one a full idea of the difficulties arising for $n \ge 3$. Not to repeat the above arguments, below, without detailed proofs the results are given concerning n-fold Fourier series of functions of bounded variation in the sense of Hardy, as well as those connected with approximating properties of n-fold Cesàro means of the series $\sigma_n[f,B]$, $B \subset M$.

In what follows, we denote by $HV(T^n)$ the class of all the functions $f \in T^n \to \mathbb{R}$ satisfying

a) $|f(x)| \le A(f)$ for $x \in T^n$;

b) all the possible limits of the form $f(x_1 \pm 0,\dots,x_n \pm 0)$ (their number is 2^n) exist;

c) for any $B \subset M$ and $h \in]0,\pi[^n$,

$$\sup_{x_j, j \in B'} \int\limits_{T^n(B)} |\Delta(f,x,h_B)|\, dx_B \le A(f,n) \prod_{i \in B} |h_i|. \tag{3.1.49}$$

In 1906, Hardy [1] introduced the class of the functions of several variables with bounded variation. Hardy actually proved that any function of the class satisfy the above conditions a)–c).

Theorem 3.1.8. a) *Let* $f \in HV(T^n)$ *and* $\beta \in]-1,0]^n$. *Then at any point* $x \in T^n$,

$$\Big\{\sigma_m^{-\beta}(x,f) - 2^{-n}\Big[f(x_1+0,x_2+0,\dots,x_n+0) + \\ + f(x_1-0,x_2+0,\dots,x_n+0) + \\ + \dots + f(x_1-0,x_2-0,\dots,x_n-0)\Big]\Big\} \to 0 \quad as \quad n \to \infty.$$

b) *Let* f *and* β *satisfy the conditions of* a). *If, in addition,* f *is continuous on a nondegenerate* n*-dimensional rectangle* $I_0^n \subset T^n$, *then the series* $\sigma_n[f]$ *is* $(C,-\beta)$*-summable to* f *uniformly on any rectangle strictly contained in* I_0^n.

In the case $n = 2$, we proved the theorem 3.1.8 in [13, Ch. IV]. Following the scheme of the proof of Theorem 3.1.1, we make sure that the theorem is true for any $n \geq 3$. It is true for $\beta \in]0,+\infty[^n$ as well.

Theorem 3.1.9. *Let* $f \in C(T^n) \cap HV(T^n)$. *Then*

$$\|\sigma_m^\beta(f) - f\|_C \leq \\ \leq A(f,n,\beta)\Big[\sum_{i=1}^n \omega_i^{1-\beta_i}\Big(\frac{1}{m_i},f\Big)_C\Big] \quad for \ \beta \in]-1,0[^n, \tag{3.1.50}$$

$$\|\sigma_m^\beta(f) - f\|_C \leq A(f,n)\Big\{\sum_{i=0}^n \omega_i\Big(\frac{1}{m_i},f\Big)_C \times \\ \times \log\Big[\omega_i\Big(\frac{1}{m_i},f\Big)_C\Big]^{-1}\Big\} \quad for \ \beta = 0, \tag{3.1.51}$$

$$\|\sigma_m^\beta(f) - f\|_c \leq A(f,n,\beta)\Big[\sum_{i=1}^n m_i^{-1}\int_{\frac{1}{m_i}}^{\pi} s_i^{-2}\omega_i(s_i,f)_C ds_i\Big] \\ for \ \ \beta \in]0,+\infty[^n. \tag{3.1.52}$$

To establish (3.1.50) and (3.1.51), the essential is to obtain the corresponding estimates for the expressions $J_{10}(x,f,B,m,\beta)$, $B \subset M$, defined by (3.1.34). But the latters follow from (3.1.49) and the schemes of the proofs of Theorem 4.1.4 from the first part and Theorems 3.1.1, 3.1.2. As to (3.1.52), it is a consequence of (3.1.17).

Taking into account the appropriate properties of the Poisson–Abel kernel, we conclude that it holds

Theorem 3.1.10. *Let $f \in L^p(T^n)$ for some $p \in [1,+\infty]$, and let $f(x,r)$ be defined by* (1.1.27) *with $r \in [0,1[^n$. Then*

$$\|f(\cdot,r)-f(\cdot)\|_p \leq A(p,n)\left[\sum_{i=1}^{n}(1-r_i)\int_{1-r_i}^{\pi} s_i^{-2}\omega_i(s_i,f)_C ds_i\right]. \qquad (3.1.53)$$

If $p=+\infty$ or $p=1$, the inequality (3.1.53) cannot be improved in the space $L^p(T^n)$. To be sure of this, it suffices to use the appropriate assertions by Natanson [2] (for $p=+\infty$) and Ul'yanov [3] (for $p=1$).

Remark 3.1.4. Using inequalities (3.1.50)–(3.1.52), we can obtain upper estimates for the expressions $\|\sigma_m^{\beta}(f)-f\|_C$ also in the case where $\beta_i \in]-1,0[$ for $i \in B_1$, $\beta_j = 0$ for $j \in B_2$, and $\beta_l \in]0,+\infty]$ for $l \in B_3$, where $B_1 \cup B_2 \cup B_3 = M$ and at most one of them is empty.

3.2. Approximating Properties of n-fold Cesàro Means of the Series $\overline{\sigma}_n[f,B]$

Note first that though Theorem 3.1.8 is not true for the series $\overline{\sigma}_n[f,B]$ with $B \subset M$, by Theorem 2.2.4 it holds

Corollary 3.2.1. *Let $f \in HV(T^n)$ and $\beta \in [0,\frac{1}{2}[^n$. Then all the series $\overline{\sigma}_n[f,B]$ are almost everywhere $(C,-\beta)$-summable.*

It is not difficult to see that for any $\beta \in]-1,+\infty[^n$, there is a function $f \in C(T^n) \cap HV(T^n)$ such that for $\lambda = 1$, we have

$$\limsup_{m_\lambda \to \infty}\left\|t_m^{\beta}(\cdot,f,M)-\overline{f}_M\left(\frac{\pi}{m}\right)\right\|_C > 0,$$

where $t_m^{\beta}(x,f,M)$ is defined by (1.1.24) for $\beta \neq 0$ and by (1.1.26) for $\beta = 0$, and $\overline{f}_M(x,\frac{\pi}{m})$ is defined by (1.1.8).

Below we will assume that for $\alpha = 0$,

$$\sigma_m^{\alpha}(x,f) \equiv S_m(x,f), \quad t_m^{\alpha}(x,f,B) \equiv \overline{S}_m(x,f,B), \ \text{ where } \ B \subset M.$$

Let $f \in L^p(T^n)$ for some $p \in]1,+\infty[$. Using the inequality (1.3.10) of Sokol–Sokolowski [1], we get

$$\|t_m^{\alpha}(\cdot,f,B)-\overline{f}_B(\cdot)\|_p \leq A(p,n)\|\sigma_m^{\alpha}(f)-f\|_p \ \text{ for } \ B \subset M.$$

Therefore the upper estimates for $\|\sigma_m^{\alpha}(f)-f\|_p$ are valid for

$$\|t_m^{\alpha}(\cdot,f,B)-\overline{f}_B(\cdot)\|_p \ \text{ with } \ B \subset M \qquad (3.2.1)$$

as well. Naturally, for (3.2.1) sharper upper estimates are possible.

Now suppose that $f \in L^p(T^n)$ with $p=+\infty$ or $p=1$. Then, as compared with the case $n=1$ which we have considered in Chapter 4 of the first part, the situation is more complex. As we have already noted, some of the assertions are given in our work [13, Ch. IV]. The theorem presented below which makes more precise those assertions is an analogue of Theorem 4.2.1 of the first part in the multidimensional case.

For the sake of brevity, we will use the notation

$$\overline{f}_{B,m}(x) \equiv \overline{f}_{B}\left(x, \frac{\pi}{m}\right). \tag{3.2.2}$$

Further, if $f \in L^p(T^n)$ for some $p \in [1, +\infty]$, then we denote

$$\begin{aligned} g(f,m,\alpha) \equiv g(f,m,\alpha,B) = \sum_{B_1 \subset B} \prod_{i \in B_1} \lambda(m_i, \alpha_i) \omega_{B_1}\left(\frac{1}{m}, f\right)_p + \\ + \sum_{B_1 \subset B} \sum_{B_2 \subset M \setminus B_1} \prod_{i \in B_2} \lambda(m_i, \alpha_i) \omega_{B_2}\left(\frac{1}{m}, \overline{f}_{B_1,m}\right)_p + \\ + \sum_{i \in B'} m_i^{-1} \int_{\frac{1}{m_i}}^{\pi} s_i^{-2} \omega_i(s_i, \overline{f}_{B,m})_p + \sum_{i=1}^{n} m_i^{-1} \int_{\frac{1}{m_i}}^{\pi} s_i^{-2} \omega_i(s_i, f)_p ds_i, \end{aligned} \tag{3.2.3}$$

where $\lambda(m_i, \alpha_i)$ are defined by (3.1.9).

Theorem 3.2.1. *Let $f \in L^p(T^n)$ for some $p \in [1, +\infty]$ and let $g(f, m, \alpha)$ be defined by* (3.2.3). *Then for any $B \subset M$,*

$$\|t_m^\alpha(\cdot, f, B) - \overline{f}_{B,m}(\cdot)\|_p \le A(p, n, \alpha) g(f, m, \alpha) \quad \textit{for } m \in \mathbb{N}^n. \tag{3.2.4}$$

The theorem can by proved by the same method as Theorem 3.1.1. To be more precise, the sets H, H_i, $1 \le i \le 5$, defined at the beginning of this chapter and relations (1; 2.2.9)–(1; 2.2.16) should be used, and the fact is to be taken into account that in the corresponding estimates, the expressions $\varphi_{m_i}^{\alpha_i}(s_i)$ with $i \in B'$ and $\psi_{m_j}^{\alpha_j}(s_j)$ with $j \in B$ behave equally.

In the case $B = M$, the theorem was proved by Leladze [1].

Some of other important results concerning approximating properties of n-fold Cesàro and Poisson–Abel means can be found in our works [35], [46].

3.3. Unsolved Problems

3.3.1. Let $\chi(n) \subset C(T^n)$ be a proper subclass (for instance, $Lip\,\alpha$, $H(\omega, C)$, and so on), and let $\beta \in]-1, +\infty[^n$. Investigate the asymptotic behaviour of the expressions

$$\sup_{f \in \chi(n)} \|\sigma_m^\beta(f) - f\|_C.$$

Some results connected with this problem for $\alpha = 0$ are given in Stepanets [1, Ch. III].

This problem can be considered for the space $L(T^n)$, as well as for the conjugate series $\overline{\sigma}_n[f, B]$.

3.2.2. Let $n \ge 2$, $\alpha \in]0,1[$, $f \in C(T^n) \cap Lip(\alpha, p)$ and $\alpha p \in [1, n]$. Investigate the problem of (C, β)-summability with $\beta \in]-1, 0]^n$ of the series $\sigma_n[f]$ in the metric of $C(T^n)$.

CHAPTER IV

Convergence and Summability of Multiple Trigonometric Fourier Series and Their Conjugates in the Spaces $L^p(T^n)$, $p \in]0, +\infty]$

4.1. Convergence and Summability of the Series $\sigma_n[f]$ and $\overline{\sigma}_n[f, B]$ in the Spaces $L^p(T^n)$, $p \in]0, 1]$

In 1925, Kolmogorov [3] for the first time used the method of representation of partial sums of the series $\sigma[f]$ and $\overline{\sigma}[f]$ in terms of conjugate functions, and established the convergence of the series $\sigma[f]$ in $L^p(T)$ for any $p \in [0, 1[$. Later on, this method was widely used in the works of various mathemeticians (see, e.g., Zygmund [15, Ch. VII], Lozinskiĭ [1], etc.). In particular, for partial sums of $\sigma[f]$ and $\overline{\sigma}[f]$, there hold (in appropriate Lebesgue classes) the inequalities of types of Kolmogorov [3] (see (1; 1.1.21)) and Zygmund [4] (see (1;1.1.5)), [15, Ch. VII] (see (1.1.47)). To be more precise,

a) if $f \in L(T)$, then for any $p \in]0, 1[$,

$$\|S_m(f)\|_p \le A(p)\|f\|_1, \quad \|\overline{S}_m(f)\|_p \le A(p)\|f\|_1; \tag{4.1.1}$$

b) if $f \in L(\log^+ L)^\eta(T)$ for some $\eta \in]0, +\infty[$, then

$$\begin{aligned} &\int_T |S_m(x, f)| \Big[\log(|S_m(x, f)| + 2)\Big]^{\eta-1} dx \le \\ &\le A(\eta)\Big[\int_T |f(x)| \Big[\log^+ |f(x)|\Big]^\eta dx + 1\Big], \end{aligned} \tag{4.1.2}$$

$$\begin{aligned} &\int_T |\overline{S}_m(x, f)| \Big[\log(|\overline{S}_m(x, f)| + 2)\Big]^{\eta-1} dx \le \\ &\le A(\eta)\Big[\int_T |f(x)| \Big[\log^+ |f(x)|\Big]^\eta dx + 1\Big]. \end{aligned} \tag{4.1.3}$$

Using (1.1.23), (1.1.24), and (4.1.1)–(4.1.3), we can conclude that the following statements are true.

1. Let $f \in L(\log^+ L)^{n-1}(T^n)$. Then for any $p \in]0,1[$,

$$\|S_m(f) - f\|_p \to 0, \quad \|S_m(\cdot, f, B) - \overline{f}_B(\cdot)\|_p \to 0 \quad \text{as } m \to \infty, \tag{4.1.4}$$

2. Let $f \in L(\log^+ L)^n(T^n)$. Then for any $B \subset M$, $B \neq \varnothing$,

$$\|S_m(f) - f\|_1 \to 0, \quad \|S_m(\cdot, f, B) - \overline{f}_B(\cdot)\|_1 \to 0 \quad \text{as } m \to \infty. \tag{4.1.5}$$

Recent investigations of Getsadze [1], [2] imply

3. Let a function $\Phi : [0,+\infty[\to [0,+\infty[$ satisfy $\Phi \uparrow \infty$ as $t \uparrow +\infty$, and

$$\Phi(t)\left[t(\log t)^{n-1}\right]^{-1} \to 0 \quad \text{as} \quad t \to +\infty.$$

Then there is a measurable function f such that

$$\int_{T^n} \Phi(|f(x)|)\, dx < +\infty,$$

yet $\sigma_n[f]$ does not converge in measure.

This result shows that the restriction $f \in L(\log^+ L)^{n-1}(T)$ in the statement 1 is essential.

Analogously, using results of Tkebuchava [2], we conclude that this restriction is likewise essential in the statement 2.

Now let $\alpha \in]0,+\infty[^n$. According to Theorem 3.1.1, if $f \in L(T^n)$, then

$$\|\sigma_m^\alpha(f) - f\|_1 \to 0 \quad \text{as} \quad m \to \infty.$$

Moreover,

$$\|\sigma_m^\alpha(f) - f\|_1 \leq$$
$$\leq A(n,\alpha)\left[\sum_{i=1}^n m_i^{-1} \int_{\frac{1}{m_i}}^{\pi} s_i^{-2}\omega_i(s_i,f)_1\, ds_i\right] \quad \text{for} \quad m \in \mathbb{N}^n. \tag{4.1.6}$$

A similar estimate (see (3.1.52)) is valid for n-fold Poisson–Abel means as well. Using (1.1.26) and Theorem 1.2.3 (see (1.2.52)), we get

4. If $f \in L(\log^+ L)^{n-1}(T^n)$, then for any $p \in]0,1[$ and $B \neq \varnothing$,

$$\|t_m^\alpha(\cdot, f, B) - \overline{f}_B(\cdot)\|_p \to 0 \quad \text{as} \quad m \to \infty. \tag{4.1.7}$$

Analogously, from Theorem 1.3.1 we get

5. If $f \in L(\log^+ L)^n(T^n)$, then

$$\|t_m^\alpha(\cdot, f, B) - \overline{f}_B(\cdot)\|_1 \to 0 \quad \text{as} \quad m \to \infty. \tag{4.1.8}$$

Analogous statements are also valid for n-fold Poisson–Abel means of the series $\overline{\sigma}_n[f,B]$. We should add that in the statements 4 and 5, the classes $L(\log^+ L)^{n-1}(T^n)$ and $L(\log^+ L)^n(T^n)$ are exact.

6. a) Let $n \geq 2$ and $\beta \in]1,n[$. If $f \in L(\log^+ L)^\beta(T^n)$, then for any $B \subset M$ with $\overline{k}(B) \in [1,\beta[$, the series $\overline{\sigma}_n[f,B]$ is summable in the norm of $L(T^n)$ to $\overline{f}_B$ by n-fold Cesàro (of positive order) and Poisson–Abel methods.

b) Let $n \geq 2$, $\alpha \in]0,+\infty[^n$, and $\beta \in [0,n[$. If $B \subset M$ with $\overline{k}(B) \in]\beta, n]$, then in the class $L(\log^+ L)^\beta(T^n)$, there are the functions f and g such that for $\lambda = 1$,

$$\limsup_{m_\lambda \to \infty} \|t_m^\alpha(\cdot, f, B)\|_1 = +\infty, \quad \limsup_{r_\lambda \to 1-} \|\overline{g}(\cdot, r, B)\|_1 = +\infty.$$

This statement follows from the scheme of the proof of Theorem 1.3.1. In the work [43], we pointed out that the statements 4 and 5 (see (4.1.7) and (4.1.8)) can be improved in a certain sense. To be more precise, it holds

Theorem 4.1.1. a) *Let $f \in L(\log^+ L)^{n-2}(T^n)$ and $\alpha \in]0,+\infty[^n$. Then for any $p \in]0,1[$ and $B \subset M$ $(B \neq \varnothing)$,*

$$\|\ell_m^\alpha(\cdot, f, B) - \overline{f}_{B,m}(\cdot)\|_p \to 0 \quad \textit{as} \quad m \to \infty. \tag{4.1.9}$$

b) *Let $f \in L(\log^+ L)^{n-1}(T^n)$ and $\alpha \in]0,+\infty[^n$. Then for any $B \subset M$ $(B \neq \varnothing)$,*

$$\|t_m^\alpha(\cdot, f, B) - \overline{f}_{B,m}(\cdot)\|_1 \to 0 \quad \textit{as} \quad m \to \infty, \tag{4.1.10}$$

where $\overline{f}_{B,m}$ are defined by (3.2.2).

Proof. It suffices to prove that under the conditions of the assertions a) and b) the inequalities

$$\|t_m^\alpha(\cdot, f, B) - \overline{f}_{B,m}(\cdot)\|_p \leq$$
$$\leq A(p,n,\alpha)\left[\int_{T^n} |f(x)|\left[\log^+|f(x)|\right]^{n-2} dx + 1\right], \quad B \subset M, \tag{4.1.11}$$

and

$$\|\ell_m^\alpha(\cdot, f, B) - \overline{f}_{B,m}(\cdot)\|_1 \leq$$
$$\leq A(p,n,\alpha)\left[\int_{T^n} |f(x)|\left[\log^+|f(x)|\right]^{n-1} dx + 1\right], \quad B \subset M, \tag{4.1.12}$$

hold, respectively.

We will assume that $B = M$, the scheme of the proof being simpler to carry out in the case $B \neq M$.

We have

$$t_m^\alpha(x, f, M) - \overline{f}_{M,m}(x) = (-1)^n\left[\int_{H(M)} \overline{\Delta}(f,x,s)\prod_{i=1}^n \tau_{m_i}^{\alpha_i}(s_i)\,ds + \right.$$
$$+ \int_{H_1(M)} \overline{\Delta}(f,x,s)\prod_{i=1}^n H_{m_i}^{\alpha_i}(s_i)\,ds +$$
$$+ \sum_{B \subset M} 2^{-k} \int_{H_1(B)\times H_1(B')} \overline{\Delta}(f,x,s)\prod_{i\in B}\operatorname{ctg}\frac{s_i}{2}\prod_{j\in B'} H_{m_j}^{\alpha_j}(s_j)\,ds +$$

$$+\sum_{B\subset M}\int_{H(B)\times H_1(B')}\overline{\Delta}(f,x,s)\prod_{i=1}^{n}\tau_{m_i}^{\alpha_i}(s_i)\,ds\Bigg]\equiv$$

$$\equiv\sum_{i=1}^{4}\delta_i(x,f,m,\alpha)\quad\text{where } k\equiv\overline{k}(B). \tag{4.1.13}$$

By (1; 2.2.14), we find

$$\|\delta_1(\cdot,f,m,\alpha)\|_p\le A(p,n,\alpha)\omega_M\Big(\frac{1}{m},f\Big)_1\quad\text{for } m\in\mathbb{N}^n, \tag{4.1.14}$$

and by (2.2.44), we get

$$\|\delta_2(\cdot,f,m,\alpha)\|_p\le$$

$$\le A(p,n,\alpha)\left[\sum_{i=1}^{n}m_i^{-1}\int_{\frac{1}{n}}^{\pi}s_i^{-2}\omega_i(s_i,f)_1ds_i\right]\quad\text{for } m\in\mathbb{N}^n. \tag{4.1.15}$$

The inequalities (4.1.14) and (4.1.15) remain valid in the case $p=1$.

Consider $\delta_3(x,f,m,\alpha)$ in (4.1.13). Each of its addends contains an expression of the type $\overline{\Delta}(\overline{f}_{B,m},x,s_{B'})$ with $B\in U_{n-k}$, $k\in[1,n-1]$, and in each of it for at least one coordinate s_j, the function $H_{m_j}^{\alpha_j}(s_j)$ is integrated over $[\frac{\pi}{m_j},\pi]$, $j\in B'$. Therefore, using (1.1.52) with respect to appropriate s_i, $i\in B$, and taking note of (2.2.44) (approximately as while proving Lemma 1.2.1), we find

$$\|\delta_3(\cdot,f,m,\alpha)\|_p\le A(p,n,\alpha)\|f\|_1\quad\text{for } n=2, \tag{4.1.16}$$

$$\|\delta_3(\cdot,f,m,\alpha)\|_p\le A(p,n,\alpha)\|\overline{f}_{B_1,m}\|_1,$$
$$\text{for }\ \overline{k}(B_1)\in[1,n-2],\quad n\ge 3. \tag{4.1.17}$$

Using Lemma 1.2.11, by virtue of (4.1.17) we have

$$\|\delta_3(\cdot,f,m,\alpha)\|_p\le A(p,n,\alpha)\left[\int_{T^n}|f(x)|\big[\log^+|f(x)|\big]^{n-2}dx+1\right]. \tag{4.1.18}$$

Now consider $\delta_4(x,f,m,\alpha)$. In each of its addends, for at least one of the coordinates s_i, the function $\tau_{m_1}^{\alpha_i}(s_i)$ is integrated over $[0,\frac{\pi}{m_i}]$, $i\in B$, and for at least one of the coordinates s_j, the function $\tau_{m_j}^{\alpha_j}(s_j)$ is integrated over $[\frac{\pi}{m_j},\pi]$, $j\in B'$. Here too, using (1.1.52) with respect to some s_j, $j\in B'$, as well as relations (1; 2.2.14), (2.2.43) and (2.2.44), we conclude that

$$\|\delta_4(\cdot,f,m,\alpha)\|_p\le A(p,n,\alpha)\|f\|_1\quad\text{for } n=2, \tag{4.1.19}$$

and

$$\|\delta_4(\cdot,f,m,\alpha)\|_p\le$$

$$\le A(p,n,\alpha)\left[\int_{T^n}|f(x)|\big[\log^+|f(x)|\big]^{n-2}dx+1\right]\quad\text{for } n\ge 3. \tag{4.1.20}$$

The above argument and Lemma 1.2.14 lead us to the statement: if $f \in L(\log^+ L)^{n-1}(T^n)$, then

$$\|\delta_i(\cdot, f, m, \alpha)\|_1 \leq$$
$$\leq A(n,\alpha)\Bigg[\int_{T^n} |f(x)|\Big[\log^+|f(x)|\Big]^{n-1} dx + 1\Bigg] \quad \text{for } i = 3,4, \quad n \geq 2. \quad (4.1.21)$$

Thus, basing on (4.1.13)–(4.1.16) and (4.1.18)–(4.1.21), we make sure that (4.1.9) and (4.1.10) are fulfilled. ■

Remark 4.1.1. In the proof of Theorem 4.1.1, we have noted that the case $B \neq M$ is simpler. But we should add that when $B \neq M$ and $n \geq 3$, the restrictions $f \in L(\log^+ L)^{n-1}(T^n)$ and $f \in L(\log^+ L)^{n-1}(T^n)$ in the assertions a) and b), respectively, become unnecessary. However, they are necessary for Theorem 4.1.1 to be true.

Since the Poisson–Abel kernel and its conjugate have all the properties of the kernels $K_{m_i}^{\alpha_i}(s_i)$ and $\tau_{m_i}^{\alpha_i}(s_i)$ we made use of while proving Theorem 4.1.1, we can conclude that it likewise holds

Theorem 4.1.2. a) *Let $f \in L(\log^+ L)^{n-2}(T^n)$, $\overline{f}(x,r,B)$ and $\overline{f}_B(x,\varepsilon)$ be defined, respectively, by* (1.1.28) *and* (1.1.8) *with $r_i \in [0,1[$, $\varepsilon_i = 1 - r_i$, $1 \leq i \leq n$. Then for any $p \in]0,1[$ and $B \subset M$ $(B \neq \varnothing)$, the relation*

$$\|\overline{f}(\cdot,r,B) - \overline{f}_B(\cdot,\varepsilon)\|_p \to 0 \quad as \quad r \to 1-,$$

is valid.

b) *If $f \in L(\log^+ L)^{n-1}(T^n)$, then for any $B \subset M$ $(B \neq \varnothing)$, we have*

$$\|\overline{f}(\cdot,r,B) - \overline{f}_B(\cdot,\varepsilon)\|_1 \to 0 \quad as \quad r \to 1- .$$

In 1972, at the seminar of Men'shov and Ul'yanov, Men'shov posed the question: if $n \geq 2$, $\beta \in [0, n-1[$ and $f \in L(\log^+ L)^{\beta}(T^n)$, then for which $B \subset M$ $(B \neq \varnothing)$ is the series $\overline{\sigma}_n[f, B]$ summable by Cesàro (of positive order) and Poisson–Abel methods in the sense of convergence in measure?

Below we give statements which, in particular, answer this question (see also our work [44]).

Theorem 4.1.3. a) *Let $n \geq 2$ and $\beta \in [0, n-1]$. If $f \in L(\log^+ L)^{\beta}(T^n)$, then for any $p \in]0,1[$ and $B \subset M$ with $\overline{k}(B) \in [1, \beta+1]$, the series $\overline{\sigma}_n[f, B]$ is summable to $\overline{f}_B$ in the metric of $L^p(T^n)$ both by Cesàro (of positive order) and Poisson–Abel methods.*

b) *Let $n \geq 2$ and $\beta \in [0, n-1[$. If $B \subset M$ with $\overline{k}(B) \in]\beta+1, n]$, then in the class $L(\log^+ L)^{\beta}$, there are functions f and g such that the expressions $t_m^{\alpha}(x, f, B)$ and $\overline{g}(x, r, B)$ do not converge in measure as $m_\lambda \to \infty$ and $r_\lambda \to 1-$, respectively, where $\lambda = 1$ and $\alpha \in]0, +\infty[^n$.*

The assertion a) of the theorem follows from a result of Kolmogorov [3] (see (1.1.47)) in the case $\overline{k}(B) = 1$, and from Theorem 1.2.3 in the case $\overline{k}(B) \geq 2$.

As to the assertion b), it can be verified on the basis of the results presented in the proof of Theorem 1.2.4 b), as well as in the assertions of Section 1.2. Note that f depends on α.

We should add that the class $L(\log^+ L)^{n-1}(T^n)$ is best possible for the convergence in measure of Cesàro (of positive order) and Poisson–Abel means of $\overline{\sigma}_n[f, M]$. Tkebuchava [3] showed that the class is best possible for their λ-convergence ($\lambda = 1$) as well.

4.2. Convergence and Summability of the Series $\sigma_n[f]$ and $\overline{\sigma}_n[f, B]$ in the Spaces $L^p(T^n)$, $p \in [1, +\infty]$

Let $f \in L^p(T^n)$ for some $p \in]1, +\infty[$. Using (3.1.25) and a theorem of Sokol–Sokolowski [1], we conclude that if $\alpha \in]0, +\infty[^n$, then

$$\|\sigma_m^\alpha(f) - f\|_p \to 0 \quad \text{as} \quad m \to \infty, \tag{4.2.1}$$

and for any $B \subset M$, $B \neq \varnothing$,

$$\|t_m^\alpha(\cdot, f, B) - f_B\|_p \to 0 \quad \text{as} \quad m \to \infty. \tag{4.2.2}$$

Relations (4.2.1) and (4.2.2) are also true for reflexive Orlicz spaces, as well as for n-fold Poisson–Abel means.

Remark 4.2.1. Some weighted estimates conserning n-fold Cesàro (of positive order) and Poisson–Abel means follow from the paper of Nakhman and Osilenker [1] (see also Dyn'kin and Osilenker [1]).

Now we consider more thoroughly the problems of convergence of the series $\sigma_n[f]$ and $\overline{\sigma}_n[f, B]$ in the spaces $L(T^n)$ and $C(T^n)$, as well as their summability by n-fold Cesàro method. We have pointed out some of the assertions stated below in the works [4], [5], [13, Ch. IV], [14], [15], [16], [20], [24], [31], [33], [40]. They, in particular, solve a problem posed by Ul'yanov in 1966: Find, in terms of partial moduli of continuity of f, a nonimprovable analogue of the well-known Dini–Lipschitz test (see, e.g., Zygmund [15, Ch. II]) for convergence in $C(T^n)$ by rectangles of the series $\sigma_n[f]$. The analogous problem for $L(T^n)$ is also solved.

Theorem 4.2.1. a) *Let $f \in L(T^n)$ and $\eta \in]0, +\infty[$. If*

$$\omega_i(\delta, f)_1 \left(\log \frac{2\pi}{\delta}\right)^{n+\eta} \le A(f, n), \quad \delta \in]0, \pi], \quad 1 \le i \le n, \tag{4.2.3}$$

then

$$\|S_m(f) - f\|_1 \to 0 \quad \text{as} \quad m \to \infty, \tag{4.2.4}$$

and

$$\|\overline{S}_m(\cdot, f, B) - \overline{f}_B(\cdot)\|_1 \to 0 \quad \text{as} \quad m \to \infty \quad (B \subset M, \quad B \neq \varnothing). \tag{4.2.5}$$

b) *Let $f \in L(T^n)$ and for some i_0, let us have*

$$\omega_{i_0}(\delta, f)_1 \left(\log \frac{2\pi}{\delta}\right)^n \to 0 \quad \text{as} \quad \delta \to 0+ \tag{4.2.6}$$

and

$$\omega_i(\delta, f)_1\Big(\log\frac{2\pi}{\delta}\Big)^n \le A(f,n) \quad \textit{for } \delta \in]0,\pi], \; 1 \le i \le n, \; i \ne i_0. \tag{4.2.7}$$

Then (4.2.4) *holds.*

c) *Given* $n \ge 2$, *there is a function* $f \in L(T^n)$ *such that*

$$\omega_i(\delta, f)_1\Big(\log\frac{2\pi}{\delta}\Big)^n \le A(f,n) \quad \textit{for } \delta \in]0,\pi], \tag{4.2.8}$$

for any $B \subset M$, $B \ne \varnothing$, *the function* $\overline{f}_B \in L(T^n)$ *and*

$$\omega_i(\delta, \overline{f}_B)_1\Big(\log\frac{2\pi}{\delta}\Big)^n \le A(f,n) \quad \textit{for } \delta \in]0,\pi], \tag{4.2.9}$$

yet for $\lambda = 1$,

$$\limsup_{m_\lambda\to\infty} \|S_m(f) - f\|_1 > 0.$$

Proof. It is not difficult to verify that (4.2.3) ensures the almost everywhere existence of all the functions $\overline{f}_B$. Using a theorem of Panjikidze [1], we can conclude that $f \in L(\log^+ L)^n(T^n)$. Then (4.1.5) implies the validity of (4.2.3) and (4.2.4). This proves the assertion a) of the theorem.

It can easily be verified that (4.2.6) and (4.2.7) guarantee

$$\omega_B\Big(\frac{1}{m}, f\Big)_1 \prod_{i\in B} \log(m_i + 2) \to 0 \quad \text{as } m \to \infty, \; B \subset M.$$

Then, using the appropriate estimate (with $p = 1$ and $\alpha = 0$) from Theorem 3.1.1, we get

$$\|S_m(f) - f\|_1 \to 0 \quad \text{as } m \to \infty.$$

That is, the assertion b) is also proved.

Now we turn to the proof of c).

Assume that $V(n,k,x)$ and $\overline{V}(n,k,x)$ are given by (1; 2.3.1) and (1; 2.3.2), respectively, and p_i is given by (2.1.41), i.e.,

$$p_i \equiv p_i(j) = 2^{2^j} \quad (j = 1, 2, \dots, \; 1 \le i \le n).$$

Let the function $f : T^n \to [-\infty, +\infty]$ be defined as

$$f(x) = \sum_{j=1}^\infty (2^{-j})^n \prod_{i=1}^n V(p_i - 1, p_i, x_i). \tag{4.2.10}$$

Using (1; 2.3.3), we conclude

$$\sum_{j=1}^\infty (2^{-j})^n \int_{T^n} \prod_{i=1}^n |V(p_i - 1, p_i, x_i)|\, dx < +\infty,$$

so that $f \in L(T^n)$.

It can easily be seen that for different j, the products

$$\prod_{i=1}^{n} V(p_i - 1, p_i, x_i).$$

do not contain similar terms. Therefore the series in the right-hand side of (4.2.10) can be represented as an n-fold trigonometric series of the form

$$\sum_{p \geq 1} a_p \prod_{i=1}^{n} \cos p_i x_i. \tag{4.2.11}$$

It is easy to observe that n-fold rectangular sums of (4.2.11) with the indices $3p_i - 2$, $1 \leq i \leq n$, converge in the metric of $L(T^n)$. Therefore this series is an n-fold Fourier series of the function f.

Now we show that for this function, (4.2.8) are fulfilled. We use the method we have applied while proving Theorem 2.1.7.

Suppose that $h_i \in]0, 10^{-2}[$, and $j_0 \equiv j_0(h_i)$ is a largest natural number for which

$$\left(\exp_2 \exp_2 j\right)^2 \leq h_i^{-1}, \tag{4.2.12}$$

and so

$$\left(\exp_2 \exp_2 j\right)^2 > h_i^{-1} \quad (j = j_0 + 1, \dots).$$

Then

$$\left(2^{-j_0 - 1} \log h_i^{-1}\right)^n \leq A(f, n). \tag{4.2.13}$$

In view of (1; 2.3.1), we have

$$\|V(p_l^{-1}, p_l, \cdot)\|_C \leq A 2^{2^l}, \quad 1 \leq l \leq n.$$

Then

$$\|V'(p_l^{-1} l - 1, p_l, \cdot)\|_C \leq A 2^{2^l}, \quad 1 \leq l \leq n. \tag{4.2.14}$$

Furthermore, we find

$$\begin{aligned} &\|f(\cdot + h_i) - f(\cdot)\|_1 \leq \\ &\leq A(n) \Bigg\{ \sum_{j=1}^{j_0} (2^{-j})^n \int_{T^n} \Bigg| V(p_i - 1, p_i, x_i + h_i) \prod_{\substack{l=1 \\ l \neq i}}^{n} V(p_l - 1, p_l, x_l) - \\ &\qquad - \prod_{l=1}^{n} V(p_l - 1, p_l, x_l) \Bigg| \, dx + \\ &+ \sum_{j=j_0+1}^{\infty} (2^{-j})^n \int_{T^n} \Bigg| V(p_i - 1, p_i, x_i + h_i) \prod_{\substack{l=1 \\ l \neq i}}^{n} V(p_l - 1, p_l, x_l) - \\ &- \prod_{l=1}^{n} V(p_l - 1, p_l, x_l) \Bigg| \, dx \Bigg\} \equiv B_1^{(0)}(f, n, h_i) + B_2^{(0)}(f, n, h_i). \end{aligned} \tag{4.2.15}$$

The expression $B_1^{(0)}(f,n,h_i)$ can be estimated like $B_1(f,n,h_i)$ in the proof of Theorem 2.1.7. To be more precise, Lagrange's theorem, Bernstein's inequality [2], and inequalities (4.2.12), (4.2.14) should be applied for the coordinate x_i, while for coordinates x_l $(1 \le l \le n,\ l \ne i)$, the inequalities (1; 2.3.3) should be used. Therefore

$$B_1^{(0)}(f,n,h_i)(2^{j_0+1})^n \le A(f,n). \tag{4.2.16}$$

By (1;2.3.3) and (4.2.15), we obtain

$$B_1^{(0)}(f,n,h_i)(2^{j_0+1})^n \le A(f,n). \tag{4.2.17}$$

Thus, on the basis of (4.1.2) and (4.2.15)–(4.2.17), we get

$$\omega_i(\delta,f)_1\Big(\log\frac{2\pi}{\delta}\Big)^n \le A(f,n),\ \ \delta \in]0,\pi],\ \ 1 \le i \le n,$$

that is, (4.2.8) is fulfilled.

Using a theorem of Panjikidze [1], we conclude that $f \in L(\log^+ L)^{n-1}(T^n)$. Therefore all the functions $\overline{f}_B$, $B \subset M$, exist almost everywhere. By definition,

$$\overline{\sigma}_n[f,B] = \sum_{p\ge 1} a_p \prod_{i\in B'} \cos p_i x_i \prod_{l\in B} \sin p_l x_l. \tag{4.2.18}$$

On the other hand, applying (1; 2.3.3), we find

$$\sum_{j=1}^{\infty}(2^{-j})^n \int_{T^n}\Big|\prod_{i\in B'} V(p_i-1,p_i,x_i)\prod_{l\in B}\overline{V}(p_l-1,p_l,x_l)\Big|\,dx < +\infty.$$

Thus the function $\varphi_B : T^n \to \mathbb{R}$ defined as

$$\varphi_B(x) = \sum_{j=1}^{\infty}(2^{-j})^n \prod_{i\in B'} V(p_i-1,p_i,x_i)\prod_{l\in B}\overline{V}(p_l-1,p_l,x_l), \tag{4.2.19}$$

is summable on T^n. Moreover, the series in the right-hand side of (4.2.19) has the form of the n-fold trigonometric series appearing in the right-hand side of (4.2.18). Hence we can easily deduce that $\varphi_B = \overline{f}_B$ almost everywhere, and $\overline{\sigma}_n[f,B] = \sigma_n[\overline{f}_B]$, $B \subset M$. Since the properties of $V(p_i-1,p_i,x_i)$ and $\overline{V}(p_i-1,p_i,x_i)$ we have used while proving (4.2.8) are the same, we conclude that (4.2.9) also holds.

It remains to show that the series $\sigma_n[f]$ and $\sigma_n[\overline{f}_B]$ do not converge (boundedly diverge) in the metric of $L(T^n)$.

It is not difficult to observe that

$$\begin{aligned} S_m(x,f) &= \sum_{j=1}^{q}(2^{-j})^n \prod_{i=1}^{n} V(p_i-1,p_i,x_i) + \\ &+ (2^{-q-1})^n g\Big(\prod_{i=1}^{n} V(p_i-1,p_i,x_i)\Big), \end{aligned} \tag{4.2.20}$$

where, as in the one-dimensional case, q is a natural number depending on m, and the function g denotes a certain part (in some cases it is absent) of the n-fold polynomial

$$\prod_{i=1}^{n} V(p_i - 1, p_i, x_i), \quad p_i \equiv p_i(q+1), \quad 1 \le i \le n.$$

Put

$$m^{(1)} = (m_1^{(1)}, \dots, m_n^{(1)}), \quad m_i^{(1)} = 2^{2^j} - 1,$$

$$m^{(2)} = (m_1^{(2)}, \dots, m_n^{(2)}), \quad m_i^{(2)} = 3 \cdot 2^{2^{j-1}} - 2, \quad (j = 1, 2, \dots; \ 1 \le i \le n).$$

By (4.2.20), we get

$$S_{m^{(1)}}(x, f) - S_{m^{(2)}}(x, f) =$$
$$= \left(\frac{1}{2^j}\right)^n \prod_{i=1}^{n} \frac{1}{p_i(j)} \sum_{\nu_i=1}^{p_i(j)-1} \nu_i \cos(\nu_i + p_i(j)).$$

Hence applying (1; 3.2.29), we find

$$\limsup_{m_\lambda \to \infty} \|S_m(f) - f\|_1 > 0. \quad \blacksquare$$

Remark 4.2.2. Analyzing the scheme of the proof of the assertion b) of Theorem 3.2.3 from the first part (see (3.2.25) and (3.2.29)), we can conclude that

$$\int_T |\overline{V}_2(n, x)|\, dx > A \log n,$$

where

$$\overline{V}_2(n, x) = \frac{1}{n} \sum_{\nu=1}^{n-1} \nu \sin(\nu + n)x$$

with $n \equiv n(j) = 2^{2^j}$ $(j = 1, 2, \dots)$. Hence it follows that for the function f defined by (4.2.10), the relations

$$\limsup_{m_\lambda \to \infty} \|\overline{S}_m(\cdot, f, B) - \overline{f}_B(\cdot)\|_1 > 0, \quad B \subset M,$$

also hold.

Theorem 4.2.1 implies

Corollary 4.2.1. *Let $f \in L(T^n)$ and for any $B \subset M$, $B \neq \varnothing$,*

$$\omega_B(\delta, f)_1 \prod_{i \in B} \log \frac{2\pi}{\delta_i} \to 0 \quad \text{as } \delta_i \to 0+, \ i \in B. \tag{4.2.21}$$

Then $\sigma_n[f]$ converges in the metric of $L(T^n)$.

Condition (4.2.21) is essential in a certain sense. To be more precise, using Theorem 3.2.1 from the first part and point a) of Theorem 4.2.1, we arrive at

Theorem 4.2.2. *Given $B_0 \subset M$ with $\overline{k}(B_0) \in [1, n]$, there is a function $f \in L(T^n)$ such that*

$$\omega_{B_0}(\delta, f)_1 \prod_{i \in B_0} \log \frac{2\pi}{\delta_i} \leq A(f, n), \quad \delta_i \in]0, \pi], \quad i \in B_0,$$

for any $B \subset M$ with $B \neq B_0$, $B \neq \varnothing$, it holds

$$\omega_B(\delta, f)_1 \prod_{i \in B} \log \frac{2\pi}{\delta_i} \to 0, \quad \delta_i \to 0+, \quad i \in B,$$

yet

$$\limsup_{m_\lambda \to \infty} \|S_m(f) - f\|_1 > 0.$$

Theorem 4.2.3. *Given $n \geq 2$, there is a function $f \in L(T^n)$ for which all the functions $\overline{f}_B \in L(T^n)$, (4.2.8) and (4.2.9) are fulfilled, yet for $\lambda = 1$,*

$$\limsup_{m_\lambda \to \infty} \|S_m(\cdot, f, M) - \overline{f}_{M,m}(\cdot)\|_1 > 0. \tag{4.2.22}$$

Proof. Let the function $f : T^n \to \mathbb{R}$ be defined by (4.2.10). Then, according to corresponding assertions of b) of Theorem 4.2.1, the inequalities (4.2.8) and (4.2.9) hold. We have

$$\begin{aligned} &\|\overline{S}_m(\cdot, f, M) - \overline{f}_{M,m}(\cdot)\|_1 \geq \\ &\geq \|\overline{S}_m(\cdot, f, M) - \overline{f}_M(\cdot)\|_1 - \|t_m^1(\cdot, f, M) - \overline{f}_M(\cdot)\|_1 - \\ &- \|t_m^1(\cdot, f, M) - \overline{f}_{M,m}(\cdot)\|_1. \end{aligned} \tag{4.2.23}$$

Using the statement of Remark 4.2.2 (for $B = M$), we conclude that

$$\limsup_{m_\lambda \to \infty} \|\overline{S}_m(\cdot, f, M) - \overline{f}_M(\cdot)\|_1 > 0. \tag{4.2.24}$$

On the other hand, according to Theorem 3.1.1,

$$\limsup_{m_\lambda \to \infty} \|t_m^1(\cdot, f, M) - \overline{f}_M(\cdot)\|_1 = 0, \tag{4.2.25}$$

while according to Theorem 4.1.1,

$$\limsup_{m \to \infty} \|t_m^1(\cdot, f, M) - \overline{f}_{M,m}(\cdot)\|_1 = 0. \tag{4.2.26}$$

Then by (4.2.23)–(4.2.26), we obtain (4.2.22). ∎

Using Theorem 3.2.1 (see (3.2.3) and (3.2.4)), we see that it holds

Theorem 4.2.4. *Let $f \in L(T^n)$ and*

$$\omega(\delta, f)_1 \left(\log \frac{2\pi}{\delta} \right)^n \to 0 \quad \text{as } \delta \to 0+. \tag{4.2.27}$$

Then for any $B \subset M$, $B \neq \varnothing$, we have

$$\limsup_{m_\lambda \to \infty} \|S_m(\cdot, f, B) - \overline{f}_{B,m}(\cdot)\|_1 = 0. \tag{4.2.28}$$

Remark 4.2.3. In the relation (4.2.28), the expressions $\overline{f}_{B,m}(x)$, in general, cannot be substituted by $\overline{f}_B(x)$. The point is that (4.2.27) does not ensure the integrability of f_M on T^n.

Now we give some assertions for the space $C(T^n)$.

Theorem 4.2.5. a) *Let* $f \in C(T^n)$ *and* $\eta \in]0, +\infty[$. *If*

$$\omega_i(\delta, f)_C \left(\log \frac{2\pi}{\delta} \right)^{n+\eta} \leq A(f,n) \quad for \ \delta \in]0, \pi], \tag{4.2.29}$$

then

$$\|S_m(f) - f\|_C \to 0 \quad as \ m \to \infty, \tag{4.2.30}$$

and

$$\|\overline{S}_m(\cdot, f, B) - \overline{f}_B(\cdot)\|_C \to 0 \quad as \ m \to \infty. \tag{4.2.31}$$

b) *Let* $f \in C(T^n)$ *and for some* i_0,

$$\omega_{i_0}(\delta, f)_C \left(\log \frac{2\pi}{\delta} \right)^n \to 0 \quad as \ \delta \to 0+ \tag{4.2.32}$$

and

$$\omega_i(\delta, f)_C \left(\log \frac{2\pi}{\delta} \right)^n \leq A(f,n) \quad for \ \delta \in]0, \pi], \ i \neq i_0. \tag{4.2.33}$$

Then (4.2.30) *holds.*

c) *Given* $n \geq 2$, *there is a function* $f \in C(T^n)$ *such that*

$$\omega_i(\delta, f)_C \left(\log \frac{2\pi}{\delta} \right)^n \leq A(f,n), \ \ \delta \in]0, \pi], \ \ 1 \leq i \leq n, \tag{4.2.34}$$

for any $B \subset M$, $B \neq \varnothing$, *the function* $\overline{f}_B \in C(T^n)$ *and*

$$\omega_i(\delta, \overline{f}_B)_C \left(\log \frac{2\pi}{\delta} \right)^n \leq A(f,n), \ \ \delta \in]0, \pi], \ \ 1 \leq i \leq n, \tag{4.2.35}$$

yet for $\lambda = 1$,

$$\limsup_{m_\lambda \to \infty} \|S_m(f) - f\|_C > 0,$$
$$\limsup_{m_\lambda \to \infty} \|\overline{S}_m(\cdot, f, B) - \overline{f}_B(\cdot)\|_C > 0, \quad B \subset M.$$

Proof. It is easy to verify that conditions (4.2.29) guarantee the everywhere existence and continuity of all the functions $\overline{f}_B$. By Theorem 3.1.1, we can conclude that ($p = +\infty$, $\alpha = 0$)

$$\|S_m(f) - f\|_C \to 0 \quad \text{as } m \to \infty.$$

Analogously, using (4.2.29), according to Theorem 3.2.1 we find

$$\|\overline{S}_m(\cdot, f, B) - \overline{f}_B(\cdot)\|_C \to 0 \quad \text{as } m \to \infty.$$

Therefore (4.2.30) and (4.2.31) are fulfilled. Thus a) is proved.

By (4.2.32) and (4.2.33), we find

$$\omega_B\left(\frac{1}{m}, f\right)_C \prod_{i\in B} \log(m_i+2) \to 0 \quad \text{as } m \to \infty.$$

Using (for $p = +\infty$ and $\alpha = 0$) Theorem 3.1.1, we conclude that (4.2.30) holds. Therefore b) is also valid.

Now we turn to the proof of c). We will use two trigonometric polynomials of Fejér [2] (see also Bari [2, Ch. I], Zygmund [15, Ch. VIII]):

$$\begin{aligned} F(x,p) = \frac{1}{\pi+8}\Big[\frac{\cos px}{p} + \frac{\cos(p+1)x}{p-1} + \cdots + \frac{\cos(2p-1)x}{1} - \\ - \frac{\cos(2p+1)x}{1} - \frac{\cos(2p+2)x}{2} - \cdots - \frac{\cos 3px}{p}\Big], \end{aligned} \tag{4.2.36}$$

$$\begin{aligned} \overline{F}(x,p) = \frac{1}{\pi+8}\Big[\frac{\sin px}{p} + \frac{\sin(p+1)x}{p-1} + \cdots + \frac{\sin(2p-1)x}{1} - \\ - \frac{\sin(2p+1)x}{1} - \frac{\sin(2p+2)x}{2} - \cdots - \frac{\sin 3px}{p}\Big] \quad \text{for } x \in T. \end{aligned} \tag{4.2.37}$$

It is well-known that (see, e.g., Bari [2, Ch. I], Pak [1])

$$\|F(\cdot,p)\|_C < 1, \quad \|\overline{F}(\cdot,p)\|_C < 1 \quad \text{for } p \in \mathbb{N}. \tag{4.2.38}$$

Define now the function $f : T^n \to \mathbb{R}$ by

$$f(x) = \sum_{j=1}^{\infty} (2^{-j})^n \prod_{i=1}^{n} F(x_i, 2^{2^j}). \tag{4.2.39}$$

In view of (4.2.38), we get $f \in C(T^n)$. If for any $B \neq \varnothing$ we put

$$\psi_B(x) = \sum_{j=1}^{\infty} (2^{-j})^n \prod_{i\in B'} F(x_i, 2^{2^j}) \prod_{l\in B} \overline{F}(x_l, 2^{2^j}), \tag{4.2.40}$$

then again using (4.2.38), we conclude $\psi_B \in C(T^n)$ and $\psi_B = \overline{f}_B$ everywhere.

Since different $F(x_j, 2^{2^j})$ and $\overline{F}(x_l, 2^{2^j})$ ($i \in B'$, $l \in B$) do not contain similar terms, the series in the right-hand sides of (4.2.39) and (4.2.40) can be represented as

$$\sigma_n[f] = \sum_{p\geq 1} a_p \prod_{i=1}^{n} \cos p_i x_i, \tag{4.2.41}$$

$$\sigma_n[\overline{f}_B] = \sum_{p\geq 1} a_p \prod_{i\in B'} \cos p_i x_i \prod_{l\in B} \sin p_l x_l, \tag{4.2.42}$$

respectively.

Let (as in the proof of the assertion c) of Theorem 4.2.1) $h_i \in]0, 10^{-2}[$, $1 \leq i \leq n$, and $j_0(h_i)$ be the largest natural number for which the inequality (4.2.12) (i.e, the inequality (4.2.13) as well) is fulfilled. Using Lagrange's theorem and Bernstein's inequality, by (4.2.36) and (4.2.37) we conclude that

$$\|F(\cdot+h_i,2^{2^j})-F(\cdot,2^{2^j})\|_C\le A2^{2^j}h_i, \tag{4.2.43}$$

$$\|\overline{F}(\cdot+h_i,2^{2^j})-\overline{F}(\cdot,2^{2^j})\|_C\le A2^{2^j}h_i. \tag{4.2.44}$$

Then, basing on (4.2.12), (4.2.13), (4.2.39), and (4.2.43), we get

$$|f(x+h_i)-f(x)|\le$$

$$\le\sum_{j=1}^{j_0}(2^{-j})^n\Big|\prod_{\substack{l=1\\ l\ne i}}F(x_l,2^{2^j})F(x_i+h_i,2^{2^j})-\prod_{l=1}^{n}F(x_l,2^{2^j})\Big|+$$

$$+\sum_{j=j_0+1}^{\infty}(2^{-j})^n\Big|\prod_{\substack{l=1\\ l\ne i}}F(x_l,2^{2^j})F(x_i+h_i,2^{2^j})-\prod_{l=1}^{n}F(x_l,2^{2^j})\Big|\le$$

$$\le A(f,n)\Big(\log\frac{1}{h_i}\Big)^{-n}.$$

Therefore

$$\omega_i(\delta,f)_C\Big(\log\frac{2\pi}{\delta}\Big)^n\le A(f,n)\quad\text{for }\ \delta\in]0,\pi].$$

Quite analogously, on the basis of (4.2.12), (4.2.13), (4.2.40), and (4.2.44), we find

$$\omega_i(\delta,\overline{f}_B)_C\Big(\log\frac{2\pi}{\delta}\Big)^n\le A(f,n)\quad\text{for }\ \delta\in]0,\pi].$$

Therefore (4.2.34) and (4.2.35) are valid.

By virtue of (4.2.39)–(4.2.42), we conclude that partial sums of the series $\sigma_n[f]$ and $\sigma_n[\overline{f}_B]$ can be represented as

$$S_m(x,f)=\sum_{j=1}^{q_1}(2^{-j})^n\prod_{i=1}^{n}F(x_i,2^{2^j})+$$

$$+2^{-(q_1+1)n}g_1\Big(\prod_{i=1}^{n}F(x_i,2^{2^j})\Big), \tag{4.2.45}$$

$$\overline{S}_m(x,f,B)=\sum_{j=1}^{q_2}(2^{-j})^n\prod_{i\in B'}F(x_i,2^{2^j})\prod_{l\in B}F(x_l,2^{2^j})+$$

$$+g_2\Big(\prod_{i\in B'}F(x_i,2^{2^j})\prod_{l\in B}\overline{F}(x_l,2^{2^j})\Big), \tag{4.2.46}$$

respectively, where q_i $(i=1,2)$ are natural numbers, and g_i $(i=1,2)$ are some parts (sometimes absent) of the polynomials

$$\prod_{i=1}^{n}F(x_i,2^{2^j}),\quad \prod_{i\in B'}F(x_i,2^{2^j})\prod_{l\in B}\overline{F}(x_l,2^{2^j}),$$

respectively.

Let now $x = 0$ and

$$m^{(1)} = (m_1^{(1)}, \ldots, m_n^{(1)}), \quad m_i^{(1)} = 2^{2^j} - 1,$$
$$m^{(2)} = (m_1^{(2)}, \ldots, m_n^{(2)}), \quad m_i^{(2)} = 3 \cdot 2^{2^j - 1}, \quad j \in \mathbb{N}, \quad 1 \leq i \leq n.$$

Then, according to (4.2.45), we find

$$\|S_{m^{(1)}}(f) - S_{m^{(2)}}(f)\|_C \geq |S_{m^{(1)}}(0, f) - S_{m^{(2)}}(0, f)| > A.$$

Therefore

$$\limsup_{m_\lambda \to \infty} \|S_m(f) - f\|_C > 0.$$

Analogously, if $x' \equiv x'(B, j)$ is a point of T^n such that

$$x_j' = 0, \quad i \in B, \quad x_l' = 4\left(\exp_2 \exp_2 j\right)^{-1} \pi, \quad l \in B,$$

we have

$$|\overline{S}_{m^{(2)}}(x', f, B) - \overline{S}_{m^{(1)}}(x', f, B)| > A,$$

so by (4.3.46) we conclude that

$$\limsup_{m_\lambda \to \infty} \|\overline{S}_m(\cdot, f, B) - \overline{f}_B(\cdot)\|_C > 0.$$

Therefore c) is valid. ■

Analyzing the scheme of the proof of Theorem 4.2.5 (for $B = M$), we arrive at

Theorem 4.2.6. *Given $n \geq 2$, there is a function $f \in C(T^n)$ for which* (4.2.34) *holds, the series $\sigma_n[f]$ everywhere converges, yet the convergence is not uniform.*

Theorem 4.2.7. a) *Let $f \in C(T^n)$, and for any $B \neq \varnothing$, let*

$$\omega_B(\delta, f)_C \prod_{i \in B} \log \frac{2\pi}{\delta_i} \to 0 \quad \text{as } \delta_i \to 0+, \quad i \in B.$$

Then $\sigma_n[f]$ converges in the metric of $C(T^n)$.

b) *Given $B_0 \subset M$ with $\overline{k}(B_0) \in [1, n]$, there is a function $f \in C(T^n)$ such that*

$$\omega_{B_0}(\delta, f)_C \prod_{i \in B_0} \log \frac{2\pi}{\delta_i} \leq A(f, n), \quad \delta_i \in]0, \pi], \quad i \in B_0,$$

for any $B \neq B_0$, $B \neq \varnothing$, it holds

$$\omega_B(\delta, f)_C \prod_{i \in B} \log \frac{2\pi}{\delta_i} \to 0, \quad \delta_i \to 0+, \quad i \in B,$$

yet for $\lambda = 1$,

$$\limsup_{m_\lambda \to \infty} \|S_m(f) - f\|_C > 0.$$

c) *Given $n \geq 2$, there is a function $f \in C(T^n)$ for which all $\overline{f}_B$ belong to $C(T^n)$,* (4.2.34) *and* (4.2.35) *are fulfilled, yet for $\lambda = 1$,*

$$\limsup_{m_\lambda \to \infty} \|\overline{S}_m(\cdot, f, M) - \overline{f}_{M,m}(\cdot)\|_C > 0.$$

Proof. The assertion a) follows (for $p = +\infty$ and $\alpha = 1$) from Theorem 3.1.1. Using an example of Fejér [3] and the assertion c) of Theorem 4.2.5, we make sure that b) is likewise true. Furthermore (as while proving Theorem 4.2.3), we have

$$\begin{aligned}\|\overline{S}_m(\cdot,f,M)-\overline{f}_{M,m}(\cdot)\|_C \ge \\ \ge \|\overline{S}_m(\cdot,f,M)-\overline{f}_M(\cdot)\|_C - \|t_m^1(\cdot,f,M)-\overline{f}_M(\cdot)\|_C - \\ -\|t_m^1(\cdot,f,M)-\overline{f}_{M,m}(\cdot)\|_C.\end{aligned} \tag{4.2.47}$$

By Theorem 3.1.1, we see that

$$\|t_m^1(\cdot,f,M)-\overline{f}_m(\cdot)\|_C \to 0 \quad \text{as } m\to\infty. \tag{4.2.48}$$

Applying Theorem 3.2.1, it is not difficult to verify that

$$\|t_m^1(\cdot,f,M)-\overline{f}_{M,m}(\cdot)\|_C \to 0 \quad \text{as } m\to\infty. \tag{4.2.49}$$

Therefore, taking note of the corresponding assertion of Theorem 4.2.5 and (4.2.27)–(4.2.29), we make sure that c) is valid too. ■

Using Theorem 3.2.1 (for $p=\infty$ and $\alpha=0$), we can verify that it holds

Theorem 4.2.8. *Let $f\in C(T^n)$ and*

$$\omega(\delta,f)_C\Big(\log\frac{2\pi}{\delta}\Big)^n \to 0 \quad \textit{as } \delta\to 0+. \tag{4.2.50}$$

Then for any $B\subset M$ $(B\neq\varnothing)$, we have

$$\limsup_{m\to\infty}\|\overline{S}_m(\cdot,f,B)-\overline{f}_{B,m}(\cdot)\|_C = 0.$$

Remark 4.2.4. If $f\in C(T^n)$, then condition (4.2.50), in general, does not ensure the continuity of $\overline{f}_M$.

Using Theorem 3.3.3 (for $p=+\infty$), we arrive at

Theorem 4.2.9. *Let $\alpha\in]0,1[$, $k\in[1,n]$ and $B\subset M$ with $\overline{k}(B)=k$. Suppose that $f\in C(T^n)$ and*

$$\omega(\delta,f)_C\delta^{-\alpha}\Big(\log\frac{2\pi}{\delta}\Big)^{n-k} \to 0 \quad \textit{as } \delta\to 0+. \tag{4.2.51}$$

If for $\beta=(\beta_1,\beta_2,\ldots,\beta_n)$ the conditions

$$\beta_i\in]0,\alpha[\ \textit{ for } i\in B,\quad \beta_j=0 \ \textit{ for } j\in B',\quad \beta_1+\beta_2+\cdots+\beta_n=\alpha$$

are fulfilled, then

$$\|\sigma_m^{-\beta}(f)-f\|_C\to 0 \quad \textit{as } m\to\infty.$$

We should add that the condition (4.2.51) is essential for Theorem 4.2.9 to be true. An analogous theorem is also true for the space $L(T^n)$.

Remark 4.2.5. As in the case $n=1$ (see, e.g., Zygmund [15, Ch. VIII]), the assertion a) of Theorem 4.2.5 can be improved. To be more precise, one can attain that $\sigma_n[f]$ be divergent on a dense subset E of $\mathbb{R}^n$ with $\mu_n(E)=0$.

Remark 4.2.6. The results of this chapter can be used in various fields. In particular, they are applicable to the theory of functions of several complex variables. In [38], we pointed out some assertions in this direction.

Remark 4.2.7. In Chapters 1–4 of the second part of the book, various theorems have been established concerning the almost everywhere existence and integrability on T^n of conjugate functions of several variables. The questions connected with almost everywhere summability by n-fold Cesàro (of positive order) and Poisson–Abel methods of the series $\overline{\sigma}_n[f, B]$, $B \subset M$ $(B \neq \varnothing)$ in the spaces $L^p(T^n)$, $p \in]0, +\infty]$, have also been investigated, as well as some regularities connected with the dimension. Some of the above results remain true for more general n-fold singular integrals and n-fold linear means of the series $\overline{\sigma}_n[f, B]$. But we would not dwell on this since they do not reveal new regularities connected with the dimension.

4.3. Unsolved Problems

4.3.1. Let $E \subset T^n$ $(n \geq 1)$ be an arbitrary set with $\mu_n(E) = 0$. Is there a function $f \in C(T^n)$ such that

$$\omega_i(\delta, f)_C \left(\log \frac{2\pi}{\delta} \right)^n \leq A(f, n) \quad \text{for } \delta \in]0, \pi], \; 1 \leq i \leq n, \tag{4.3.1}$$

all the functions $\overline{f}_B$ belong to $C(T^n)$, and

$$\omega_i(\delta, \overline{f}_B)_C \left(\log \frac{2\pi}{\delta} \right)^n \leq A(f, n) \quad \text{for } \delta \in]0, \pi], \; 1 \leq i \leq n, \tag{4.3.2}$$

yet the series $\sigma_n[f]$ diverges (λ-diverges for some $\lambda \in [1, +\infty[$) on E?

4.3.2. Let $E \subset T^n$ $(n \geq 1)$ be a set of type G_δ with $\mu_n(E) = 0$. Is there a function $f \in C(T^n)$ such that all $\overline{f}_B$ belong to $C(T^n)$, (4.3.1) and (4.3.2) are fulfilled, yet the series $\sigma_n[f]$ diverges (λ-diverges for some $\lambda \in [1, +\infty[$) on E and converges (λ-converges) on $T^n \setminus E$?

We should note that in Problems 4.3.1 and 4.3.2, the phrases "λ-diverges", "λ-converges" are meaningfull for $n \geq 2$.

In connection with Problems 4.3.1 and 4.3.2, the following should be added.

Ul'yanov posed the question: is it true that for any set $E \subset T$ with $\mu_1(E) = 0$, there is a function $f \in C(T)$ whose Fourier series unboundedly diverges on E? The answer is contained in Buzdalin [1].

The following problem is also due to Ul'yanov. Let $E \subset T$ be an arbitrary set of type C_δ with $\mu_1(E) = 0$. Is there a function $f \in C(T)$ such that the series $\sigma[f]$ diverges unboundedly on E, and converges on $T \setminus E$? As far as we know, the problem is still unsolved. In the case where E is closed, the answer is contained in Buzdalin [2].

4.3.3. Let $\alpha \in]0, n[$. Is there a function $f \in L(\log^+ L)^\alpha(T^n)$ such that the series $\sigma_n[f]$ λ-diverges (for $\lambda = 1$) on a set $E \subset T^n$ with $\mu_n(E) > 0$ (almost everywhere, everywhere), and converges in the metric of $L(T^n)$, $n \geq 2$?

4.3.4. Let $\alpha \in]0, n[$. Is there a function $f \in L(T^n)$ such that

$$\omega_i(\delta, f)_1 \left(\log \frac{2\pi}{\delta} \right)^\alpha \leq A(f, n) \quad \text{for } \delta \in]0, \pi], \ 1 \leq i \leq n,$$

the series $\sigma_n[f]$ λ-deverges (for $\lambda = 1$) on a set $E \subset T$ with $\mu_n(E) > 0$ (almost everywhere, everywhere), and

$$\lim_{m_\lambda \to \infty} \|S_m(f) - f\|_1 = 0\,?$$

4.3.5. Let $f : T^n \to [-\infty, +\infty]$ be measurable and almost everywhere finite. Suppose that $\beta \in]-1, 0[^n$, $p \in]0, +\infty[$, and $\varepsilon \in]0, 1[$. Is there a function $g \in C(T^n)$ such that

$$\mu_n \Big[\big\{ x \in T^n : f(x) \neq g(x) \big\} \Big] < \varepsilon,$$

and the series $\sigma_n[g]$ is (C, β)-summable in the metric of $L^p(T^n)$?

Some of the unsolved problems presented here were discussed earlier in this chapter.

In Chapter 4 of the first part (see Section 3.3), some results have been given concerning the behavior of the series $\sigma[f]$ and $\overline{\sigma}[f]$ of even and odd functions. In the multidimensional case, a little is done in this direction. In the work [13, Ch. III], we presented some assertions in the case $n = 2$, but they give no important information about $\sigma_n[f]$ and $\overline{\sigma}_n[f, B]$.

Remark 4.3.1. Problems can also be posed on other types of convergence and summability of $\sigma_n[f]$ and $\overline{\sigma}_n[f, B]$ (including spherical convergence).

CHAPTER V

Summability of Series $\sigma_2[\mathrm{f}]$ and $\overline{\sigma}_2[\mathrm{f}, \mathrm{B}]$ by a Method of the Marcinkiewicz Type

5.1. Summability Almost Everywhere of Series $\sigma_2[f]$ and $\overline{\sigma}_2[f, B]$ by a Method of the Marcinkiewicz Type

This chapter basically deals with the case $n = 2$.

It will be assumed that $x = (x_1, x_2)$, $y = (y_1, y_2), \ldots, s = (s_1, s_2)$; $M = \{1, 2\} \equiv B_3$, $B_1 \equiv \{1\}$, $B_2 \equiv \{2\}$, and m is a natural number.

Let $\alpha \in]-1, 0[\cup]0, +\infty[$. If $f \in L(T^2)$, then by $S_{j,j}(x, f)$ and $\overline{S}_{j,j}(x, f, B_l)$ ($j = 0, 1, \ldots,$ $l = 1, 2, 3$) we denote the quadratic sums of the series $\sigma_2[f]$ and $\overline{\sigma}_2[f, B]$, respectively ($l = 1, 2, 3$). We will call the (C, α)-means of these sums, i.e., the expressions

$$\sigma_m^\alpha(x, f) = (A_m^\alpha)^{-1} \sum_{j=0}^{m} A_{m-j}^{\alpha-1} S_{j,j}(x, f) \tag{5.1.1}$$

and

$$t_m^\alpha(x, f, B_l) = (A_m^\alpha)^{-1} \sum_{j=0}^{m} A_{m-j}^{\alpha-1} \overline{S}_{j,j}(x, f, B_l) \quad (l = 1, 2, 3) \tag{5.1.2}$$

Marcinkiewicz type means of $\sigma_2[f]$ and $\overline{\sigma}_2[f, B]$, respectively.

In 1939, Marcinkiewicz [3] proved that if $f \in L(\log^+ L)(T^2)$, then for $\alpha = 1$, we have

$$\lim_{m \to \infty} \sigma_m^\alpha(x, f) = f(x)$$

almost everywhere.

In the works [4], [11], [13, Ch. IV], we set forth the assertions generalizing Marcinkiewicz's result in two senses: weakening of the condition and generalizing of the summation method. In [22], the assertions are given concerning the (C, α)-summability of $\overline{\sigma}_2[f, B_l]$ ($l = 1, 2, 3$) for $\alpha \in]0, +\infty[$.

Later in this section, we present the results from the above mentioned works of the author.

In what follows, we will need some auxiliary results.

Lemma 5.1.1. *Let* $\alpha \in]-1,0[\cup]0,+\infty[$ *and*

$$k_m^\alpha(x) = (A_m^\alpha)^{-1} \sum_{j=0}^{m} A_{m-j}^{\alpha-1} D_j(x_1) D_j(x_2). \tag{5.1.3}$$

Then

$$\|k_m^\alpha\|_C \le A(\alpha) m^2, \quad \alpha \in]0,+\infty[, \tag{5.1.4}$$

$$\|k_m^\alpha\|_C \le A(\alpha) m^{2-\alpha}, \quad \alpha \in]-1,0[. \tag{5.1.5}$$

Proof. Since

$$\|D_j\|_C \le Aj \quad \text{for } j \in \mathbb{N},$$

by (1; 2.2.3) and (5.1.3) we have

$$\|k_m^\alpha\|_C \le A (A_m^\alpha)^{-1} \sum_{j=0}^{m} |A_{m-j}^{\alpha-1}| (j+1)^2 \le$$

$$\le \begin{cases} A(\alpha) n^2 & \text{for } \alpha \in]0,+\infty[, \\ A(\alpha) n^{2-\alpha} & \text{for } \alpha \in]-1,0[. \end{cases}$$

Therefore (5.1.4) and (5.1.5) hold. ■

Using the inequalities

$$|D_j(x_l)| \le A(|x_l|)^{-1} \quad \text{for } |x_l| \in]0,\pi] \ (l=1,2),$$

we can quite analogously verify

Lemma 5.1.2. *If* $k_m^\alpha(x)$ *is defined by* (5.1.3), *then*

$$|k_m^\alpha(x)| \le \begin{cases} A(\alpha)(|x_1 x_2|)^{-1} & \text{for } \alpha \in]0,+\infty[, \\ A(\alpha) m^{-\alpha} (|x_1 x_2|)^{-1} & \text{for } \alpha \in]-1,0[, \ |x_l| \in]0,\pi[\ (l=1,2). \end{cases}$$

Lemma 5.1.3. *Let* $x_l \in]0,\pi[$ $(l=1,2)$, $\eta \in]0,2\pi[$, *and* $\alpha \in]-1,0[\cup]0,+\infty[$. *If* $0 < x_1 \pm x_2 \le 2\pi - \eta$, *then*

$$k_m^\alpha(x) = \varphi_m^\alpha(x) + r_m^\alpha(x), \tag{5.1.6}$$

where

$$\varphi_m^\alpha(x) = \left(8A_m^\alpha \sin\frac{x_1}{2} \sin\frac{x_2}{2}\right)^{-1} \left\{ \left(2\sin\frac{x_1 - x_2}{2}\right)^{-\alpha} \times \right.$$
$$\times \cos\left[\left(m + \frac{1}{2} + \frac{\alpha}{2}\right)(x_1 - x_2) - \frac{\pi\alpha}{2}\right] - \left(2\sin\frac{x_1 + x_2}{2}\right)^{-\alpha} \times$$
$$\left. \times \cos\left[\left(m + \frac{1}{2} + \frac{\alpha}{2}\right)(x_1 + x_2) - \frac{\pi\alpha}{2}\right]\right\}, \tag{5.1.7}$$

$$|r_m^\alpha(x)| \le A(\alpha)(m^2 x_1 x_2)^{-1} \left[(x_1 - x_2)^{-2} + (2\pi - x_1 - x_2)^{-2}\right]. \tag{5.1.8}$$

Proof. We have

$$k_m^\alpha(x) = \left(8A_m^\alpha \sin\frac{x_1}{2}\sin\frac{x_2}{2}\right)^{-1} \operatorname{Re}\left\{\sum_{j=0}^{m} A_{m-j}^{\alpha-1}\left[e^{i(j+\frac{1}{2})(x_1-x_2)} - \right.\right.$$

$$\left.\left. - e^{i(j+\frac{1}{2})(x_1+x_2)}\right]\right\} = \left(8A_m^\alpha \sin\frac{x_1}{2}\sin\frac{x_2}{2}\right)^{-1} \times$$

$$\times \operatorname{Re}\left\{e^{i(m+\frac{1}{2})(x_1-x_2)}\sum_{j=0}^{m} A_j^{\alpha-1}e^{-ij(x_1-x_2)} - \right.$$

$$\left. - e^{i(m+\frac{1}{2})(x_1+x_2)}\sum_{j=0}^{m} A_j^{\alpha-1}e^{-ij(x_1+x_2)}\right\} \equiv$$

$$\equiv \psi_m^\alpha(x) + r_m^\alpha(x), \tag{5.1.9}$$

where

$$\varphi_m^\alpha(x) = \left(8A_m^\alpha \sin\frac{x_1}{2}\sin\frac{x_2}{2}\right)^{-1} \operatorname{Re}\left\{e^{i(m+\frac{1}{2})(x_1-x_2)} \times \right.$$

$$\left. \times \left[1-e^{-i(x_1-x_2)}\right]^{-\alpha} - [e^{i(m+\frac{1}{2})(x_1+x_2)}\left[1-e^{-i(x_1+x_2)}\right]^{-\alpha}\right\} =$$

$$= \left(8A_m^\alpha \sin\frac{x_1}{2}\sin\frac{x_2}{2}\right)^{-1}\left\{\left(2\sin\frac{x_1-x_2}{2}\right)^{-\alpha} \times \right.$$

$$\times \cos\left[\left(m+\frac{1}{2}+\frac{\alpha}{2}\right)(x_1-x_2) - \frac{\pi\alpha}{2}\right] - \left(2\sin\frac{x_1+x_2}{2}\right)^{-\alpha} \times$$

$$\left. \times \cos\left[\left(m+\frac{1}{2}+\frac{\alpha}{2}\right)(x_1+x_2) - \frac{\pi\alpha}{2}\right]\right\}, \tag{5.1.10}$$

$$r_m^\alpha(x) = \left(8A_m^\alpha \sin\frac{x_1}{2}\sin\frac{x_2}{2}\right)^{-1} \times$$

$$\times \operatorname{Re}\left\{e^{i(m+\frac{1}{2})(x_1+x_2)}\sum_{j=m+1}^{\infty} A_j^{\alpha-1}e^{-ij(x_1+x_2)} - \right.$$

$$\left. - e^{i(m+\frac{1}{2})(x_1-x_2)}\sum_{j=m+1}^{\infty} A_j^{\alpha-1}e^{-ij(x_1-x_2)}\right\}. \tag{5.1.11}$$

Using twice Abel's transform, we easily find

$$e^{i(m+\frac{1}{2})(x_1+x_2)}\sum_{j=m+1}^{\infty} A_j^{\alpha-1}e^{-ij(x_1+x_2)} =$$

$$= A_{m+1}^{\alpha-1}e^{-\frac{1}{2}(x_1+x_2)}\left[1-e^{-i(x_1+x_2)}\right]^{-1} +$$

$$+ A_{m+2}^{\alpha-2}e^{-\frac{3}{2}(x_1+x_2)}\left[1-e^{-i(x_1+x_2)}\right]^{-2} +$$

$$+ \left[\sum_{j=m+1}^{\infty} A_{m+2+j}^{\alpha-2}e^{-i(j+\frac{3}{2})(x_1+x_2)}\right]\left[1-e^{-i(x_1+x_2)}\right]^{-2}, \tag{5.1.12}$$

$$e^{i(m+\frac{1}{2})(x_1-x_2)}\sum_{j=m+1}^{\infty}A_j^{\alpha-1}e^{-ij(x_1-x_2)}=$$

$$=A_{m+1}^{\alpha-1}e^{-\frac{1}{2}(x_1-x_2)}\Big[1-e^{-i(x_1-x_2)}\Big]^{-1}+$$

$$+A_{m+2}^{\alpha-2}e^{-\frac{3}{2}(x_1-x_2)}\Big[1-e^{-i(x_1-x_2)}\Big]^{-2}+$$

$$+\Big[\sum_{j=m+1}^{\infty}A_{m+2+j}^{\alpha-3}e^{-i(j+\frac{3}{2})(x_1-x_2)}\Big]\Big[1-e^{-i(x_1-x_2)}\Big]^{-2}. \tag{5.1.13}$$

Then, by (5.1.11), (5.1.12), and (5.1.13), we get

$$r_m^\alpha(x)=\Big(8A_m^\alpha\sin\frac{x_1}{2}\sin\frac{x_2}{2}\Big)^{-1}\Big\{\Big(2\sin\frac{x_1+x_2}{2}\Big)^{-2}\times$$

$$\times\Big[A_{m+2}^{\alpha-2}\cos\frac{x_1+x_2}{2}+\sum_{j=1}^{\infty}A_{m+2+j}^{\alpha-3}\cos\Big(j+\frac{1}{2}\Big)\Big(x_1+x_2\Big)\Big]-$$

$$-\Big(2\sin\frac{x_1-x_2}{2}\Big)^{-2}\Big[A_{m+2}^{\alpha-2}\cos\frac{x_1-x_2}{2}+$$

$$+\sum_{j=1}^{\infty}A_{m+2+j}^{\alpha-3}\cos\Big(j+\frac{1}{2}\Big)\Big(x_1-x_2\Big)\Big]\Big\}. \tag{5.1.14}$$

Hence, by virtue of (1; 2.2.5), it follows (5.1.12). Then, on the basis of (5.1.9), (5.1.10), and (5.1.11), we conclude that lemma is true. ■

Corollary 5.1.1. *Let $k_m^\alpha(x)$ be defined by* (5.1.3). *Suppose that $x_l\in]0,\pi[$ $(l=1,2)$, $\eta\in]m^{-1},2\pi[$, and $\alpha\in[-1,0[\cup]0,1[$. If $m^{-1}\le x_1\pm x_2\le 2\pi-\eta$, then*

$$|k_m^\alpha(x)|\le A(\alpha)(m^\alpha x_1x_2)^{-1}\Big[(x_1-x_2)^{-\alpha}+(2\pi-x_1-x_2)^\alpha\Big]. \tag{5.1.15}$$

Lemma 5.1.4. *Suppose that $x_1\in[2m^{-1},\pi]$ and $x_2\in]0,m^{-1}]$. If $|\alpha|\in]0,1[$, then*

$$|k_m^\alpha(x)|\le A(\alpha)m^{1-\alpha}(x_1-x_2)^{-1-\alpha}. \tag{5.1.16}$$

Proof. Taking note of (5.1.7) and the conditions of the lemma, and using the Lagrange formula, we get

$$|\varphi_m^\alpha(x)|\le A(\alpha)\Big[m^{1-\alpha}(x_1-x_2)^{-1-\alpha}+m^{-\alpha}(x_1-x_2)^{-2-\alpha}\Big]. \tag{5.1.17}$$

Analogously, by (5.1.14), we find

$$|r_m^\alpha(x)|\le A(\alpha)\Big[m^{1-\alpha}(x_1-x_2)^{-1-\alpha}+$$

$$+m^{-\alpha}(x_1-x_2)^{-2-\alpha}+m^{-1}(x_1-x_2)^{-3}\Big]. \tag{5.1.18}$$

Thus, on the basis of (5.1.6), (5.1.17), and (5.1.18), the inequality (5.1.16) is valid. ■

Put

$$k_m^{(1,\alpha)}(x)=(A_m^\alpha)^{-1}\sum_{j=0}^{m}A_{m-j}^{\alpha-1}\overline{D}_j(x_1)D_j(x_2), \tag{5.1.19}$$

$$k_m^{(2,\alpha)}(x) = (A_m^\alpha)^{-1}\sum_{j=0}^{m} A_{m-j}^{\alpha-1} D_j(x_1)\overline{D}_j(x_2), \tag{5.1.20}$$

$$k_m^{(3,\alpha)}(x) = (A_m^\alpha)^{-1}\sum_{j=0}^{m} A_{m-j}^{\alpha-1} \overline{D}_j(x_1)\overline{D}_j(x_2), \tag{5.1.21}$$

where $\overline{D}_j$ is the conjugate Dirichlet kernel, $j \in \mathbb{N}_0$.

It is not difficult to observe that the assertions of Lemmas 5.1.1 and 5.1.2 are likewise true for the expression $k_m^{(\nu,\alpha)}$ $(\nu = 1,2,3)$. To be more precise, it holds

Lemma 5.1.5. a) *Let* $\alpha \in]-1,0[\cup]0,+\infty[$. *Then*

$$\|k_m^{(\nu,\alpha)}\|_C \le A(\alpha)m^2 \quad \textit{for } \alpha \in]0,+\infty[\ (\nu = 1,2,3),$$

$$\|k_m^{(\nu,\alpha)}\|_C \le A(\alpha)m^{2-\alpha} \quad \textit{for } \alpha \in]-1,0[\ (\nu = 1,2,3).$$

b) *Let* $x_l \in]0,\pi]$ $(l = 1,2)$ *and* $\alpha \in]-1,0[\cup]0,+\infty[$. *Then*

$$|k_m^{(\nu,\alpha)}(x)| \le \begin{cases} A(\alpha)(|x_1x_2|)^{-1} & \textit{for } \alpha \in]0,+\infty[, \\ A(\alpha)m^{-\alpha}(|x_1x_2|)^{-1} & \textit{for } \alpha \in]-1,0[\quad (\nu = 1,2,3). \end{cases}$$

Lemma 5.1.6. *Let* $x_l \in]0,\pi]$ $(l = 1,2)$, $\eta \in]0,2\pi[$, *and* $\alpha \in]-1,0[\cup]0,+\infty[$. *Let, moreover,* $1 < x_1 \pm x_2 \le 2\pi - \eta$. *Then*

$$k_m^{(2,\alpha)}(x) = \frac{1}{2}\operatorname{ctg}\frac{x_2}{2}\, k_m^\alpha(x_1) + \varphi_m^{(2,\alpha)}(x) + r_m^{(2,\alpha)}(x), \tag{5.1.22}$$

$$\begin{aligned} k_m^{(3,\alpha)}(x) &= \frac{1}{4}\operatorname{ctg}\frac{x_1}{2}\operatorname{ctg}\frac{x_2}{2} + \frac{1}{2}\operatorname{ctg}\frac{x_1}{2}\, H_m^\alpha(x_2) + \\ &+ \frac{1}{2}\operatorname{ctg}\frac{x_2}{2}\, H_m^\alpha(x_1) + \psi_m^{(3,\alpha)}(x) + r_m^{(3,\alpha)}(x), \end{aligned} \tag{5.1.23}$$

where K_m^α *is the Cesàro kernel, and*

$$H_m^\alpha(x_l) = -(A_m^\alpha)^{-1}\sum_{j=0}^{m} A_{m-j}^{\alpha-1}\left(2\sin\frac{x_l}{2}\right)^{-1}\cos\left(j+\frac{1}{2}\right)x_l \quad (l = 1,2),$$

$$\begin{aligned} \varphi_m^{(2,\alpha)}(x) &= \left(8A_m^\alpha \sin\frac{x_1}{2}\sin\frac{x_2}{2}\right)^{-1}\left\{\left(2\sin\frac{x_1+x_2}{2}\right)^{-\alpha}\times\right. \\ &\times \sin\left[\left(m+\frac{1}{2}+\frac{\alpha}{2}\right)(x_1+x_2) - \frac{\alpha\pi}{2}\right] - \left(2\sin\frac{x_1-x_2}{2}\right)^{-\alpha}\times \\ &\left.\times \sin\left[\left(m+\frac{1}{2}+\frac{\alpha}{2}\right)(x_1-x_2) - \frac{\alpha\pi}{2}\right]\right\}, \end{aligned} \tag{5.1.24}$$

$$\begin{aligned} \varphi_m^{(3,\alpha)}(x) &= \left(8A_m^\alpha \sin\frac{x_1}{2}\sin\frac{x_2}{2}\right)^{-1}\left\{\left(2\sin\frac{x_1+x_2}{2}\right)^{-\alpha}\times\right. \\ &\times \cos\left[\left(m+\frac{1}{2}+\frac{\alpha}{2}\right)(x_1+x_2) - \frac{\alpha\pi}{2}\right] + \left(2\sin\frac{x_1-x_2}{2}\right)^{-\alpha}\times \\ &\left.\times \cos\left[\left(m+\frac{1}{2}+\frac{\alpha}{2}\right)(x_1-x_2) - \frac{\alpha\pi}{2}\right]\right\}, \end{aligned} \tag{5.1.25}$$

$$|r_m^{(2,\alpha)}(x)| \le A(\alpha)(mx_1x_2)^{-1}\left[(x_1-x_2)^{-1} + (2\pi - x_1 - x_2)^{-1}\right], \tag{5.1.26}$$

$$|r_m^{(3,\alpha)}(x)| \le A(\alpha)(m^2 x_1 x_2)^{-1}\Big[(x_1-x_2)^2+(2\pi-x_1-x_2)^2\Big]. \tag{5.1.27}$$

The lemma can be verified by the same method as Lemma 5.1.3. We should only add that according to (5.1.29) and (5.1.21), the functions $k_m^{(2,\alpha)}(x)$ and $k_m^{(3,\alpha)}$ can be represented as

$$\begin{aligned}
k_m^{(2,\alpha)}(x) &= \frac{1}{2}\operatorname{ctg}\frac{x_2}{2}k_m^\alpha(x_1) - \left(4A_m^\alpha \sin\frac{x_1}{2}\sin\frac{x_2}{2}\right)^{-1}\times\\
&\times \sum_{j=0}^{m} A_{m-j}^{\alpha-1}\sin\left(j+\frac{1}{2}\right)x_1\cos\left(j+\frac{1}{2}\right)x_2 =\\
&= \frac{1}{2}\operatorname{ctg}\frac{x_2}{2}k_m^\alpha(x_1) + \left(8A_m^\alpha \sin\frac{x_1}{2}\sin\frac{x_2}{2}\right)^{-1}\times\\
&\times \operatorname{Im}\left\{\sum_{j=0}^{m} A_{m-j}^{\alpha-1}\left[e^{-i(j+\frac{1}{2}(x_1+x_2)}-e^{-i(j+\frac{1}{2}(x_1-x_2)}\right]\right\},\\
k_m^{(3,\alpha)}(x) &= \frac{1}{4}\operatorname{ctg}\frac{x_1}{2}\operatorname{ctg}\frac{x_2}{2}+\frac{1}{2}\operatorname{ctg}\frac{x_1}{2}H_m^\alpha(x_2)+\\
&+\frac{1}{2}\operatorname{ctg}\frac{x_2}{2}H_m^\alpha(x_1)+\left(4A_m^\alpha \sin\frac{x_1}{2}\sin\frac{x_2}{2}\right)^{-1}\times\\
&\times \sum_{i=0}^{m} A_{m-j}^{\alpha-1}\cos\left(j+\frac{1}{2}\right)x_1\cos\left(j+\frac{1}{2}\right)x_2 =\\
&= \frac{1}{4}\operatorname{ctg}\frac{x_1}{2}\operatorname{ctg}\frac{x_2}{2}+\frac{1}{2}\operatorname{ctg}\frac{x_1}{2}H_m^\alpha(x_2)+\\
&+\frac{1}{2}\operatorname{ctg}\frac{x_2}{2}H_m^\alpha(x_1)+\left(8A_m^\alpha \sin\frac{x_1}{2}\sin\frac{x_2}{2}\right)^{-1}\times\\
&\times \operatorname{Re}\left\{\sum_{j=0}^{m} A_{m-j}^{\alpha-1}\left[e^{-i(j+\frac{1}{2}(x_1+x_2)}-e^{-i(j+\frac{1}{2}(x_1-x_2)}\right]\right\}.
\end{aligned}$$

Lemma 5.1.6 implies

Lemma 5.1.7. *Let $x_l \in [0,\pi]$ $(l=1,2)$, $\eta \in]0,2\pi[$, and $\alpha \in]-1,0[\cup]0,1[$. If $0 < x_1 \pm x_2 \le 2\pi-\eta$, then*

$$k_m^{(1,\alpha)}(x) = \frac{1}{2}\operatorname{ctg}\frac{x_1}{2}\,k_m^\alpha(x_2)+\varphi_m^{(1,\alpha)}(x)+r_m^{(1,\alpha)}, \tag{5.1.28}$$

where

$$\begin{aligned}
\varphi_m^{(1,\alpha)}(x) &= \left(8A_m^\alpha \sin\frac{x_1}{2}\sin\frac{x_2}{2}\right)^{-1}\left\{\left(2\sin\frac{x_1+x_2}{2}\right)^{-\alpha}\times\right.\\
&\times\sin\left[\left(m+\frac{1}{2}+\frac{\alpha}{2}\right)(x_1+x_2)-\frac{\alpha\pi}{2}\right]-\left(2\sin\frac{x_1-x_2}{2}\right)^{-\alpha}\times\\
&\left.\times\sin\left[\left(m+\frac{1}{2}+\frac{\alpha}{2}\right)(x_1-x_2)-\frac{\alpha\pi}{2}\right]\right\},
\end{aligned} \tag{5.1.29}$$

$$|r_m^{(1,\alpha)}(x)| \le A(\alpha)(m x_1 x_2)^{-1}\Big[(x_1-x_2)^{-1}+(2\pi-x_1-x_2)^{-1}\Big]. \tag{5.1.30}$$

Lemma 5.1.8. *Let $x_2 \in [2m^{-1}, \pi]$ and $x_1 \in]0, m^{-1}]$. If $|\alpha| \in]0,1[$, then*

$$(A_m^\alpha)^{-1}\left|\sum_{j=0}^{m} A_{m-j}^{\alpha-1}\overline{D}_j(x_1)\Big(2\sin\frac{x_2}{2}\Big)^{-1}\cos\Big(j+\frac{1}{2}\Big)x_2\right| \le$$

$$\le A(\alpha)m^{1-\alpha}(x_2-x_1)^{-1-\alpha}. \tag{5.1.31}$$

The lemma can be proved by the same method as Lemma 5.1.4.

Lemma 5.1.9. *Let functions $h_l : [0,+\infty[\to [0,+\infty[$ $(l=1,2)$ be continuous, strictly increasing up to $+\infty$, and let $h_l(0)=0$ $(l=1,2)$. Suppose that T_1^2 is the square $|x_l| \le 3\pi$ $(l=1,2)$ and $f \in L(T_1^2)$. For any point $x \in T^2$, put*

$$f_*(x) = \sup_{\tau\le\tau_0}\frac{1}{4h_1(\tau)h_2(\tau)}\int_{-h_1(\tau)}^{h_1(\tau)}\int_{-h_1(\tau)}^{h_2(\tau)}|f(x+s)|\,ds. \tag{5.1.32}$$

Let, moreover, $f_*^{\alpha,\beta}$ denote the function f_* with αh_1 and βh_2 instead of h_1 and h_2, respectively, where α and β are positive numbers. If $\lambda \in]0,+\infty[$,

$$f_\lambda^*(x) = \sup_{i,j\in\mathbb{N}_0}\left\{f_*^{2^i,2^j}(x)2^{-\lambda(i+j)}\right\} \tag{5.1.33}$$

and

$$E_\lambda(\xi) = \left\{x : f_\lambda^*(x) > \xi > 0,\ \ x\in T^2\right\},$$

then

$$\mu_2[E_\lambda(\xi)] \le A_\xi^{-1}\int_{T_1^2}|f(x)|\,dx. \tag{5.1.34}$$

The lemma was proved by Marcinkiewicz and Zygmund [1] in the case $\lambda = \frac{1}{2}$ (see also Zygmund [16,Ch. XVII]), but the method of the proof works for any $\lambda \in]0,+\infty[$ (see also Zygmund [10]).

Lemma 5.1.10. *Suppose that h_l $(l=1,2)$ and f satisfy the conditions of Lemma* 5.1.9, *and for any point $x \in T^2$, put*

$$\varphi_*(x) = \sup_{\tau\le i_0}\frac{1}{4h_1(\tau)h_2(\tau)}\int_{-h_1(\tau)}^{h_1(\tau)}\int_{s_1-h_1(\tau)}^{s_1+h_2(\tau)}|f(x+s)|\,ds.$$

Let $\varphi_*^{\alpha,\beta}$ denote the function φ_* with αh_1 and βh_2 instead of h_1 and h_2, respectively, where α and β are positive numbers. If $\lambda \in]0,+\infty[$,

$$\varphi_\lambda^*(x) = \sup_{i,j\in\mathbb{N}_0}\left\{f_*^{2^i,2^j}(x)2^{-\lambda(i+j)}\right\}, \tag{5.1.35}$$

and

$$E_\lambda^{(1)}(\xi) = \left\{x : \varphi_\lambda^*(x) > \xi > 0,\ \ x\in T^2\right\},$$

then

$$\mu_2[E_\lambda^{(1)}(\xi)] \le A\xi^{-1} \int_{T_1^2} |f(x)|\,dx. \tag{5.1.36}$$

The lemma is proved quite analogously to Lemma 5.1.9.

Theorem 5.1.1. *Let* $f \in L(T^2)$ *and* $\alpha \in]0+\infty[$. *Then*

$$\lim_{m\to\infty} \sigma_m^\alpha(x,f) = f(x)$$

almost everywhere.

Proof. It suffices to confine ourselves to the case $\alpha \in]0,1[$. By (5.1.1) and (5.1.3), we obtain

$$\sigma_m^\alpha(x,f) = \frac{1}{\pi^2}\int_0^\pi\int_0^\pi f(x+s)k_m^\alpha(s)\,ds =$$

$$= \frac{1}{\pi^2}\int_0^\pi\int_0^\pi \Big[f(x_2+s_1,x_2+s_2) + f(x_1-s_1,x_2+s_2) +$$

$$+ f(x_1+s_1,x_2-s_2) + f(x_1-s_1,x_2-s_2)\Big]\,k_m^\alpha(s)\,ds. \tag{5.1.37}$$

Define

$$V_m(x) \equiv V_m(x,\alpha) = \frac{1}{\pi^2}\int_0^\pi\int_0^\pi f(x+s)k_m^\alpha(s)\,ds, \tag{5.1.38}$$

and consider for $m \in \mathbb{N}$ the following subsets of $[0,\pi]^2$:

$$U_1 = U_1(m) = \Big\{s : s_l \in [0,2m^{-1}];\ l=1,2\Big\},$$

$$U_2 = U_2(m) = \Big\{s : s_1 \in \Big[2m^{-1},\frac{3\pi}{4}\Big],\ s_2 \in [m^{-1}, s_1 - m^{-1}]\Big\},$$

$$U_3 = U_3(m) = \Big\{s : s_1 \in [2m^{-1},\pi];\ s_2 \in [0,m^{-1}]\Big\},$$

$$U_4 = U_4(m) = \Big\{s : s_1 \in [m^{-1}, s_2 - m^{-1}]\ s_2 \in \Big[2m^{-1},\frac{3\pi}{4}\Big]\Big\},$$

$$U_5 = U_5(m) = \Big\{s : s_1 \in [0,m^{-1}];\ s_2 \in [2m^{-1},\pi]\Big\},$$

$$U_6 = U_6(m) = \Big\{s : s_1 \in \Big[2m^{-1},\frac{3\pi}{4}\Big],\ s_2 \in [s_1 - m^{-1}, s_1]\Big\},$$

$$U_7 = U_7(m) = \Big\{s : s_1 \in [s_2 - m^{-1}, s_2]\ s_2 \in \Big[2m^{-1},\frac{3\pi}{4}\Big]\Big\},$$

$$U_8 = U_8(m) = \Big\{s : s_1 \in \Big[\frac{3\pi}{4},\pi\Big],\ s_2 \in [m^{-1}, s_1 - m^{-1}]\Big\},$$

$$U_9 = U_9(m) = \Big\{s : s_1 \in \Big[\frac{3\pi}{4},\pi\Big],\ s_2 \in [s_1 - m^{-1}, s_1]\Big\},$$

$$U_{10} = U_{10}(m) = \Big\{s : s_1 \in [m^{-1}, s_2 - m^{-1}]\ s_2 \in \Big[\frac{3\pi}{4},\pi\Big]\Big\},$$

$$U_{11} = U_{11}(m) = \left\{s : s_1 \in [s_2 - m^{-1}, s_2] \;\; s_2 \in \left[\frac{3\pi}{4}, \pi\right]\right\}.$$

Then, by (5.1.38), we will have

$$|V_m(x)| \le \sum_{\nu=1}^{11} \frac{1}{\pi^2}\left|\int_{U_0} f(x+s)k_m^\alpha(s)\,ds\right|. \tag{5.1.39}$$

Put

$$V_m^{(\nu)}(x) \equiv V_m^{(\nu)}(x,\alpha) = \frac{1}{\pi^2}\int_{U_0} f(x+s)k_m^\alpha(s)\,ds, \;\; 1 \le \nu \le 11, \tag{5.1.40}$$

and

$$\Gamma_\nu(x) \equiv \Gamma_\nu(x,f,\alpha) = \sup_m |V_m^{(\nu)}(x,f,\alpha)|, \quad 1 \le \nu \le 11. \tag{5.1.41}$$

Show that

$$\Gamma_\nu(x) \le A(\alpha)\Big[f^*_{\frac{\alpha}{4}}(x) + \varphi^*_{\frac{\alpha}{4}}(x)\Big], \quad 1 \le \nu \le 11, \tag{5.1.42}$$

where $f^*_{\frac{\alpha}{4}}$ and $\varphi^*_{\frac{\alpha}{4}}$ are defined by (5.1.33) and (5.2.35), respectively, with $\lambda = \frac{\alpha}{4}$ (see also (5.1.32)). Since the sets U_ν, $2 \le \nu \le 11$, are symmetric with respect to coordinates of s, it suffices to establish (5.1.42) in the cases $\nu = 1, 2, 3, 6, 8, 9$.

According to (5.1.4) and (5.1.40), we get

$$|V_m^{(1)}(x)| \le A(\alpha)m^2 \int_0^{2/m}\int_0^{2/m} |f(x+s)|\,ds.$$

Hence, using Lemma 5.1.9 (see (5.1.33) and (5.1.41)), we obtain

$$\Gamma_1(x) \le A(\alpha) f^*_{\frac{\alpha}{4}}(x). \tag{5.1.43}$$

Now consider the expression $V_m^{(2)}(x)$ in (5.1.40). We have

$$|V_m^{(2)}(x)| \le \frac{1}{\pi^2}\int_{1/m}^{3\pi/4}\int_{s_2+1/m}^{3\pi/4} |f(x+s)|\,|k_m^\alpha(s)|\,ds. \tag{5.1.44}$$

But, in view of (5.1.15), it is easy to conclude that

$$|k_m^\alpha(x)| \le \begin{cases} A(\alpha)m^{-\alpha}s_1^{-1-\frac{\alpha}{2}}s_2^{-1-\frac{\alpha}{2}}, & s_2 \in]0, 2^{-1}s_1], \\ A(\alpha)m^{-\alpha}s_1^{-1-\frac{\alpha}{2}}(s_1 - s_2)^{-1-\frac{\alpha}{2}}, & s_2 > 2^{-1}s_1,\; s \in U_2. \end{cases} \tag{5.1.45}$$

Therefore, by (5.1.44) and (5.1.45), we find

$$|V_m^{(2)}(x)| \le A(\alpha)\Big\{m^{-\alpha}\int_{1/m}^{3\pi/4}\int_{s_2+1/m}^{3\pi/4} |f(x+s)|s_1^{-1-\frac{\alpha}{2}}s_2^{-1-\frac{\alpha}{2}}\,ds +$$

$$+m^{-\alpha}\int\limits_{1/m}^{3\pi/4}\int\limits_{s_2+1/m}^{3\pi/4}|f(x+s)|s_1^{-1-\frac{\alpha}{2}}(s_1-s_2)^{-1-\frac{\alpha}{2}}ds\Big\}\equiv$$

$$\equiv V_m^{(12)}(x,f,\alpha)+V_m^{(13)}(x,f,\alpha). \tag{5.1.46}$$

Choose a natural q such that $\frac{3\pi}{4}<\frac{2^q}{m}\le\pi$. Then, according to Lemma 5.1.9, we get

$$V_m^{(12)}(x,f,\alpha)\le$$

$$\le A(\alpha)\Big\{\sum_{i=1}^{q}\sum_{j=1}^{q}m^{-\alpha}\int\limits_{2^{i-1}/m}^{2^i/m}\int\limits_{2^{j-1}/m}^{2^j/m}|f(x+s)|s_1^{-1-\frac{\alpha}{2}}s_2^{-1-\frac{\alpha}{2}}ds\Big\}\le$$

$$\le A_1(\alpha)\Big\{\sum_{i=1}^{q}\sum_{j=1}^{q}m^{-\alpha}2^{-(1+\frac{\alpha}{2})i}2^{-(1+\frac{\alpha}{2})j}m^2\int\limits_{-2^{i-1}/m}^{2^i/m}\int\limits_{-2^{j-1}/m}^{2^j/m}|f(x+s)|\,ds\Big\}\le$$

$$\le A_2(\alpha)\Big\{\sum_{i=1}^{q}\sum_{j=1}^{q}m^{-\alpha}2^{-\frac{\alpha}{2}i}2^{-\frac{\alpha}{2}j}f_*^{2^i,2^j}(x)\Big\}\le A_3(\alpha)f_{\frac{\alpha}{4}}^*(x). \tag{5.1.47}$$

Analogously, using Lemma 5.1.10 (see (5.1.35) and (5.1.46)), we easily see that

$$V_m^{(13)}(x,f,\alpha)\le$$

$$\le A(\alpha)\Big\{\sum_{i=1}^{q}\sum_{j=1}^{q}m^{-\alpha}2^{-(1+\frac{\alpha}{2})i}2^{-(1+\frac{\alpha}{2})j}m^2\int\limits_{-2^i/m}^{2^i/m}\int\limits_{s_2-2^j/m}^{s_2+2^j/m}|f(x+s)|\,ds\Big\}\le$$

$$\le A_1(\alpha)\Big\{\sum_{i=1}^{q}\sum_{j=1}^{q}m^{-\alpha}2^{-\frac{\alpha}{2}i}2^{-\frac{\alpha}{2}j}\varphi_*^{2^i,2^j}(x)\Big\}\le A(\alpha)\varphi_{\frac{\alpha}{4}}^*(x). \tag{5.1.48}$$

Then, by (5.1.40), (5.1.41), and (5.1.46)–(5.1.48), we have

$$\Gamma_2(x)\le A(\alpha)\big[f_{\frac{\alpha}{4}}^*(x)+\varphi_{\frac{\alpha}{4}}^*(x)\big]. \tag{5.1.49}$$

On the basis of Lemmas 5.1.4 and 5.1.9, we find

$$|V_m^{(3)}(x)|\le A(\alpha)m^{1-\alpha}\int\limits_0^{1/m}\int\limits_{2/m}^{\pi}|f(x+s)|s_1^{-1-\alpha}ds\le$$

$$\le A_1(\alpha)\Big\{\sum_{i=1}^{q}2^{-(1+\alpha)i}m^2\int\limits_0^{1/m}\int\limits_{2^{i-1}/m}^{2^i/m}|f(x+s)|\,ds\Big\}\le$$

$$\le A_2(\alpha)f_{\frac{\alpha}{4}}^*(x),$$

so that

$$\Gamma_3(x)\le A(\alpha)f_{\frac{\alpha}{4}}^*(x). \tag{5.1.50}$$

Further, using Lemma 5.1.2 and (5.1.4), we get

$$|V_m^{(6)}(x)| \le A(\alpha) \int_{2/m}^{3\pi/4} \int_{s_1+1/m}^{3\pi/4} |f(x+s)|(s_1 s_2)^{-1} ds.$$

Hence, applying Lemma 5.1.10, we conclude

$$|V_m^{(6)}(x)| \le A(\alpha) m^{1-\alpha} \int_{2/m}^{3\pi/4} \int_{s_1-1/m}^{s_1} |f(x+s)| s_1^{-2} ds \le$$

$$\le A_1(\alpha) \left\{ \sum_{i=2}^{q} 2^{-2i} m^2 \int_{-2^{i-1}/m}^{2^i/m} \int_{s_1-1/m}^{s_1+1/m} |f(x+s)|\, ds \right\} \le$$

$$\le A_2(\alpha) \varphi^*_{\frac{\alpha}{4}}(x).$$

Therefore, according to (5.1.41), we obtain

$$\Gamma_6(x) \le A(\alpha) \varphi^*_{\frac{\alpha}{4}}(x). \tag{5.1.51}$$

For $V_m^{(8)}(x)$ from (5.1.40), we have

$$\pi^2 |V_m^{(8)}(x)| \le \int_{3\pi/4}^{\pi} \int_{1/m}^{\pi/2} |f(x+s)|\, k_m^{\alpha}(s)|\, ds +$$

$$+ \int_{3\pi/4}^{\pi} \int_{\pi/2}^{\pi} |f(x+s)|\, k_m^{\alpha}(s)|\, ds. \tag{5.1.52}$$

But, using (5.1.15) as while proving (5.1.49), we see that

$$\int_{3\pi/4}^{\pi} \int_{1/m}^{\pi/2} |f(x+s)|\, k_m^{\alpha}(s)|\, ds \le A(\alpha) f^*_{\frac{\alpha}{4}}(x),$$

and on the basis of Lemma 5.1.2, we find

$$\int_{3\pi/4}^{\pi} \int_{\pi/2}^{\pi} |f(x+s)|\, k_m^{\alpha}(s)|\, ds \le A(\alpha) f^*_{\frac{\alpha}{4}}(x).$$

Hence, by (5.1.41) and (5.1.52), we get

$$\Gamma_8(x) \le A(\alpha) f^*_{\frac{\alpha}{4}}(x). \tag{5.1.53}$$

Now it is clear that for $V_m^{(9)}(x)$, the relation

$$\Gamma_9(x) \le A(\alpha) f^*_{\frac{\alpha}{4}}(x) \tag{5.1.54}$$

is likewise true.

Thus, adding term by term the inequalities (5.1.43), (5.1.49), (5.1.50), (5.1.51), (5.1.53) and (5.1.54), we have

$$\sup_m |\sigma_m^\alpha(x,f)| \le A(\alpha)\left[f^*_{\frac{\alpha}{4}}(x) + \varphi^*_{\frac{\alpha}{4}}(x)\right], \tag{5.1.55}$$

whence by Lemmas 5.1.9 and 5.1.10 (see (5.1.34) and (5.1.36)), we can conclude that

$$\lim_{m\to\infty} \sigma_m^\alpha(x) = f(x)$$

almost everywhere. ■

Corollary 5.1.2. *Let $f \in L(T^2)$ and the series $\sigma_2[f]$ λ-converge (for some $\lambda \in [1,+\infty[$) to ψ on a set $E \subset T^2$ with $\mu_2(E) > 0$. Then $f(x) = \psi(x)$ almost everywhere on E.*

In particular, $\sigma_2[f]$ can not converge to $+\infty$ or $-\infty$ on a set of positive Lebesgue measure.

For double trigonometric series conjugate to $\sigma_2[f]$, it holds

Theorem 5.1.2. *Let $f \in \log^+ L(T^2)$ and $\alpha \in]0,+\infty[$. Then*

$$\lim_{m\to\infty} t_m^\alpha(x,f,B_l) = \overline{f}_{B_l}(x) \quad (l = 1,2,3)$$

almost everywhere.

We are not going to give a detailed proof of this theorem since otherwise we would have to repeat the above proofs (see the schemes of the proofs of Theorems 2.2.7 and 5.1.1). However, we should note the following.

Since under the conditions of the theorem $\overline{\sigma}_2[f,B_l] = \sigma_1[\overline{f}_{B_l}]$ $(l = 1,2)$, its validity for $l = 1,2$ follows from Theorem 5.1.1. Analyzing the scheme of the proof of Theorem 5.1.1 and taking into account Lemmas 5.1.5, 5.1.6, and 5.1.8 (see (5.1.23), (5.1.25), (5.1.37), and (5.1.31), respectively), we conclude that

$$\lim_{m\to\infty} t_m^\alpha(x,f,B_3) = \overline{f}_{B_3}(x)$$

almost everywhere.

In [22], we set down a detailed proof of Theorem 5.1.2. Therein the matter whether it is nonimprovable is also discussed. In this direction, we should note that analyzing the proofs of the assertion b) of Theorem 1.2.4 (as well as consequent remarks), the assertion a) of Theorem 2.2.7, Theorem 5.1.1, a theorem of Tkebuchava [3], and basing on relations (5.1.2), (5.1.19) and (5.1.28)-(5.1.31), we arrive to

Theorem 5.1.3. *Let a function $\Phi : [0,+\infty[\to [0,+\infty[$ satisfy $\Phi \uparrow +\infty$ as $t \uparrow +\infty$ and*

$$\liminf_{m\to\infty} \Phi(t)(t\log t)^{-1} = 0.$$

If $\alpha \in]0,+\infty[$, then there are functions $f_l \in L(T^2)$ $(l = 1,2,3)$ such that

$$\int_{T^2} \Phi(|f_l(x)|)\,dx < +\infty \quad (l = 1,2,3),$$

yet

$$\limsup_{m\to\infty} t_m^{\alpha}(x, f_l, B_l)| = +\infty \quad (l = 1, 2, 3)$$

almost everywhere.

Analyzing the results of Chapter IV of the second part of the book and the scheme of the proof of Theorem 5.1.1, and using (5.1.2), (5.1.19), and (5.1.28)–(5.1.31), we conclude that it holds

Theorem 5.1.4. a) *Let $f \in L(T^2)$ and $\alpha \in]0, +\infty[$. Then for any $p \in]0, 1[$,*

$$\lim_{m\to\infty} \|t_m^{\alpha}(\cdot, f, B_l) - \overline{f}_{B_l}(\cdot)\|_p = 0 \quad (l = 1, 2).$$

b) *Let $f \in L\log^+ L(T^2)$ and $\alpha \in]0, +\infty[$. Then for any $p \in]0, 1[$,*

$$\lim_{m\to\infty} \|t_m^{\alpha}(\cdot, f, B_3) - \overline{f}_{B_3}(\cdot)\|_p = 0.$$

c) *Let a function $\Phi : [0, +\infty[\to [0, +\infty[$ satisfy $\Phi \uparrow +\infty$ as $t \uparrow +\infty$ and*

$$\liminf_{m\to\infty} \Phi(t)(t\log t)^{-1} = 0.$$

If $\alpha \in]0, +\infty[$, then there is a function $f \in L(T^2)$ such that

$$\int_{T^2} \Phi(|f(x)|)\, dx < +\infty,$$

yet the sequence $\left(t_m^{\alpha}(x, f, B_3)\right)_{m\geq 1}$ does not converge in measure on T^2.

5.2. Some Approximating Properties of Marcikiewicz Type Means of the Series $\sigma_2[f]$ in the Spaces $L^p(T^2)$, $p \in [1, +\infty]$

In the works [4], [9], [13, Ch. IV], we pointed out the assertions connected with approximating properties of Marcinkiewicz type means of $\sigma_2[f]$ in the spaces $L^p(T^2)$, where $p \in [1, +\infty]$ and $\alpha \in [-1, 0[\cup]0, +\infty[$. Below we present them in a more precise form.

Theorem 5.2.1. *Let $\alpha \in]-1, 0[\cup]0, 1[$, and suppose that $f \in L^p(T^2)$ for some $p \in [1, +\infty]$.*

a) *If $\alpha \in]0, +\infty[$, then*

$$\|\sigma_m^{\alpha}(f) - f\|_p \leq$$
$$\leq A(p, \alpha)\left\{\frac{1}{m}\left[\int_{1/m}^{\pi} s_1^{-2}\omega_1(s_1, f)_p ds_1 + \int_{1/m}^{\pi} s_2^{-2}\omega_2(s_2, f)_p ds_2\right]\right\}; \tag{5.2.1}$$

b) *If $\alpha \in]-1, 0[$, then*

$$\|\sigma_m^{\alpha}(f) - f\|_p \leq$$
$$\leq A(p, \alpha)\left\{m^{-\alpha}\log(m+2)\left[\omega_1\left(\frac{1}{m}, f\right)_p + \omega_2\left(\frac{1}{m}, f\right)_p\right]\right\}; \tag{5.2.2}$$

Proof. We will assume that $|\alpha| \in]0,1[$. According to (5.1.3), we have

$$\frac{1}{\pi^2}\int_{T^2} k_m^\alpha(s)\,ds = 1.$$

Therefore, by (5.1.1),

$$\pi^2\Big[\sigma_m^\alpha(x,f) - f(x)\Big] =$$

$$= \int_0^\pi\int_0^\pi \Big[f(x_1+s_1,x_2+s_2) - f(x_1+s_1,x_2) - f(x_1,x_2+s_2) + f(x)\Big]k_m^\alpha(s)\,ds +$$

$$+ \int_0^\pi\int_0^\pi \Big[f(x_1-s_1,x_2+s_2) - f(x_1-s_1,x_2) - f(x_1,x_2+s_2) + f(x)\Big]k_m^\alpha(s)\,ds +$$

$$+ \int_0^\pi\int_0^\pi \Big[f(x_1+s_1,x_2-s_2) - f(x_1+s_1,x_2) - f(x_1,x_2-s_2) + f(x)\Big]k_m^\alpha(s)\,ds +$$

$$+ \int_0^\pi\int_0^\pi \Big[f(x_1-s_1,x_2-s_2) - f(x_1-s_1,x_2) - f(x_1,x_2-s_2) + f(x)\Big]k_m^\alpha(s)\,ds +$$

$$+2\int_0^\pi\int_0^\pi \Big[f(x_1+s_1,x_2) - f(x_1-s_1,x_2) - 2f(x)\Big]k_m^\alpha(s)\,ds +$$

$$+2\int_0^\pi\int_0^\pi \Big[f(x_1,x_2+s_2) - f(x_1,x_2-s_2) - 2f(x)\Big]k_m^\alpha(s)\,ds \equiv$$

$$\equiv \sum_{j=1}^{6} I_m^{(j)}(x,f,\alpha). \tag{5.2.3}$$

The last equality shows that to prove (5.2.1) and (5.2.2), it suffices to obtain the appropriate estimates for $I_m^{(1)}(x,f,\alpha)$ and $I_m^{(5)}(x,f,\alpha)$.

Consider first $I_m^{(1)}(x,f,\alpha)$. According to our notation, we have

$$I_m^{(1)}(x,f,\alpha) \equiv \int_0^\pi\int_0^\pi \Delta(f,x,s)k_m^\alpha(s)\,ds.$$

Using the sets U_ν, $1 \le \nu \le 11$, defined in the proof of Theorem 5.1.1, we get

$$\|I_m^{(1)}(\cdot,f,\alpha)\|_p \le \sum_{\nu=1}^{11}\left\|\int_{U_\nu}\Delta(f,\cdot,s)k_m^\alpha(s)\,ds\right\|_p. \tag{5.2.4}$$

Put

$$B_m^{(\nu)}(x) \equiv B_m^{(\nu)}(x,f\alpha) = \int_{U_\nu}\Delta(f,x,s)k_m^\alpha(s)\,ds. \tag{5.2.5}$$

In view of (5.1.4) and (5.1.5), we obtain

$$\|B_m^{(1)}\|_p \le \begin{cases} A(p,\alpha)\Big[\omega_1\Big(\frac{1}{m},f\Big)_p + \omega_2\Big(\frac{1}{m},f\Big)_p\Big], & \alpha \in]0,1[, \\ A(p,\alpha)m^{-\alpha}\Big[\omega_1\Big(\frac{1}{m},f\Big)_p + \omega_2\Big(\frac{1}{m},f\Big)_p\Big], & \alpha \in]-1,0[. \end{cases} \tag{5.2.6}$$

Further, according to (5.1.6), the expression (5.2.5) can be rewritten as

$$\begin{aligned} B_m^{(2)}(x) = & \int_{2/m}^{4\pi/m}\int_{1/m}^{s_1-1/m} \Delta(f,x,s)k_m^\alpha(s)\,ds + \\ & + \int_{4\pi/m}^{3\pi/4}\int_{1/m}^{2\pi/m} \Delta(f,x,s)k_m^\alpha(s)\,ds + \\ & + \int_{4\pi/m}^{3\pi/4}\int_{2\pi/m}^{s_1-1/m} \Delta(f,x,s)\varphi_m^\alpha(s)\,ds + \\ & + \int_{4\pi/m}^{3\pi/4}\int_{2\pi/m}^{s_1-1/m} \Delta(f,x,s)r_m^\alpha(s)\,ds \equiv \\ & \equiv \sum_{j=1}^{4} g_M^{(j)}(x,f,\alpha). \end{aligned} \tag{5.2.7}$$

Hence, using (5.1.15), we get

$$\begin{aligned} & \|g_M^{(1)}(\cdot,f,\alpha)\|_p \le \\ & \le A(p,\alpha)m^{-\alpha}\omega_1\Big(\frac{1}{m},f\Big)_p \int_{2/m}^{4\pi/m}\int_{1/m}^{s_1-1/m} (s_1s_2)^{-1}(s_1-s_2)^{-\alpha}ds \le \\ & \le A_1(p,\alpha)\omega_1\Big(\frac{1}{m},f\Big)_p. \end{aligned} \tag{5.2.8}$$

Analogously, we find that

$$\|g_M^{(2)}(\cdot,f,\alpha)\|_p \le \begin{cases} A(p,\alpha)\omega_2\Big(\frac{1}{m},f\Big)_p, & \alpha \in]0,1[, \\ A(p,\alpha)m^{-\alpha}\omega_2\Big(\frac{1}{m},f\Big)_p, & \alpha \in]-1,0[. \end{cases} \tag{5.2.9}$$

By virtue of (5.1.8) and (5.2.7), we have

$$\begin{aligned} & \|g_M^{(4)}(\cdot,f,\alpha)\|_p \le \\ & \le A(p,\alpha)\Big\{m^{-2}\int_{2/m}^{\pi}\int_{1/m}^{s_1-1/m} (s_1-s_2)^{-1}(s_1-s_2)^2\omega_1(s_1,f)_p ds\Big\} \le \end{aligned}$$

$$\leq A(p,\alpha)\left\{m^{-2}\int\limits_{1/m}^{\pi} s_1^{-2}\omega_1(s_1,f)_p ds_1\right\}. \tag{5.2.10}$$

Put

$$b_\alpha(s)\equiv b_\alpha(s_1,s_2)=\left(\sin\frac{s_1}{2}\sin\frac{s_2}{2}\right)^{-1}\left[\left(\sin\frac{s_1-s_2}{2}\right)^{-\alpha}\pm\right.$$
$$\left.\pm\left(\sin\frac{s_1+s_2}{2}\right)^{-\alpha}\right]. \tag{5.2.11}$$

Then by (5.1.7) and (5.2.7), we can conclude that to estimate $g_M^{(3)}(x,f,\alpha)$, it suffices to evaluate

$$g_M^{(5)}(x,f,\alpha)=m^{-\alpha}\int\limits_{4\pi/m}^{3\pi/4}\int\limits_{2\pi/m}^{s_1-1/m}\Delta(f,x,s)b_\alpha(s)\cos ms_1\cos ms_2\,ds.$$

But $g_M^{(5)}(x,f,\alpha)$ can be represented as

$$2g_M^{(5)}(x,f,\alpha)=$$
$$=m^{-\alpha}\left\{\int\limits_{4\pi/m}^{3\pi/4}\int\limits_{2\pi/m}^{s_1-1/m}\left[\Delta(f,x,s)-\Delta\left(f,x,s_1,s_2+\frac{\pi}{m}\right)\right]b_\alpha(s)\cos ms_1\cos ms_2\,ds+\right.$$
$$+\int\limits_{4\pi/m}^{3\pi/4}\int\limits_{2\pi/m}^{s_1-1/m}\Delta\left(f,x,s_1,s_2+\frac{\pi}{m}\right)\left[b_\alpha(s)-b_\alpha\left(s_1,s_2+\frac{\pi}{m}\right)\right]\cos ms_1\cos ms_2\,ds-$$
$$-\int\limits_{4\pi/m}^{3\pi/4}\int\limits_{s_1-1/m-\pi/m}^{s_1-1/m}\Delta\left(f,x,s_1,s_2+\frac{\pi}{m}\right)b_\alpha\left(s_1,s_2+\frac{\pi}{m}\right)\cos ms_1\cos ms_2\,ds-$$
$$\left.-\int\limits_{4\pi/m}^{3\pi/4}\int\limits_{\pi/m}^{2\pi/m}\Delta\left(f,x,s_1,s_2+\frac{\pi}{m}\right)b_\alpha\left(s_1,s_2+\frac{\pi}{m}\right)\cos ms_1\cos ms_2\,ds\right\}\equiv$$
$$\equiv\sum_{j=6}^{9}g_M^{(j)}(x,f,\alpha), \tag{5.2.12}$$

whence, by (5.2.11), we find

$$\|g_M^{(6)}(\cdot,f,\alpha)\|_p\leq$$
$$\leq A(p,\alpha)\left\{m^{-\alpha}\omega_2\left(\frac{1}{m},f\right)_p\int\limits_{2/m}^{\pi}\int\limits_{1/m}^{s_1-1/m}(s_1s_2)^{-1}(s_1-s_2)^{-\alpha}ds\right\}\leq$$
$$\leq A(p,\alpha)\left\{m^{-\alpha}\omega_2\left(\frac{1}{m},f\right)_p\int\limits_{2/m}^{\pi}\int\limits_{1/m}^{s_1-1/m}\left[s_1^{-1-\alpha}s_2^{-1}+s_1^{-2}(s_1-s_2)^{-\alpha}\right]ds\right\}\leq$$

$$\leq \begin{cases} A(p,\alpha)\omega_2\left(\frac{1}{m},f\right)_p, & \alpha\in]0,1[,\\ A(p,\alpha)m^{-\alpha}\log(m+2)\omega_2\left(\frac{1}{m},f\right)_p, & \alpha\in]-1,0[. \end{cases} \tag{5.2.13}$$

On the other hand, with the help of (5.2.11), we obtain

$$\left|b_\alpha(s)-b_\alpha\left(s_1,s_2+\frac{\pi}{m}\right)\right|\leq$$

$$\leq A(\alpha)m^{-1}\left[s_1^{-1}s_2^{-2}(s_1-s_2)^{-\alpha}+(s_1s_2)^{-1}(s_1-s_2)^{-1-\alpha}\right]. \tag{5.2.14}$$

Then, according to (5.2.12), we will have

$$\|g_M^{(7)}(\cdot,f,\alpha)\|_p\leq$$

$$\leq A(p,\alpha)\Bigg\{m^{-1-\alpha}\int\limits_{4/m}^{3\pi/4}\int\limits_{2/m}^{s_1-1/m}\left[s_1^{-1}s_2^{-2}(s_1-s_2)^{-\alpha}+\right.$$

$$\left.+(s_1s_2)^{-1}(s_1-s_2)^{-1-\alpha}\omega_2(s_2,f)_pds\right\}\leq$$

$$\leq A(p,\alpha)\Bigg\{m^{-1-\alpha}\int\limits_{4/m}^{3\pi/4}\int\limits_{2/m}^{s_1-2/m}\left[s_1^{-1-\alpha}s_2^{-2}+s_1^{-2}(s_1-s_2)^{-1-\alpha}\right]\omega_2(s_2,f)_pds\Bigg\}\leq$$

$$\leq \begin{cases} A_2(p,\alpha)\omega_2\left(\frac{1}{m},f\right)_p & \text{for } \alpha\in]0,1[,\\ A_2(p,\alpha)m^{-\alpha}\log(m+2)\omega_2\left(\frac{1}{m},f\right)_p & \text{for } \alpha\in]-1,0[. \end{cases} \tag{5.2.15}$$

Taking note of (5.2.11) and (5.2.12), we get

$$\|g_M^{(8)}(\cdot,f,\alpha)\|_p\leq A(p,\alpha)\left[m^{-1}\int\limits_{1/m}^{\pi}s_1^{-2}\omega_1(s_1,f)_pds_1\right], \tag{5.2.16}$$

$$\|g_M^{(9)}(\cdot,f,\alpha)\|_p\leq$$

$$\leq \begin{cases} A(p,\alpha)\omega_2\left(\frac{1}{m},f\right)_p & \text{for } \alpha\in]0,1[,\\ A(p,\alpha)m^{-\alpha}\log(m+2)\omega_2\left(\frac{1}{m},f\right)_p & \text{for } \alpha\in]-1,0[. \end{cases} \tag{5.2.17}$$

Therefore (5.2.12), (5.2.13), and (5.2.15)–(5.2.17) imply

$$\|g_M^{(5)}(\cdot,f,\alpha)\|_p\leq A(p,\alpha)\Bigg\{m^{-1}\Bigg[\int\limits_{1/m}^{\pi}s_1^{-2}\omega_1(s_1,f)_pds_1+$$

$$+\int\limits_{1/m}^{\pi}s_2^{-2}\omega_2(s_2,f)_pds_2\Bigg]\Bigg\},\quad \alpha\in]0,1[,$$

$$\|g_M^{(5)}(\cdot,f,\alpha)\|_p \le A(p,\alpha)\Big\{m^{-\alpha}\log(m+2)\Big[\omega_1\Big(\frac{1}{m},f\Big)_p+$$
$$+\omega_2\Big(\frac{1}{m},f\Big)_p\Big]\Big\} \quad \text{for } \alpha\in]-1,0[.$$

For $g_M^{(3)}(x,f,\alpha)$ from (5.2.7), a similar estimate can be obtained.

Thus, using (5.2.7)–(5.2.10), we conclude that

$$\|B_m^{(2)}\|_p \le \begin{cases} A(p,\alpha)\Big\{m^{-1}\Big[\int\limits_{1/m}^{\pi} s_1^{-2}\omega_1(s_1,f)_p ds_1 + \\ \quad + \int\limits_{1/m}^{\pi} s_2^{-2}\omega_2(s_2,f)_p ds_2\Big]\Big\} & \text{for } \alpha\in]0,1[, \\ A(p,\alpha)\Big\{m^{-\alpha}\log(m+2)\Big[\omega_1\Big(\frac{1}{m},f\Big)_p + \\ \quad +\omega_2\Big(\frac{1}{m},f\Big)_p\Big]\Big\} & \text{for } \alpha\in]-1,0[. \end{cases} \tag{5.2.18}$$

Using Lemma 5.1.4 and (5.2.5), we have

$$\|B_m^{(3)}\|_p \le A(p,\alpha)\omega_2\Big(\frac{1}{m},f\Big)_p m^{1-\alpha}\int\limits_{2/m}^{\pi}\int\limits_{0}^{1/m}(s_1-s_2)^{-1-\alpha}ds \le$$
$$\le \begin{cases} A_1(p,\alpha)\omega_2\Big(\frac{1}{m},f\Big)_p & \text{for } \alpha\in]0,1[, \\ A_1(p,\alpha)m^{-\alpha}\omega_2\Big(\frac{1}{m},f\Big)_p & \text{for } \alpha\in]-1,0[, \end{cases} \tag{5.2.19}$$

and by Lemma 5.1.2 and (5.1.5), we find

$$\|B_m^{(6)}\|_p \le A(p,\alpha)\Big\{m^{-1}(1+m^{-\alpha})\int\limits_{1/m}^{\pi} s_1^{-2}\omega_1(s_1,f)ds_1\Big\} \le$$
$$\le \begin{cases} A(p,\alpha)m^{-1}\int\limits_{1/m}^{\pi} s_1^{-2}\omega_1(s_1,f)ds_1 & \text{for } \alpha\in]0,1[, \\ A(p,\alpha)m^{-\alpha}\omega_1\Big(\frac{1}{m},f\Big)_p & \text{for } \alpha\in]-1,0[, \end{cases} \tag{5.2.20}$$

$$\|B_m^{(9)}\|_p \le \begin{cases} A(p,\alpha)\omega_1\Big(\frac{1}{m},f\Big)_p & \text{for } \alpha\in]0,1[, \\ A_1(p,\alpha)m^{-\alpha}\omega_1\Big(\frac{1}{m},f\Big)_p & \text{for } \alpha\in]-1,0[. \end{cases} \tag{5.2.21}$$

Now consider the expression $B_m^{(2)}(x)$ from (5.2.5). With the help of (5.1.6), it can be rewritten as

$$B_m^{(8)}(x) = \int_{3\pi/4}^{\pi}\int_{1/m}^{s_1-1/m} \Delta(f,x,s)\varphi_m^{\alpha}(s)\,ds + \\ + \int_{3\pi/4}^{\pi}\int_{1/m}^{s_1-1/m} \Delta(f,x,s)r_m^{\alpha}(s)\,ds \equiv \\ \equiv g_M^{(10)}(x,f,\alpha) + g_M^{(11)}(x,f,\alpha). \tag{5.2.22}$$

To estimate $g_M^{(10)}(x,f,\alpha)$, it suffices to evaluate appropriately the expression

$$g_M^{(12)}(x,f,\alpha) = m^{-\alpha}\int_{3\pi/4}^{\pi}\int_{1/m}^{s_1-1/m} \Delta(f,x,s)b_\alpha(s)\cos ms_1\cos ms_2 ds,$$

where $b_\alpha(x)$ is defined by (5.2.11).

As while estimating $g_M^{(5)}(x,f,\alpha)$ (see (5.2.12)), $g_M^{(12)}(x,f,\alpha)$ can be represented as

$$2g_M^{(12)}(x,f,\alpha) =$$

$$= m^{-\alpha}\Bigg\{\int_{3\pi/4}^{\pi}\int_{2\pi/m}^{s_1-1/m}\Big[\Delta(f,x,s) - \Delta\Big(f,x,s_1,s_2+\frac{\pi}{m}\Big)\Big]b_\alpha(s)\cos ms_1\cos ms_2\,ds + \\ + \int_{3\pi/4}^{\pi}\int_{2\pi/m}^{s_1-1/m}\Delta\Big(f,x,s_1,s_2+\frac{\pi}{m}\Big)\Big[b_\alpha(s) - b_\alpha\Big(s_1,s_2+\frac{\pi}{m}\Big)\Big]\cos ms_1\cos ms_2\,ds + \\ + \int_{3\pi/4}^{\pi}\int_{s_1-1/m-\pi/m}^{s_1-1/m}\Delta\Big(f,x,s_1,s_2+\frac{\pi}{m}\Big)b_\alpha\Big(s_1,s_2+\frac{\pi}{m}\Big)\cos ms_1\cos ms_2\,ds - \\ - \int_{3\pi/4}^{3\pi/4}\int_{\pi/m}^{2\pi/m}\Delta\Big(f,x,s_1,s_2+\frac{\pi}{m}\Big)b_\alpha\Big(s_1,s_2+\frac{\pi}{m}\Big)\cos ms_1\cos ms_2\,ds + \\ +2\int_{3\pi/4}^{\pi}\int_{1/m}^{2\pi/m}\Delta(f,x,s)b_\alpha(s)\cos ms_1\cos ms_2 ds\Bigg\}.$$

whence in the same way as while evaluating $g_M^{(j)}(x,f,\alpha)$, $6 \le j \le 9$, from (5.2.12) (taking also note of the behavior of the expression $[\sin \frac{s_1 \pm s_2}{2}]^{-\alpha}$ in $b_\alpha(s)$), we find

$$\|g_M^{(12)}\|_p \le \begin{cases} A(p,\alpha)\Big\{m^{-1}\Big[\int\limits_{1/m}^{\pi} s_1^{-2}\omega_1(s_1,f)_p ds_1 + \\ \qquad + \int\limits_{1/m}^{\pi} s_2^{-2}\omega_2(s_2,f)_p ds_2\Big]\Big\} & \text{for } \alpha \in]0,1[, \\ A(p,\alpha)\Big\{m^{-\alpha}\log(m+2)\Big[\omega_1\Big(\frac{1}{m},f\Big)_p + \\ \qquad +\omega_2\Big(\frac{1}{m},f\Big)_p\Big]\Big\} & \text{for } \alpha \in]-1,0[. \end{cases} \tag{5.2.23}$$

Therefore the same estimate is also valid for $g_M^{(10)}(x,f,\alpha)$ from (5.2.22).

Rewrite $g_M^{(11)}(x,f,\alpha)$ from (5.2.22) as

$$g_M^{(11)}(x,f,\alpha) = \int\limits_{3\pi/4}^{\pi}\int\limits_{1/m}^{\pi/2} \Delta(f,x,s) r_m^\alpha(s)\,ds + \int\limits_{3\pi/4}^{\pi}\int\limits_{\pi/2}^{s_1-1/m} \Delta(f,x,s) r_m^\alpha(s)\,ds.$$

Hence, using (5.1.8), we get

$$\|g_M^{(11)}\|_p \le A(p,\alpha)\Big\{m^{-1}\Big[\int\limits_{1/m}^{\pi} s_1^{-2}\omega_1(s_1,f)_p ds_1 + \\ + \int\limits_{1/m}^{\pi} s_2^{-2}\omega_2(s_2,f)_p ds_2\Big]\Big\}. \tag{5.2.24}$$

Then, by (5.2.22)–(5.2.24), we will have

$$\|B_m^{(8)}\|_p \le \begin{cases} A(p,\alpha)\Big\{m^{-1}\Big[\int\limits_{1/m}^{\pi} s_1^{-2}\omega_1(s_1,f)_p ds_1 + \\ \qquad + \int\limits_{1/m}^{\pi} s_2^{-2}\omega_2(s_2,f)_p ds_2\Big]\Big\} & \text{for } \alpha \in]0,1[, \\ A(p,\alpha)\Big\{m^{-\alpha}\log(m+2)\Big[\omega_1\Big(\frac{1}{m},f\Big)_p + \\ \qquad +\omega_2\Big(\frac{1}{m},f\Big)_p\Big]\Big\} & \text{for } \alpha \in]-1,0[. \end{cases} \tag{5.2.25}$$

Thus, taking note of (5.2.4), (5.2.6), (5.2.18), (5.2.19), (5.2.20), (5.2.21), and (5.2.25), we conclude

$$\|I_m^{(1)}(\cdot,f,\alpha)\|_p \le \begin{cases} A(p,\alpha)\Big\{m^{-1}\Big[\int\limits_{1/m}^{\pi} s_1^{-2}\omega_1(s_1,f)_p ds_1 + \\ \quad + \int\limits_{1/m}^{\pi} s_2^{-2}\omega_2(s_2,f)_p ds_2\Big]\Big\} & \text{for } \alpha\in]0,1[, \\ A(p,\alpha)\Big\{m^{-\alpha}\log(m+2)\Big[\omega_1\Big(\frac{1}{m},f\Big)_p + \\ \quad +\omega_2\Big(\frac{1}{m},f\Big)_p\Big]\Big\} & \text{for } \alpha\in]-1,0[. \end{cases} \tag{5.2.26}$$

Therefore, according to (5.2.3),

$$\|I_m^{(j)}(\cdot,f,\alpha)\|_p \le$$

$$\le \begin{cases} A(p,\alpha)\Big\{m^{-1}\Big[\int\limits_{1/m}^{\pi} s_1^{-2}\omega_1(s_1,f)_p ds_1 + \\ \quad + \int\limits_{1/m}^{\pi} s_2^{-2}\omega_2(s_2,f)_p ds_2\Big]\Big\} & \text{for } \alpha\in]0,1[, \\ A(p,\alpha)\Big\{m^{-\alpha}\log(m+2)\Big[\omega_1\Big(\frac{1}{m},f\Big)_p + \\ \quad +\omega_2\Big(\frac{1}{m},f\Big)_p\Big]\Big\} & \text{for } \alpha\in]-1,0[\ \ (j=2,3,4). \end{cases} \tag{5.2.27}$$

On the other hand, the expression $I_m^{(5)}(x,f,\alpha)$ can be represented as

$$I_m^{(5)}(x,f,\alpha) = \frac{1}{\pi}\int\limits_0^{\pi}\Big[f(x_1+s_1,x_2)+f(x_1-s_1,x_2)-2f(x)\Big]k_m^{\alpha}(s_1)\,ds_1.$$

So, using Theorem 4.1.1 from the first part, we obtain

$$\|I_m^{(5)}(\cdot,f,\alpha)\|_p \le A(p,\alpha)\Big\{(1+m^{-\alpha})\omega_1\Big(\frac{1}{m},f\Big)_p + \\ + m^{-1}\int\limits_{1/m}^{\pi} s_1^{-2}\omega_1(s_1,f)_p ds_1\Big\}. \tag{5.2.28}$$

For $I_m^{(6)}(x,f,\alpha)$ from (5.2.3), a similar estimate is valid. To be more precise,

$$\|I_m^{(6)}(\cdot,f,\alpha)\|_p \le A(p,\alpha)\Big\{(1+m^{-\alpha})\omega_2\Big(\frac{1}{m},f\Big)_p + \\ + m^{-1}\int\limits_{1/m}^{\pi} s_2^{-2}\omega_2(s_2,f)_p ds_2\Big\}. \tag{5.2.29}$$

Thus, basing on (5.2.3) and (5.2.26)–(5.2.29), we conclude that (5.2.1) and (5.2.2) are true. ■

Remark 5.2.1. When $p = +\infty$ or $p = 1$, the inequality (5.2.1) is precise in the sense of order for all the space $L^p(T^2)$. As to (5.2.2) for $p = +\infty$ or $p = 1$, it remains an open problem whether it is best possible in the same sense. In the case where $f \in L^p(T^2)$ for some $p \in]1, +\infty[$, (5.2.1) and (5.2.2) can be improved. For instance, in this case the inequality

$$\|\sigma_m^\alpha(f) - f\|_p \leq A(p,\alpha)\left\{(1+m^{-\alpha})E_{[\frac{m}{2}],[\frac{m}{2}]}(f)_p + m^{-1}\sum_{j=1}^{[\frac{m}{2}]} E_{j,j}(f)_p\right\}$$

is valid.

A number of corollaries can be obtained from Theorem 5.2.1. We formulate two of them which are characteristic.

Corollary 5.2.1. *Let $f \in Lip\,\alpha$ on T^2 for some $\alpha \in]0,1[$. Then for any $\beta \in]0,\alpha[$, it holds*

$$\|\sigma_m^{-\beta}(f) - f\|_p \to 0 \quad \text{as } m \to \infty.$$

Corollary 5.2.2. *Let $f \in C(T)$ and $\alpha \in]0,1[$. Let, moreover,*

$$R_m^\alpha(x,f) = (A_m^\alpha)^{-1}\sum_{j=0}^{m} A_{m-j}^{\alpha-1} S_j^2(x,f) \quad \text{for } m \in \mathbb{N}.$$

Then

$$\|R_m^\alpha(f) - f^2\|_C \leq A(\alpha)m^{-1}\int_{1/m}^{\pi} t^{-2}\omega(t,f)_C dt.$$

Remark 5.2.2. We could bring theorems concerning approximating properties of the expressions $t_m^\alpha(x,f,B_l)$ $(l = 1,2,3)$ in the spaces $L^p(T^2)$ for $p = +\infty$ and $p = 1$. But, as we have seen in Chapter 3 (Theorem 3.2.1), their formulation is cumbersome enough. For this reason, we do not give them here.

Now we would like to say a few words on Poisson–Abel means of series $\sigma_2[f]$ and $\overline{\sigma}_2[f, B_l]$ $(l = 1,2,3)$.

Let

$$f(x,r) = (1-r)\sum_{j=0}^{\infty} r^j S_{j,j}(x,f), \tag{5.2.30}$$

and

$$f(x,r,B_l) = (1-r)\sum_{j=0}^{\infty} r^j S_{j,j}(x,f,B_l) \quad (l=1,2,3), \quad r \in [0,1[. \tag{5.2.31}$$

Analogues of Theorems 5.1.1–5.1.4, 5.2.1 are valid for Poisson–Abel means as well. We are not going to repeat everything, but regard as appropriate to formulate the analogue of Theorem 5.2.1.

Theorem 5.2.2. *Let $f \in L^p(T^2)$ for some $p \in [1, +\infty]$. Then*

$$\|f(\cdot, r) - f(\cdot)\|_p \leq$$

$$\leq A(p)\Big\{(1-r)\Big[\int_{1-r}^{\pi} s_1^{-2}\omega_1(s_1, f)_p ds_1 + \int_{1-r}^{\pi} s_2^{-2}\omega_2(s_2, f)_p ds_2\Big]\Big\}.$$

By now, various questions concerning linear means of $\sigma_n[f]$ $(n \geq 2)$ of Marcinkiewicz type have been investigated by a number of researchers. We have set them down in [35, Ch. II], [46]. We think it to be appropriate to dwell here on the works of D'yachenko [1], [2].

Let $n \geq 3$ and $f \in L(T^n)$, and let the expressions $S_{j,\dots,j}(x, f)$ and $S_{j,\dots,j}(x, f, B)$, $j \in \mathbb{N}_0$, $B \subset M$ $(B \neq \varnothing)$ denote the cubic partial sums of the series $\sigma_n[f]$ and $\overline{\sigma}_n[f, B]$, respectively. For $\alpha \in]0, +\infty[$, define

$$\sigma_m^\alpha(x, f) = (A_m^\alpha)^{-1} \sum_{j=0}^{m} A_{m-j}^{\alpha-1} S_{j,\dots,j}(x, f) \quad \text{for } m \in \mathbb{N}, \tag{5.2.32}$$

and

$$t_m^\alpha(x, f, B) = (A_m^\alpha)^{-1} \sum_{j=0}^{m} A_{m-j}^{\alpha-1} S_{j,\dots,j}(x, f, B) \quad \text{for } m \in \mathbb{N}. \tag{5.2.33}$$

If $r \in [0, 1[$, we will assume that

$$f(x, r) = (1-r)\sum_{j=0}^{m} r^j S_{j,\dots,j}(x, f), \tag{5.2.34}$$

and

$$f(x, r, B) = (1-r)\sum_{j=0}^{m} r^j S_{j,\dots,j}(x, f, B). \tag{5.2.35}$$

The results of D'yachenko [1], [2] show, in particular, that Theorem 5.1.1 is valid in the case $n \geq 3$ as well, i.e., for the expression (5.2.32). It can be shown that the same conclusion is also true for $f(x, r)$ defined by (5.2.34).

Remark 5.2.3. Theorems 2.2.1, 2.2.2, 2.2.7, 2.2.8, and 2.2.9 remain true if we substitute n-fold Cesàro (of positive order) and Poisson–Abel means by the means (5.2.35) (in Theorems 2.2.1 and 2.2.2) and the means (5.2.33) (in Theorems 2.2.7, 2.2.8 and 2.2.9).

A similar remark is valid for Theorems 4.1.1, 4.1.2, and 4.1.3.

An analogue of Theorem 5.1.3 for $n \geq 3$ should also be pointed out. We give it for the means (5.2.33) since for the means (5.2.35), its formulation is quite analogous.

Theorem 5.2.3. *Let $n \geq 3$, and let a function $\Phi : [0, +\infty[\to [0, +\infty[$ satisfy $\Phi \uparrow +\infty$ as $t \uparrow +\infty$. If*

$$\liminf_{t\to\infty} \Phi(t)\left[t(\log t)^{n-1}\right]^{-1} = 0$$

and $\alpha \in]0,+\infty[$, *then there is a function* $f \in L(T^n)$ *such that*

$$\int_{T^n} \Phi(|f(x)|)\,dx < +\infty,$$

yet almost everywhere,

$$\limsup_{m\to\infty} |t_m^\alpha(x,f,M)| = +\infty.$$

Remark 5.2.4. In Theorem 5.1.3, the conclusion concerns all the conjugate series $\overline{\sigma}_2[f,B_l]$ $(l=1,2,3)$. As to the just formulated theorem, the conclusion is valid only for $\overline{\sigma}_n[f,M]$.

5.3. Unsolved Problems

5.3.1. Let $\alpha \in]0,+\infty[$, functions $\nu_i : \mathbb{N}_0 \to \mathbb{N}_0$ be such that $\nu_i \uparrow +\infty$, $1 \le i \le n$, and let $\nu = (\nu_1, \nu_2, \ldots, \nu_n)$. Put

$$\sigma_m^\alpha(x,f,\nu) = (A_m^\alpha)^{-1} \sum_{j=0}^{m} A_{m-j}^{\alpha-1} S_{\nu_1(j),\ldots,\nu_n(j)}(x,f). \tag{5.3.1}$$

Find necessary and sufficient conditions which should be imposed on ν_i for Theorems 5.1.1 and 5.2.1 to remain valid (for appropriate classes of functions f). If the problem is solved, there arise similar questions (with appropriate changes) for the means of the type (5.2.36) of $\overline{\sigma}_n[f,B]$, $B \subset M$ $(B \neq \varnothing)$, as well as for the means of the Poisson–Abel type.

5.3.2. Let $p = +\infty$ or $p = 1$, and let $f \in L^p(T^n)$ $(n \ge 2)$. Suppose that $\alpha \in]-1,0[\cup]0,+\infty[$. In terms of $L^p(T^n)$-moduli of continuity, establish an exact (for all the space $L^p(T^n)$) upper estimate for the expression

$$\|\sigma_m^\alpha(f) - f\|_p.$$

5.3.3. Let $p = +\infty$ or $p = 1$, and let $W(T^n) \subset L^p(T^n)$ be a subclass (say, $Lip\,\alpha$, $H(\omega, L^p), \ldots$). Investigate the asymptotic behavior of

$$\sup_{f\in W(T^n)} \|\sigma_m^\beta(f) - f\|_p.$$

A similar problem arises for

$$\sup_{f\in W(T^n)} \|f(\cdot,r) - f(\cdot)\|_p$$

as well.

Bibliography

Aĭzenberg, L. A.

[1] *Multivariate logarithmic residue and its applications (Russian),* In: Itogi Nauki i Tekhniki, Sovrem. Probl. Mat., Fund. Napravl., vol. **8**, Akad. Nauk SSSR, Inst. Nauchn. i Tekhn. Inform., Moscow, 1985, pp. 29–45.

Akhiezer, D. N.

[1] *Homogeneous complex manifolds (Russian),* In: Itogi Nauki i Tekhniki, Sovrem. Probl. Mat., Fund. Napravl., vol. **10**, Akad. Nauk SSSR, Inst. Nauchn. i Tekhn. Inform., Moscow, 1986, pp. 223–275.

Akhiezer, N. I.

[1] *Lectures in approximation theory (Russian),* Nauka, Moscow, (1965).

Akhobadze, T. I.

[1] *On the convergence and summability of simple trigonometric Fourier series (Russian),* In the collection: Some problems of the theory of functions, vol. **I**, Tbilisi State Univ., Tbilisi, (1979), pp. 5–66.

[2] *Classes of functions and convergence of Fourier series (Russian),* Acta Math. Acad. Sci. Hung., **37** (1981) , No. 1–3, pp. 95-119.

[3] *On convergence and summability of Fourier series,* Analysis Math., **8** (1982), No. 2, pp. 79–102.

[4] *On uniform convergence and Cesàro (C,α)-summability of trigonometric Fourier series (Russian),* Bull. Acad. Sci. Georgian SSR, **128** (1987), No. 2, pp. 241–244.

[5] *On a problem of convergence and (C,α)-summability of trigonometric Fourier series (Russian),* Acta Math. Acad. Sci. Hung., **55** (1990), No. 1–2, pp. 3–31.

Aleksandrov, A. B.

[1] *Function theory in the ball (Russian),* In: Itogi Nauki i Tekhniki, Sovrem. Probl. Mat., Fund. Napravl., vol. **8**, Akad. Nauk SSSR, Inst. Nauchn. i Tekhn. Inform., Moscow, 1985, pp. 115–190.

Alexits, G.

[1] *Sur l'ordre de grandeur de l'approximation d'une fonction par les moyennes de la série de Fourier,* Mat. Fiz. Lapok., **48** (1941), No. 3–4, pp. 410–422.

[2] *Konvergenzprobleme der Orthogonalreihen,* Verlag der Ungarischen Akademie der Wissenschaften, Budapest, (1960).

Alimov, Sh. A., Ashurov, R. R., Pulatov, A. K.

[1] *Multiple Fourier series and integrals (Russian)*, In: Itogi Nauki i Tekhniki, Sovrem. Probl. Mat., Fund. Napravl., vol. **42**, Akad. Nauk SSSR, Inst. Nauchn. i Tekhn. Inform., Moscow, 1989, pp. 7–104.

Alimov, Sh. A., Il'in V. A., Nikishin, E. M.

[1] *Problems of convergence of multiple trigonometric series and spectral decompositions,* I *(Russian)*, Uspekhi Mat. Nauk. **31** (1976), No. 6, pp. 28–83.

[2] *Problems of convergence of multiple trigonometric series and spectral decompositions,* II *(Russian)*, Uspekhi Mat. Nauk. **32** (1977), No. 1, pp. 107–130.

Asatiani, V. O.

[1] *Uniform convergence of Cesàro means of negative order of Fourier series (Russian)*, Bull. Acad. Sci. Georgian SSR, **92** (1978), No. 3, pp. 549–552.

Ash, J. M., Gluck, L.

[1] *A divergent multiple Fourier series of power series type,* Stud. Math., **44** (1972), No. 5, pp. 477–491.

Babenko, K. I.

[1] *On conjugate functions (Russian)*, Dokl. Akad. Nauk. SSSR, **62** (1948), No. 2, pp. 157–160.

Bahbuh, M., Nikishin, E. M.

[1] *Convergence of double Fourier series of continuous functions,* Sibirsk. Mat. Zh., **14** (1973), No. 6, pp. 1189–1199.

Bari, N. K.

[1] *On the best approximation of two conjugate functions by trigonometric polynomials (Russian)*, Izv. Akad. Nauk SSSR, Ser. Mat., **19** (1955), No. 3, pp. 285–302.

[2] *A Treatise on Trigonometric Series, vol.* **1–2**, Pergamon Press, New York, (1964).

Bari, N. K., Stechkin, S. B.

[1] *The best approximation and differential properties of two conjugate functions (Russian)*, Trudy Moskov. Mat. Obshch., **5** (1956), pp. 483–552.

Baskakov, V. A., Telyakovskiĭ, S. A.

[1] *Approximation of differentiablc functions by Fejér sums (Russian)*, Mat. Zametki, **32** (1982), No. 2, pp. 429–440.

Bellman, R.

[1] *A note on a theorem of Hardy on Fourier constants,* Bull. Amer. Math. Soc., **50** (1944), No. 8, pp. 741–744.

Belov, A. S.

[1] *Everywhere divergent trigonometric series (Russian)*, Mat. Sbornik, **85** (1971), No. 2, pp. 224–237; English translation: Math. USSR Sb., **14** (1971), pp. 219–232.

[2] *A study of some trigonometric series (Russian),* Mat. Zametki, **13** (1973), No. 4, pp. 481–492; English translation: Math. Notes, **13** (1973), pp. 291–298.

[3] *Remarks on everywhere divergent trigonometric series (Russian),* Sibirsk. Mat. Zh., **18** (1977), No. 1, pp. 23–31; English translation: Siberian Math. J., **18** (1977), No. 1, pp. 16–22.

Berenstein, C., Struppa, D.

[1] *Complex analysis and convolution equations (Russian),* In: Itogi Nauki i Tekhniki, Sovrem. Probl. Mat., Fund. Napravl., vol. **54**, Akad. Nauk SSSR, Inst. Nauchn. i Tekhn. Inform., Moscow, 1989, pp. 5–111.

Bergman, S., Marcinkiewicz, J.

[1] *Sur les fonctions analytiques de deux variables complexes,* Fund. Math., **33** (1939), No. 1, pp. 75–94.

Bernstein, S. N.

[1] *Sur l'ordre de la meilleure approximation des fonction continues par les polynomes de degré donné,* Mem. Acad. Roy. Belgique, 2-me serie, **4** (1912), pp. 1–104.

[2] *On the best approximation of continuous functions by polynomials of given degree (Russian),* Collect. papers, vol. **1**, Akad. Nauk SSSR, Moscow, (1952), pp. 11–104.

Bers, L.

[1] *Bounded analytic functions of two complex variables,* Amer. J. Math., **64** (1942), No. 5, pp. 514–529.

Besicovitch, A. S.

[1] *Sur la nature des fonctions á carré sommable mesurables,* Fund. Math., **4** (1923), pp. 172–195.

[2] *A general property of summable functions,* Proc. Lond. Math. Soc., **1** (1926), pp. 120–128.

Bloshanskiĭ, I. L.

[1] *Convergence of double Fourier series of functions belonging to L_p, $p > 1$ (Russian),* Mat. Zametki, **21** (1977), No. 6, pp. 777–787.

[2] *Divergence almost eveverywhere of a Fourier series on a given set and convergence to zero outside of it (Russian),* Dokl. Akad. Nauk. SSSR, **280** (1985), No. 4, pp. 777–780.

[3] *The structure and geometry of the maximal sets of convergence and of unbounded divergence almost everywhere of multiple Fourier series from L, equal to zero on a given set (Russian),* Izv. Akad. Nauk. SSSR, Ser. Mat., **53** (1989), No. 4, pp. 675–707; English translation: Math. USSR Izv., **35** (1990), No. 1, pp. 1–35.

Bochner, S.

[1] *Boundary values of analytic functions in several variables and almost periodic functions,* Ann. Math., **45** (1944), No. 4, pp. 708–722.

Bochner, S., Martin, W. T.

[1] *Several complex variables,* Princeton Math. Series, vol. 10, Princeton Univ. Press, Princeton, (1948).

Bohr, H.

[1] *Über einen Satz von J. Pál,* Acta Sci. Math. (Szeged), **7** (1935), No. 1, pp. 129–135.

Bregvadze, D. V.

[1] *On summability of general orthogonal series (Russian),* Bull. Acad. Sci. Georgian SSR, **132** (1988), No. 2, pp. 241–244.

Bruckner, A. M.

[1] *Differentiation of real functions,* Lecture Notes in Mathematics, **659** (1978), Berlin–New York, Springer-Verlag.

Buadze, A. I.

[1] *On a problem of P. L. Ul'yanov (Russian),* Bull. Acad. Sci. Georgian SSR, **40** (1965), No. 3, pp. 45–550.

Burkholder, D. L., Gundy, R. F., Silverstein, M. L.

[1] *A maximal function characterization of the class H^p,* Trans. Amer. Math. Soc., **157** (1971), pp. 137–153.

Busemann, H., Feller, W.

[1] *Zur Differentation der Lebesgueschen Integrale,* Fund. Math., **22** (1934), pp. 226–256.

Buzdalin, V. V.

[1] *On unboundedly divergent trigonometric Fourier series of continuous functions (Russian),* Mat. Zametki, **7** (1970), No. 1, pp. 7–18.

[2] *Trigonometric Fourier series of continuous functions diverging on a given set (Russian),* Mat. Sbornik, **95** (1974), No. 1, pp. 84–107.

Calderón, A. P.

[1] *On theorems of M. Riesz and Zygmund,* Proc. Amer. Math. Soc., **1** (1950), pp. 533–535.

[2] *Singular integrals,* Bull. Amer. Math. Soc., **72** (1966), No. 3, pp. 426–465.

Calderón, A. P., Zygmund, A.

[1] *Note on the boundary values of functions of several complex variables,* Ann. Math. Studies, **25** (1950), pp. 144–165.

Calderón, C. P.

[1] *On the Dini test and divergence of Fourier series,* Proc. Amer. Math. Soc., **82** (1981), No. 3, pp. 382–384.

Cantor, G.

[1] *Über die Ausdehnung eines Satzes aus der Theorie der trigonometrischen Reihen,* Math. Ann., **5** (1872), pp. 123–132.

Carleson, L.

[1] *On convergence and growth of partial sums of Fourier series,* Acta Math., **116** (1966), No. 1–2, pp. 135–157.

Cesari, L.

[1] *Sulle serie di Fourier delle funzioni Lipshtziane di più variabili,* Ann. Scuola Norm. Sup. Pisa, **7** (1938), pp. 279–295.

Chelidze, V. G.

[1] *Some methods of summability of double series and double integrals (Russian),* Tbilisi State University, Tbilisi, (1977).

Chelidze, V. G., Jvarsheishvili, A. G.

[1] *The theory of the Denjoy integral and some applications,* World Scientific, Singapore, New Jersey, London, Hong Kong, (1989).

Chen, K.

[1] *On the absolute Cesàro summability of negative order for a Fourier series at a given point,* Amer. J. Math., **66** (1944), No. 2, pp. 199–312.

Chen, Y. M.

[1] *Theorems of asymptotic approximations,* Math. Ann., **140** (1960), No. 5, pp. 360–407,

[2] *A remarkable divergent Fourier series,* Proc. Jap. Acad., **38** (1962), pp. 239–344.

[3] *On conjugate functions,* Canad. J. Math., **15** (1963), No. 3, pp. 486–494.

[4] *Some asymptotic approximation methods,* Proc. Lond. Math. Soc., **15** (1965), No. 2, pp. 323–345.

Chernykh, N. I.

[1] *On behavior of partial sums of trigonometric Fourier series (Russian),* Uspekhi Mat. Nauk, **23** (1968), No. 6, pp. 3–50.

Chirka, E. M.

[1] *Complex analytic sets (Russian),* In: Itogi Nauki i Tekhniki, Sovrem. Probl. Mat., Fund. Napravl., vol. **7**, Akad. Nauk SSSR, Inst. Nauchn. i Tekhn. Inform., Moscow, 1985, pp. 125–166.

Coifman, R. R.

[1] *Remarks on weak type inequalities for operators commuting with translations,* Bull. Amer. Math. Soc., **74** (1968), No. 8, pp. 710–714.

Danilyuk, I. I.

[1] *Irregular boundary value problems on the plane (Russian),* Nauka, Moscow, (1975).

Davis, B. I.

[1] *On the weak type* $(1,1)$ *inequality for conjugate function,* Proc. Amer. Math. Soc., **44** (1974), pp. 307–311.

[2] *On the distribution of values of bounded analytic functions,* Trans. Amer. Math. Soc., **222** (1976), No. 1, pp. 179–192.

Dektyarëv, I. M.

[1] *Multidimensional theory of distribution of values (Russian),* In: Itogi Nauki i Tekhniki, Sovrem. Probl. Mat., Fund. Napravl., vol. **9**, Akad. Nauk SSSR, Inst. Nauchn. i Tekhn. Inform., Moscow, 1986, pp. 37–71.

Dolbeault, P.

[1] *General theory of residues in several variables (Russian),* In: Itogi Nauki i Tekhniki, Sovrem. Probl. Mat., Fund. Napravl., vol. **7**, Akad. Nauk SSSR, Inst. Nauchn. i Tekhn. Inform., Moscow, 1985, pp. 227–251.

Du Bois Reymond, P.

[1] *Untersuchungen über die Konvergenz und Divergenz der Fourierischen Darstellungsformen,* Abb. Akad. Wiss. München, **XII** (1876), pp. 1–103.

Dunford, N., Schwartz, J. T.

[1] *Linear operators, vol.* **1**, Interscience Publishers, New York, (1958).
[2] *Linear operators, vol.* **2**, Interscience Publishers, New York, (1963).

Duren, P. L.

[1] *Theory of H^p spaces,* Academic Press, New York, (1970).

Dvoretzky, A., Erdös, P.

[1] *On power series diverging everywhere on the circle of convergence,* Michigan Math. J., **3** (1955–1956), No. 1, pp. 31–35.

D'yachenko, M. I.

[1] *On (C,α)-summability of multiple trigonometric Fourier series (Russian),* Bull. Acad. Sci. Georgian SSR, **131** (1988), No. 2, pp. 261–263.
[2] *On some properties of multiple Fourier series and Fourier transforms (Russian),* Trudy. Mat. Inst. Akad. Nauk. SSSR, **190** (1989), pp. 88–101.
[3] *Some problems of the theory of multiple trigonometric series (Russian),* Uspekhi Mat. Nauk. **47** (1992), No. 5(287), pp. 97–162.

Dyn'kin, E. M.

[1] *Methods of theory of singular integrals (Hilbert transform and Calderón–Zygmund theory) (Russian),* In: Itogi Nauki i Tekhniki, Sovrem. Probl. Mat., Fund. Napravl., vol. **15**, Akad. Nauk SSSR, Inst. Nauchn. i Tekhn. Inform., Moscow, 1987, pp. 197–292.
[2] *Methods of theory of singular integrals. Littlewood–Paley theory and its applications (Russian),* In: Itogi Nauki i Tekhniki, Sovrem. Probl. Mat., Fund. Napravl., vol. **42**, Akad. Nauk SSSR, Inst. Nauchn. i Tekhn. Inform., Moscow, 1989, pp. 105–198.

Dyn'kin, E. M., Osilenker, B. P.

[1] *Weighted estimates of singular integrals and their applications (Russian),* In: Itogi Nauki i Tekhniki, Sovrem. Probl. Mat., Mat. Anal., vol. **21**, Akad. Nauk SSSR, Inst. Nauchn. i Tekhn. Inform., Moscow, 1983, pp. 42–129.

Dzyadyk, V. K.

[1] *Introduction to the theory of uniform approximation of functions by polynomials (Russian),* Nauka, Moscow, (1977).

Edwards, R. E.

[1] *Fourier series. A modern introduction, vol.* **1**, Springer-Verlag, (1979).

[2] *Fourier series. A modern introduction, vol.* **2**, Springer-Verlag, (1982).

Efimov, A. V.

[1] *On approximation of some classes of continuous functions by Fourier and Fejér sums (Russian),* Izv. Akad. Nauk. SSSR, Ser. Mat., **22** (1958), No. 1, pp. 81–116.

[2] *Approximation of conjugate functions by Fejér sums (Russian),* Uspekhi Mat. Nauk, **14** (1959), No. 1, pp. 183–188.

[3] *Linear methods of approximation of some classes of continuous periodic functions (Russian),* Trudy Mat. Inst. Akad. Nauk. SSSR, **62** (1961), pp. 3–47.

Faber, G.

[1] *Über stetige Funktionen* II, Math. Ann. **69** (1910), pp. 372–443.

Falaleev, A. P.

[1] *Approximation of conjugate functions by Cesàro sums (Russian),* Mat. Zametki, **28** (1980), No. 3, pp. 451–458.

[2] *Exactness of representation of conjugate functions by Cesàro sums (Russian),* Sibirsk. Mat. Zh., **25** (1984), No. 4, pp. 199–206.

Fatou, P.

[1] *Séries trigonometriques et séries de Taylor,* Acta Math., **30** (1906), pp. 335–400.

Fefferman, Ch.

[1] *On the divergence of multiple Fourier series,* Bull. Amer. Math. Soc., **77** (1971), No. 2, pp. 191–195.

[2] *On the convergence of multiple Fourier series,* Bull. Amer. Math. Soc., **77** (1971), No. 5, pp. 744–745.

[3] *Estimates of double Hilbert transform,* Stud. Math., **44** (1972), No. 1, pp. 1–15.

[4] *Pointwise convergence of Fourier series,* Ann. Math., **98** (1973), No. 3, pp. 551–571.

Fefferman, Ch., Stein, E. M.

[1] *H^p spaces of several variables,* Acta Math., **129** (1972), No. 3–4, pp. 137–193.

Fefferman, R., Stein, E. M.

[1] *Singular integrals on product spaces,* Advances in Math., **45** (1982), pp. 117–143.

Fejér, L.

[1] *Untersuchungen über Fourierschen Reihen,* Math. Ann., **58** (1904), pp. 501–569.

[2] *Sur les singularités de la serie de Fourier des fonctions continues,* Ann. Ecol. Norm. Sup., **28** (1911), pp. 63–103.

Fekete, M.

[1] *Über die Faktorenfolgen welche die "Klasse" einer Fourierschen Reihen unverandert lassen,* Acta Sci. Math. (Szeged), **1** (1923), pp. 148–166.

Flett, T. M.

[1] *On the degree of approximation by the Cesàro means of the Fourier series,* Quart. J. Math., **7** (1956), No. 1, pp. 81–95.

[2] *Some theorems on odd and even functions,* Proc. Lond. Math. Soc., **3** (1958), No. 8, pp. 135–148.

Forelli, F.

[1] *The Marcel Riesz theorem on the conjugate functions,* Trans. Amer. Math. Soc., **106** (1963), No. 3, pp. 369–390.

Fuks, B. A.

[1] *An introduction to the theory of analytic functions of several complex variables (Russian),* GIFML, Moscow, (1968).

[2] *Special chapters in the theory of analytic functions of several complex variables,* Translations of Math. Monographs, vol. **14**, Amer. Math. Soc., Providence, RI, (1965).

Galstyan, S. Sh.

[1] *On everywhere divergent trigonometric Fourier series with rapidly decreasing coefficients (Russian),* Mat. Sbornik, **122** (1983), No. 2, pp. 157–167.

[2] *On everywhere divergent trigonometric series (Russian),* Mat. Zametki, **37** (1985), No. 2, pp. 186–191.

Gamelin, T. W.

[1] *Uniform algebras and Jensen measures,* London Math. Soc., Lecture notes, **392** (1979).

Gaposhkin, V. F.

[1] *A generalization of M. Riesz's theorem on conjugate functions (Russian),* Mat. Sbornik, **46** (1958), No. 3, pp. 359–372.

Garnett, J. B.

[1] *Bounded analytic functions,* Academic Press, New York, (1980).

Garsia, A. M.

[1] *Topics in almost everywhere convergence,* Markham Publ. Co., Chicago, (1970).

Geĭt, V. É.

[1] *Embedding conditions for the classes* $H^{\omega}_{k,R}$ *and* $\widetilde{H}^{\omega}_{k,R}$ *(Russian)*, Mat. Zametki, **13** (1973), No. 2, pp. 169–178; English translation: Math. Notes, **13** (1973), pp. 101–106.

Gergen, I. I.

[1] *Convergence and summability criteria for Fourier series,* Quar. J. Math., **1** (1930), pp. 252–275.

[2] *Convergence criteria for double Fourier series,* Trans. Amer. Math. Soc., **35** (1933), No. 1, pp. 29–64.

Getsadze, R. D.

[1] *On divergence in measure of multiple Fourier series (Russian),* Bull. Acad. Sci. Georgian SSR, **122** (1986), No. 3, pp. 269–271.

[2] *On divergence in measure of multiple Fourier series (Russian),* In the collection: Some problems of the theory of functions and functional analysis , vol. IV, Tbilisi State University, Tbilisi, 1988, pp. 59–76.

Gogoladze, L. D.

[1] *On summability of double conjugate trigonometric series (Russian),* Bull. Acad. Sci. Georgian SSR, **54** (1969), No. 1, pp. 21–24.

[2] *On behavior of multiple power series in the skeleton of boundary (Russian),* Annot. Dokl. Semin. Inst. Prikl. Mat. I. N. Vekua, **7**, Tbilisi, (1973).

[3] *Boundness of converging means of multiple function series (Russian),* Mat. Zametki, **4** (1983), No. 6, pp. 845–855.

[4] *The problem of existence of conjugate functions of several variables (Russian),* Bull. Acad. Sci. Georgian SSR, **114** (1984), No. 2, pp. 477–479.

[5] *Existence of conjugate functions of several variables (Russian),* Mat. Sbornik, **225** (1984), No. 4, pp. 481–488.

[6] *On behavior of multiple conjugate trigonometric series and power series (Russian),* Tbilisi State University, Tbilisi, (1987).

Golubov, B. I.

[1] *Multiple Fourier series and integrals (Russian),* In: Itogi Nauki i Tekhniki, Matem. Analiz, vol. **19**, Akad. Nauk SSSR, Inst. Nauchn. i Tekh. Inform., Moscow, 1982, pp. 3–54.

Gosselin, R. P.

[1] *On the divergence of Fourier series,* Proc. Amer. Math. Soc., **9** (1958), No. 2, pp. 278–282.

Grigor'ev, A. I.

[1] *Convergence of Cesàro means of trigonometric Fourier series (Russian),* Mat. Zametki, **34** (1983), No. 4, pp. 501–514; English Translation: Math. Notes, **34** (1983), pp. 740–747.

Gui Zhu Rui

[1] *Approximation of a continuous function by Cesàro means of its Fourier series,* Sci. Sinica, **11** (1962), pp. 1625–1634.

Gunning, R. C., Rossi, H.

[1] *Analytic functions of several complex variables,* Prentice-Hall, Inc., Englewood Cliffs, NJ, (1965).

Gurielashvili, R. I.

[1] *The convergence and boundness in $\Psi(L)$ of the Fourier series of a monotone function (Russian),* Dokl. Akad. Nauk SSSR, **220** (1975), No. 5, pp. 1015–1018.

[2] *The convergence and boundness in L of the Fourier series of a monotone function (Russian),* Bull. Acad. Sci. Georgian SSR, **94** (1979), No. 3, pp. 537–540.

[3] *On the Hilbert transform (Russian),* Bull. Acad. Sci. Georgian SSR, **103** (1981), No. 2, pp. 273–276.

[4] *On conjugate functions (Russian),* Bull. Acad. Sci. Georgian SSR, **113** (1984), No. 3, pp. 473–476.

[5] *On the Hilbert transform,* Anal. Math., **13** (1987), No. 2, pp. 121–137.

de Guzmán, M.

[1] *Differentiation of integrals in $\mathbb{R}^n$,* Springer-Verlag, Berlin, (1975).

de Guzmán, M., Welland, G. V.

[1] *On the differentiation of integrals,* Rev. Univ. Mat. Argentina, **25** (1971), pp. 253–276.

Halmos, P. R.

[1] *Measure theory,* Van Nostrand Company, New York, (1950).

Hardy, G. H.

[1] *On double Fourier series and especially those which represent the double Zeta-function with real and incommeasurable parameters,* Quart. J. Math., **37** (1906), No. 1, pp. 53–79.

[2] *The mean values of the modulus of an analytic function,* Proc. Lond. Math. Soc., **14** (1915), pp. 269–277.

[3] *On certain criteria for the convergence of the Fourier series of a continuous function,* Mess. of Math., **47** (1918), pp. 149–159.

[4] *Notes on some points in the integral calculus,* LXVI, *The arithmetic mean of a Fourier constant,* Mess. Math., **58** (1928), pp. 50–52.

[5] *On a theorem of Paley and Wiener,* Proc. Cambr. Phil. Soc., **33** (1937), No. 1, pp. 1–5.

[6] *Divergent series,* Oxford Univ. Press, New York, (1949).

Hardy, G. H., Littlewood, J. E.

[1] *Contributions to the arithmetic theory of series,* Proc. Lond. Math. Soc., **11** (1912), No. 2, pp. 411–478.

[2] *Some problems of Diophantine approximation; A remarkable trigonometrical series,* Proc. Nat. Acad. Sci. USA, **2** (1916), pp. 583–586.

[3] *Abel's theorem and its converse,* Proc. Lond. Math. Soc., **18** (1919), pp. 205–235.

[4] *The Fourier series of a positive function,* J. Lond. Math. Soc., **1** (1926), pp. 134–138.
[5] *Some properties of fractional integral,* Math. Z., **27** (1928), pp. 565–606.
[6] *A convergence criterion for Fourier series,* Math. Z., **28** (1928), No. 4, pp. 612–634.
[7] *A maximal theorem with function-theoretic applications,* Acta Math., **54** (1930), No. 1–2, pp. 81–116.
[8] *Some more theorems concerning Fourier series and Fourier power series,* Duke Math. J., **2** (1936), No. 3, pp. 354–382.

Hardy, G. H., Littlewood, J. E., Pólya, G.
[1] *Inequalities,* Cambridge Univ. Press, New York, (1934).

Hardy, G. H., Rogosinski, W. W.
[1] *Fourier series,* Cambridge Univ. Press, New York, (1944).

Helson, H.
[1] *Conjugate series in several variables,* Pacif. J. Math., **9** (1959), No. 2, pp. 511–523.

Helson, H., Lowdenslager, D.
[1] *Prediction theory and Forier series in several variables,* Acta Math., **99** (1958), No. 3–4, pp. 165–202.

Helson, H., Szegö, G.
[1] *A problem in prediction theory,* Ann. Math. Pure Appl., **51** (1960), No. 4, pp. 107–138.

Herzog, E.
[1] *A note on a power series which diverge everywhere on the circle,* Michig. Math. J., **2** (1953–1954), No. 2, pp. 175–177.

Hille, E.
[1] *On the analytical theory of semi-groups,* Proc. Nat. Acad. USA, **28** (1942), pp. 421–424.

Hille, E., Tamarkin, J. D.
[1] *On the summability of Fourier series,* I, Trans. Amer. Math. Soc., **34** (1932), pp. 757–783.
[2] *On the summability of Fourier series,* II, Ann. Math., **34** (1933), pp. 329–344.

Hirschman, I. I.
[1] *The decomposition of Walsh and Fourier series,* Memoirs Amer. Math. Soc., **15** (1955).

Hoffman, K.
[1] *Banach spaces of analytic functions,* Prentice-Hall, Englewood Cliffs, NJ, (1962).

Hörmander, L.

[1] *An introduction to complex analysis in several variables,* Van Nostrand Company, Prinveton, NJ, (1966).

Hristov, V. H.

[1] *Tests for uniform convergence (Russian),* In the collection: Constructive theory of functions -77, Sophia, 1980, pp. 171–178.

Hsiang, F. Ch.

[1] *A local property of Fourier series,* Ann. Mat. Pura Appl., **53** (1961), No. 4, pp. 333–343.

Hua Loo-Keng

[1] *Harmonic analysis of functions of several complex variables in classical domains (Russian translation),* IL, Moscow, (1959).

Hunt, R. A.

[1] *On the convergence of Fourier series,* Not. Amer. Math. Soc., **14** (1967), No. 2, p. 537.

[2] *On the convergence of Fourier series,* Proc. Confer.: "Orthog. expression and their continuous analogues", South Illinois Univ. Press, Cardanbale, Ill., 1968, pp. 235–256.

Hunt, R. A., Muckenhoupt, B., Wheeden, R. L.

[1] *Weighted norm inequalities for the conjugate function and Hilbert transform,* Trans. Amer. Math. Soc., **176** (1972), No. 1, pp. 227–251.

Hunt, R. A., Young, W. S.

[1] *A weighted norm inequalities for Fourier series,* Bull. Amer. Math. Soc., **80** (1974), No. 2, pp. 274–277.

Igari, S.

[1] *Transformations of conjugate functions,* II, Tôhoku Math. J., **14** (1962), No. 1, pp. 121–126.

Izumi, M., Izumi, S.

[1] *On some theorems on conjugate functions,* Acta Math. Acad. Sci. Hung., **13** (1962), No. 1–2, pp. 133–143.

[2] *Modulus of continuity of functions defined by trigonometric series,* J. Math. Anal. Appl., **24** (1968), No. 3, pp. 564–581.

Izumi, S.

[1] *Some trigonometrical series,* Tôhoku Math. J., **6** (1954).

[2] *Some trigonometrical series,* Proc. Jap. Acad., **31** (1955), pp. 399–403.

Izumi, S., Matsuyama, N., Tsuchikura, T.

[1] *Some negative examples,* Tôhoku Math. J., **5** (1953), pp. 43–51.

Jackson, D.

[1] *Über Genauigkeit der Annäherung stetiger Funktionen durch ganze rationale Funktionen gegebenen Grades und trigonometrische Summengegebener Ordnung,* Diss.: Göttingen, (1911).

[2] *On approximations by trigonometrical sums and polynomials,* Trans. Amer. Math. Soc., **13** (1912), pp. 491–515.

[3] *The theory of approximation,* Amer. Math. Soc. Coll. Publ., **XI**, New York, (1930).

Jessen, B., Marcinkewicz, J., Zygmund, A.

[1] *Note on the differentiability of multiple integrals,* Fund. Math., **25** (1935), pp. 217–234.

Kaczmarz, S.

[1] *Zur Theorie der Fourierschen Doppelreihen,* Stud. Math., **2** (1930), No. 1, pp. 91–96.

Kashin, B. S., Kosheleva, G. G.

[1] *An approach to "correction" theorems (Russian),* Vestnik Moskov. Univ. Ser. I, Mat. Mekh., (1988), No. 4, pp. 6–9; English translation: Moscow Univ. Math. Bull., **43** (1988), No. 4, pp. 1–5.

Kashin, B. S., Saakyan, A. A.

[1] *Orthogonal series,* Translations of Math. Monographs, **75**, American Math. Soc., Providence, RI, (1989).

Katznelson, Y.

[1] *An introduction to harmonic analysis (Second corrected edition),* Dover Publications, New York, (1976).

Kawata, T.

[1] *On Fourier constants,* Proc. Imper. Acad. Tokyo, **20** (1944), pp. 218–222.

Khavin, V. P.

[1] *The spaces of analytic functions (Russian),* In: Matem. Analiz 1964, Inst. Nauchn. i Tekhn. Inform., Moscow, 1966, pp. 76–164.

[2] *The methods and the structure of comutative harmonic analysis (Russian),* In: Itogi Nauki i Tekhniki, Sovrem. Probl. Mat., Fund. Napravl., vol. **15**, Akad. Nauk SSSR, Inst. Nauchn. i Tekhn. Inform., Moscow, 1985, pp. 6–134.

Kheladze, Sh. V.

[1] *On everywhere divergence of Fourier series of functions of the class $L\alpha(L)$ (Russian),* Trudy Tbil. Mat. Inst. Akad. Nauk GSSR, **99** (1989), pp. 51–59.

Khenkin, G. M.

[1] *Method of integral representations in complex analysis (Russian),* In: Itogi Nauki i Tekhniki, Sovrem. Probl. Mat., Fund. Napravl., vol. **7**, Akad. Nauk SSSR, Inst. Nauchn. i Tekhn. Inform., Moscow, 1985, pp. 23–124.

Khenkin, G. M., Chirka, E. M.

[1] *Boundary properties of holomorphic functions of several complex variables (Russian),* In: Itogi Nauki i Tekhniki, Sovrem. Probl. Mat., Fund. Napravl., vol. **4**, Akad. Nauk SSSR, Inst. Nauchn. i Tekhn. Inform., Moscow, 1975, pp. 13–142.

Khvedelidze, B. V.

[1] *The method of Cauchy type integrals in discountinuous boundary value problems in the theory of holomorphic functions of a complex variable* (*Russian*), In: Itogi Nauki i Tekhniki, Sovrem. Probl. Mat., vol. **7**, Akad. Nauk SSSR, Inst. Nauchn. i Tekhn. Inform., Moscow, 1975, pp. 5–162.

Kinukawa, M., Igari, S.

[1] *Transformations of conjugate functions,* Tôhoku Math. J., **13** (1961), No. 2, pp. 277–280.

Kislyakov, S. V.

[1] *Classic problems of Fourier analysis* (*Russian*), In: Itogi Nauki i Tekhniki, Sovrem. Probl. Mat., Fund. Napravl., vol. **15**, Akad. Nauk SSSR, Inst. Nauchn. i Tekhn. Inform., Moscow, 1987, pp. 135–196.

[2] *Exclusive sets in harmonic analysis* (*Russian*), In: Itogi Nauki i Tekhniki, Sovrem. Probl. Mat., Fund. Napravl., vol. **42**, Akad. Nauk SSSR, Inst. Nauchn. i Tekhn. Inform., Moscow, 1989, pp. 199–227.

Kober, H.

[1] *A note on Hilbert's operator,* Bull. Amer. Math. Soc., **48** (1942), No. 6, pp. 421–426.

[2] *A note on Hilbert transforms,* J. Lond. Math. Soc., **18** (1943), No. 2, pp. 66–71.

Kogbetliantz, E.

[1] *Analogies entre les séries trigonometriques et les séries spheriques,* Ann. l'Ecole Norm. Super., **40** (1923), No. 3, pp. 259–323.

Kokilashvili, V. M.

[1] *Maximal functions and singular integrals in the weighted functional spaces* (*Russian*), Trudy Tbilliss. Mat. Inst. Razmadze Akad. Nauk Gruz. SSR, **80** (1985).

Kolmogorov, A. N.

[1] *Une série de Fourier–Lebesgue divergente presque partout,* Fund. Math., **4** (1923), pp. 324–328.

[2] *Sur l'ordre de grandeur des coefficients de la série de Fourier–Lebesgue,* Bull. Acad. Polon., (1923), pp. 83–85.

[3] *Sur les fonctions harmoniques et les séries de Fourier,* Fund. Math., **7** (1925), pp. 24–29.

[4] *Une série de Fourier–Lebesgue divergente partout,* C. R. Acad. Sci., Paris, **183** (1926), pp. 1327–1329.

Kolmogorov, A. N., Seliverstov, G. A.

[1] *Sur la convergence des séries de Fourier,* C. R. Acad. Sci., Paris, **178** (1924), No. 3, pp. 303–305.

[2] *Sur la convergence des séries de Fourier,* Rend. Acad. Lincei, **3** (1926), pp. 307–310.

Kominar, V. G.

[1] *The Fourier series of continuous functions of bounded variation (Russian),* In: Researches into the modern problems of the constructive theory of functions, GIFML, Moscow, 1961, pp. 197–201.

Konyagin, S. V.

[1] *Divergence with respect to measure of multiple trigonometric Fourier series (Russian),* Mat. Zametki, **44** (1988), No. 2, pp. 196–201; English translation: Math. Notes, **44** (1988), No. 1-2, pp. 589–592.

[2] *Divergence with respect to measure of multiple trigonometric Fourier series (Russian),* Dokl. Rasshir. Zased. Sem. Inst. Prikl. Mat. I. N. Vekua (Reports Enlarged Sess. of Sem. I. Vekua Inst. Appl. Math.), Tbilisi, **3** (1988), No. 2, pp. 53–56.

Konyushkov, A. A.

[1] *On the Lipschitz classes (Russian),* Izv. Akad. Nauk SSSR, Ser. Math., **21** (1957), No. 3, pp. 423–444.

Koosis, P.

[1] *Introduction to H^p spaces,* Cambridge Univ. Press, Cambridge, (1980).

Korányi, A.

[1] *A remark on boundary values of functions of several complex variables,* Lecture Notes in Mathematics, **155** (1970), Springer-Verlag, pp. 115–121.

Korneĭchuk, N. P.

[1] *Extremal problems of approximation theory (Russian),* Nauka, Moscow, (1976).

[2] *S. M. Nikolskiĭ and the development of researches in the theory of approximation of functions in the USSR (Russian),* Uspekhi Mat. Nauk, **40** (1985), No. 5, pp. 71–131.

Körner, T. W.

[1] *Sets of divergence for Fourier series,* Bull. Lond. Math. Soc., **3** (1971), No. 2, pp. 152–154.

[2] *Everywhere divergent Fourier series,* Colloq. Math., **45** (1981), No. 1, pp. 103–118.

Kuz'min, O. R.

[1] *On everywhere divergent trigonometric series (Russian),* Trudy Leningrad. Industr. Inst., Fiz. i Mat., **10** (1936), No. 3, pp. 53–56.

Lamperti, J.

[1] *A note on conjugate functions,* Proc. Amer. Math. Soc., **10** (1959), No. 1, pp. 71–76.

Lebed', G. K., Avdeenko, A. A.

[1] *Approximation of continuous functions by Fejér sums (Russian),* Izv. Akad. Nauk SSSR, Ser. Mat., **35** (1971), No. 1, pp. 83–92.

Lebesgue, H.

[1] *Intégrale, longuer, aire,* Ann. Math., **7** (1902), pp. 231–359.

[2] *Recherches sur la convergence des séries de Fourier,* Math. Ann., **61** (1905), pp. 251–280.

[3] *Leçons sur les séries trigonométriques,* Paris, (1906).

[4] *Sur la représentation trigonométrique approchée des fonctions satisfaisantes à une condition de Lipschitz,* Bull. Soc. Math. France, **38** (1910), pp. 184–210.

Leindler, L.

[1] *Strong approximation by Fourier series,* Akadémiai Kiadó, Budapest, (1985).

Leiterer, J.

[1] *Holomorphic vector bundles and the Oka–Grauert principle (Russian),* In: Itogi Nauki i Tekhniki, Sovrem. Probl. Mat., Fund. Napravl., vol. **10**, Akad. Nauk SSSR, Inst. Nauchn. i Tekhn. Inform., Moscow, 1987, pp. 75–121.

Lekishvili, M. M.

[1] *On approximation of periodic functions by (C, α) means (Russian),* Mat. Sbornik, **121** (1983), No. 4, pp. 499–509.

Leladze, D. V.

[1] *On conjugate trigonometric series (Russian),* Bull. Acad. Sci. Georgian SSR, **115** (1984), No. 3, pp. 497–500.

Lesniewicz, R.

[1] *On conjugate functions and Fourier series in Orlicz spaces,* Bull. Acad. Polon. Sci., Ser. Math., Astron. et Phys., **14** (1966), No. 11, pp. 627–636.

Littlewood, J. E.

[1] *On a theorem of Kolmogoroff,* J. Lond. Math. Soc., **1** (1926), pp. 229–231.

Littlewood, J. E., Paley, R. E.

[1] *Theorems on Fourier series and power series,* I, J. Lond. Math. Soc., **6** (1931), No. 1, pp. 330–333.

[2] *Theorems on Fourier series and power series,* II, Proc. Lond. Math. Soc., **42** (1936), No. 1, pp. 52–89.

[3] *Theorems on Fourier series and power series,* III, Proc. Lond. Math. Soc., **43** (1937), No. 2, pp. 103–107.

Livingston, A.

[1] *The space H^p $(0 < p < 1)$ is not normable,* Pacif. J. Math., **3** (1953), No. 3, pp. 613–616.

Loo Ching Tsün

[1] *Note on the properties of Fourier coefficients,* Amer. J. Math., **71** (1949), No. 2, pp. 269–282.

Loomis, L. H.

[1] *A note on the Hilbert transform,* Bull. Amer. Math. Soc., **52** (1946), pp. 1082–1086.

Lorentz, G. G.

[1] *Fourier-Koeffizienten und Funktionenklassen,* Math. Z., **51** (1948), pp. 135–149.

[2] *Approximation of function,* Holt, Rinehart and Winston, New York, (1966).

Lozinskiĭ, S. M.

[1] *On convergene and summability of Fourier series (Russian),* Mat. Sbornik, **14** (1944), No. 3, pp. 175–268.

Lukashenko, S. Yu.

[1] *Bounded divergence of trigonometric series (Russian),* Mat. Zametki, **27** (1980), No. 4, pp. 583–595.

[2] *The structure of the sets of divergence of trigonometric and Walsh series (Russian),* Dokl. Acad. Sci. USSR, **253** (1980), No. 3, pp. 528–529.

Lukashenko, T. P.

[1] *Almost everywhere convergence of Fourier series of the functions summable in square (Russian),* Moscow State Univ., Moscow, (1978).

[2] *On functions of generalized bounded variation (Russian),* Izv. Akad. Nauk SSSR, Ser. Mat., **46** (1982), No. 2, pp. 276–313.

[3] *Properties of the functions of bounded variation on a set,* Mat. Sbornik, **122** (1983), No. 1, pp. 41–63.

Luzin, N. N.

[1] *Über eine Potenzreihe,* Rendic. Circ. Mat. di Palermo, **32** (1911), pp. 386–390.

[2] *Integral and trigonometric series (Russian),* GIFTL, Moscow–Leningrad, (1951).

Magnaradze, L. G.

[1] *On the theory of conjugate Fourier series (Russian),* Dokl. Rasshir. Zased. Sem. Inst. Prikl. Mat. I. N. Vekua (Reports Enlarged Sess. of Sem. I. N. Vekua Inst. Appl. Math.), Tbilisi, **3** (1988), No. 2, pp. 108–113.

Marcinkiewicz, J.

[1] *Sur les séries de Fourier,* Fund. Math., **27** (1936), No. 1, pp. 38–69.

[2] *Sur les multiplicateurs des séries de Fourier,* Stud. Math., **8** (1939), No. 1, pp. 78–91.

[3] *Sur une méthode remarquable de sommation des séries doubles de Fourier,* Ann. Scuola Norm. Sup. Pisa, **8** (1939), No. 2, pp. 149–160.

[4] *Sur une nouvelle conditions pour la convergence presque partout de séries de Fourier,* Ann. Scuola Norm. Sup. Pisa, **8** (1939), No. 3, pp. 239–240.

[5] *Sur l'interpolation d'operations,* C. R. Acad. Sci. Paris, Ser. A–B, **208** (1939), pp. 1272–1273.

Marcinkiewicz, J., Zygmund, A.

[1] *On the summability of double Fourier series,* Fund. Math., **32** (1939), No. 1, pp. 112–132.

Men'shov, D. E.

[1] *Sur les séries de Fourier des fonctions continues,* Mat. Sbornik, **8** (1940), No. 3, pp. 493–518.

[2] *Sur la représentation des fonctions mesurables par les séries trigonométriques,* Mat. Sbornik, **51** (1941), pp. 667–692.

[3] *On convergence in measure of trigonometric series (Russian),* Trudy. Mat. Inst. Akad. Nauk SSSR, **32** (1950), pp. 3–97.

[4] *On Fourier series of continuous functions (Russian),* Uch. Zap. Moskov. Univ., **148** (1951), Mat., vol. **IV**, pp. 108–132.

[5] *Properties of Cesàro means of negative order and certain other T-means for Fourier series of continuous functions (Russian),* Mat. Sbornik, **86** (1971), No. 3, pp. 419–445.

Morozov, A. Yu., Perelomov, A. M.

[1] *Complex geometry and the theory of strings,* In: Itogi Nauki i Tekhniki, Sovrem. Probl. Mat., Fund. Napravl., vol. **54**, Akad. Nauk SSSR, Inst. Nauchn. i Tekhn. Inform., Moscow, 1989, pp. 197–279.

Morse, M., Transue, W.

[1] *A new application of the Young-Pollard convergence criteria for a Fourier series,* Duke Math. J., **18** (1951), No. 2, pp. 563–571.

Mozzochi, Ch. J.

[1] *On the pointwise convergence of Fourier series,* Lecture Notes in Mathematics, **199** (1971), No. 7, pp. 1–87.

Muckenhoupt, B.

[1] *Weighted norm inequalities for the Hardy maximal function,* Trans. Amer. Math. Soc., **165** (1972), No. 1, pp. 207–226.

[2] *Two weight function norm inequalities for the Poisson integrals,* Trans. Amer. Math. Math .Soc., **210** (1975), No. 1, pp. 225–231.

Muskhelishvili, N. I.

[1] *Singular integral equations,* Walters–Noordhoff Publishing, Groningen, (1967).

Nagy, B.

[1] *Függvenyek megközelitiése Fourier-sorok szamtani közepeivel,* Mat. Fiz. Lapok, **49** (1942), pp. 122–138.

[2] *Approximation der Funktionen durch die arithmetischen Mittel ihrer Fourierschen Reihen,* Acta Sci. Math. (Szeged), **11** (1946), No. 1, pp. 71–84.

Nakhman, A. D., Osilenker, B. P.

[1] *Estimates of weighted norms of certain operators generated by multiple trigonometric Fourier series (Russian),* Izv. Vyssh. Uchebn. Zaved., Matem., 1982, No. 4, pp. 39–50.

Natanson, I. P.

[1] *Constructive function theory, vol.* **1**: *Uniform approximation,* Frederick Ungar Publ. Co., New York, (1964); *vol.* **2**: *Approximation in mean,* Constable & Co., Ltd, London, (1965); *vol.* **3**: *Interpolation and approximation,* Frederick Ungar Publ. Co., New York, (1965).

[2] *On exactness of representation of continuous periodic functions by singular integrals (Russian),* Dokl. Akad. Nauk SSSR, **73** (1950), No. 2, pp. 273–276.

[3] *Theory of functions of a real variable,* Frederick Ungar Publ. Co., New York, (1965).

Neder, L.

[1] *Zur Theorie der trigonometrischen Reihen,* Math. Ann., **84** (1921), pp. 117–136.

Neri, U.

[1] *Singular integrals,* Lecture Notes in Mathematics, **200**, Springer-Verlag, (1971).

Nevai, G. P.

[1] *Fourier-sorok és interpoláció,* Math. Lapok, (1973), No. 3–4, pp. 261–290.

Nikishin, E. M.

[1] *Resonance theorems and superlinear operators (Russian),* Uspekhi Mat. Nauk, **25** (1970), No. 6, pp. 129–191.

Nikolskiĭ, S. M.

[1] *On asymptotic behavior of the residue while approximation of functions satisfying Lipschitz condition by Fejér means (Russian),* Izv. Akad. Nauk SSSR, Ser. Mat., **4** (1940), No. 6, pp. 501–508.

[2] *Approximation of periodic functions by trigonometric polynomials (Russian),* Trudy Mat. Inst. Akad. Nauk SSSR, **15** (1945), pp. 1–76.

[3] *Generalization of an inequality of S. N. Bernstein (Russian),* Dokl.Akad. Nauk SSSR, **60** (1948), No. 9, pp. 1507–1510.

[4] *Inequalities for entire functions of finite order and their application to the theory of differentiable functions of several variables (Russian),* Trudy Mat. Inst. Akad. Nauk SSSR, **38** (1951), pp. 244–278.

[5] *Approximation of functions of several variables and imbedding theorems (Russian),* Nauka, Moscow, (1977).

Olevskiĭ, A. M.

[1] *Change of variable and absolute convergence of Fourier series (Russian),* Dokl. Akad. Nauk SSSR, **256** (1981), No. 2, pp. 284–288.

[2] *Modification of functions and Fourier series (Russian),* Uspekhi Mat. Nauk, **40** (1985), No. 3, pp. 157–193.

Oskolkov, K. I.

[1] *The uniform modulus of continuity of summable functions on sets of positive measure (Russian),* Dokl. Akad. Nauk SSSR, **229** (1976), No. 2, pp. 304–306.

[2] *Approximative properties of integrable functions on sets of full measure (Russian),* Mat. Sbornik, **103** (1977), No. 4, pp. 563–589.

[3] *On C-property of Luzin for conjugate functions (Russian),* Trudy Mat. Inst. Akad. Nauk SSSR, **164** (1983), pp. 124–135.

[4] *Subsequences of Fourier sums of integrable functions (Russian),* Trudy Mat. Inst. Akad. Nauk SSSR, **167**(1985), pp. 239–260.

Pak, I. N.

[1] *On the sums of trigonometric series (Russian),* Uspekhi Mat. Nauk, **35** (1980), No. 2, pp. 91–144.

Paley, R.

[1] *On Fourier series with positive coefficients,* J. Lond. Math. Soc., **7** (1932), pp. 205–208.

Panjikidze, L. K.

[1] *Imbedding theorems and the connection between the best approximations in different metrics (Russian),* In the collection: Some problems of the theory of functions, vol. **I**, Tbilisi State Univ., Tbilisi, 1979, pp. 243–306.

Paplauskas, A. B.

[1] *Trigonometric series (from Euler to Lebesgue) (Russian),*Nauka, Moscow, (1969).

Pichorides, S. K.

[1] *On the best values of constants in the theorems of M. Riesz, Zygmund and Kolmogorov,* Stud. Math. (PRL), **44** (1972), No. 2, pp. 165–179.

[2] *Une proprieté de la transformée de Hilbert,* C. R. Acad. Sci. Paris, **280** (1975), No. 18, pp. A1197–A1199.

Pinchuk, S. I.

[1] *Holomorphic maps in $\mathbb{C}^n$ and the theorem on holomorphic equivalence (Russian),* In: Itogi Nauki i Tekhniki, Sovrem. Probl. Mat., Fund. Napravl., vol. **9**, Akad. Nauk SSSR, Inst. Nauchn. i Tekhn. Inform., Moscow, 1986, pp. 195–223.

Plessner, A. I.

[1] *Zur Theory der konjugierten trigonometrischen Reihen,* Dissertation: Giessen, (1923).

[2] *Über Konvergenz von trigonometrischen Reihen,* J. Reine Angew. Math., **155** (1926), pp. 15–25.

[3] *Zur Theorie der konjugierten trigonometrischen Reihen,* Mitt. Math. Sem. Univ. Giessen. **10** (1928), pp. 1–36.

[4] *On conjugate trigonometric series (Russian),* Dokl. Akad. Nauk SSSR, **4** (1935), No. 6–7, pp. 235–238.

Poletskiĭ, E. A., Shabat, B. V.

[1] *Invariant metrics,* In: Itogi Nauki i Tekhniki, Sovrem. Probl. Mat., Fund. Napravl., vol. **9**, Akad. Nauk SSSR, Inst. Nauchn. i Tekhn. Inform., Moscow, 1986, pp. 37–71.

Privalov, I. I.

[1] *Sur les fonctions conjugués,* Bull. Soc. Math. France, **44** (1916), pp. 100–103.

[2] *Cauchy Integral (Russian),* Saratov, (1919).

[3] *Boundary properties of analytic functions (Russian),* GIFTL, Moscow–Leningrad, (1950).

Prokhorenko, V. I.

[1] *On divergent Fourier series (Russian),* Mat. Sbornik, **75** (1968), No. 2, pp. 185–198.

Ramsamujh, T. I.

[1] *A comparison of the Jordan and Dini tests,* Real Anal. Exch., **12** (1986–1987), No. 2, pp. 510–515.

Riesz, M.

[1] *Sur la sommation des séries de Fourier,* Acta Sci. Math. (Szeged), **1** (1923), pp. 104–113.

[2] *Les fonctions conjuguées et les séries de Fourier,* C.R. Acad. Sci. Paris, Ser. A–B, **178** (1924), pp. 1464–1467.

[3] *Sur les fonctions conjuguées,* Math. Z., **27** (1927), No. 2, pp. 218–244.

Ronkin, L. I.

[1] *Entire functions (Russian),* In: Itogi Nauki i Tekhniki, Sovrem. Probl. Mat., Fund. Napravl., vol. **9**, Akad. Nauk SSSR, Inst. Nauchn. i Tekhn. Inform., Moscow, 1986, pp. 5–36.

Rosenblum, M.

[1] *Summability of Fourier series in $L^p(d\mu)$,* Trans. Amer. Math. Soc., **165** (1962), pp. 32–42.

Roslyĭ, A. A., Khudaverdyan, O. M., Shwarts, A. S.

[1] *Supersimmetry and complex geometry (Russian),* In: Itogi Nauki i Tekhniki, Sovrem. Probl. Mat., Fund. Napravl., vol. **9**, Akad. Nauk SSSR, Inst. Nauchn. i Tekhn. Inform., Moscow, 1986, pp. 247–284.

Rudin, W.

[1] *Function theory in polydiscs,* W. A. Benjamin, Inc., New York-Amsterdam, (1967).

[2] *Functional analysis,* McGraw-Hill Book Co., New York, (1973).

[3] *Function theory in the unit ball of $\mathbb{C}^n$,* Springer-Verlag, (1980).

Ryan, P.

[1] *Conjugate functions in Orlicz spaces,* Pacif. J. Math., **13** (1963), No. 4, pp. 1371–1377.

Sadullaev, A.

[1] *Plurisubharmonic functions (Russian),* In: Itogi Nauki i Tekhniki, Sovrem. Probl. Mat., Fund. Napravl., vol. **8**, Akad. Nauk SSSR, Inst. Nauchn. i Tekhn. Inform., Moscow, 1985, pp. 65–114.

Saks, S.

[1] *Remarks on the differentiability of the Lebesgue indefinite integral,* Fund. Math., **22** (1934), pp. 257–261.

[2] *Theory of the integral,* Dover Publications, New York, (1964).

Satô, M.

[1] *Uniform convergence of Fourier series,* Proc. Jap. Acad., **30** (1954), pp. 698–701.

[2] *Uniform convergence of Fourier series,* Proc. Jap. Acad., **31** (1955), pp. 261–263.

Sawyer, E. T.

[1] *A characterization of a two-weight norm inequality for maximal operators,* Stud. Math. (PRL), **75** (1982), No. 1, pp. 1–11.

Sawyer, S.

[1] *Maximal inequalities of weak type,* Ann. Math., **84** (1966), No. 1, pp. 157–174.

Shabat, B. V.

[1] *Introduction to comlex analysis, part* 2 *(Russian),* Nauka, Moscow, (1976).

Shapiro, V. L.

[1] *Fourier series of several variables,* Bull. Amer. Math. Soc., **70** (1964), No. 1, pp. 48–93.

Shvedenko, S. V.

[1] *Hardy classes and spaces of analytic functions connected with them in the unit circle, semicircle and ball (Russian),* In: Itogi Nauki i Tekhniki, Sovrem. Probl. Mat., Mat. Anal., vol. **23**, Akad. Nauk SSSR, Inst. Nauchn. i Tekhn. Inform., Moscow, 1985, pp. 3–124.

Sidon, S.

[1] *Reihentheoretische Sätze und ihre Anvendungen in der Theorie der Fourierschen Reihen,* Math. Z., **10** (1921), pp. 121–127.

Siobhan O'shen

[1] *On a divergent trigonometrical series given by Steinhaus,* Proc. Amer. Math. Soc., **10** (1959), No. 1, pp. 68–70.

Sjolin, P.

[1] *An inequality of Paley and convergence a.e. of Walsh–Fourier series,* Ark. for Math., **7** (1969), No. 6, pp. 551–570.

[2] *Convergence almost everywhere of certain singular integrals and multiple Fourier series,* Ark. for Math., **9** (1971), No. 1, pp. 65–90.

Smirnov, V. I.

[1] *Sur les valeurs limites des fonctions analitiques,* C. R. Acad. Sci. Paris, **188** (1929), pp. 131–133.

Sokol–Sokolowski, K.

[1] *On trigonometric series conjugate to Fourier series of two variables,* Fund. Math., **34** (1947), pp. 166–188.

Stechkin, S. B.

[1] *On approximation of continuous functions by Fourier sums (Russian),* Uspekhi Mat. Nauk, **7** (1952), No. 4, pp. 139–140.

[2] *On absolute convergence of Fourier series (Russian),* Izv. Akad. Nauk SSSR, Ser. Mat., **19** (1955), No. 4, pp. 221-246.

[3] *On the best approximation of conjugate functions by trigonometric polynomials (Russian),* Izv. Akad. Nauk SSSR, Ser. Mat., **20** (1956), No. 2, pp. 197–206.

[4] *On trigonometric series divergent at every point (Russian),* Izv. Akad. Nauk SSSR, Ser. Mat., **21** (1957), No. 5, pp. 711–728.

[5] *On approximation of periodic functions by Fejér sums (Russian),* Trudy Matem. Inst. Akad. Nauk. SSSR, **62** (1961), pp. 48–60.

Stein, E. M.

[1] *On limits of sequences of operators,* Ann. Math., **74** (1961), No. 1, pp. 140–170.

[2] *Note on the class $L(logL)$,* Stud. Math. (PRL), **31** (1969), pp. 305-310.

[3] *Boundary values of harmonic functions,* Bull. Amer. Math. Soc., **76** (1970), No. 6, pp. 1992–1996.

[4] *Singular integrals and differentiability properties of functions,* Princeton Univ. Press, (1970).

Stein, E. M., Weiss, G.

[1] *Introduction to Fourier analysis on Euclidean spaces,* Princeton Univ. Press, (1971).

Steinhaus, H.

[1] *Une série trigonométrique partout divergente,* C. R. Soc. Sci. Varsovie, **5** (1912), No. 3, pp. 219–229.

[2] *A divergent trigonometrical series,* J. Lond. Math. Soc., **4** (1929), pp. 86–88.

Stepanets, A. I.

[1] *Uniform approximation by trigonometric polynomials (Russian),* Naukova Dumka, Kiev, (1981).

Storozhenko, E. A.

[1] *Approximation of functions of the class H^p $(0 < p < 1)$,* Bull. Acad. Sci. Georgian SSR, **88** (1977), No. 1, pp. 45–48.

[2] *Approximation of functions of the class H^p $(0 < p < 1)$,* Mat. Sbornik., **105** (1978), No. 4, pp. 601–621.

Suetin, P. K.

[1] *Orthogonal polynomials in two variables (Russian),* Nauka, Moscow, (1988).

Sunouchi, G.

[1] *On Fourier constants,* Proc. Imper. Acad. Tokyo, **20** (1944), pp. 542–544.

[2] *A Fourier series which belongs to the class H diverges almost everywhere,* Kodai Math. Sem. Rep., **1** (1953), pp. 27-28.

Sunouchi, G., Yano, S.

[1] *Convergence and summability of orthogonal series (Notes of Fourier analysis, XXXIX),* Proc. Acad. Japan, **26** (1950), No. 1, pp. 10–16.

Szász, O.

[1] *Zur Konvergenztheorie der Fourierschen Reihen,* Ann. Math., **61** (1933), No. 1, pp. 185–201.

Szegö, G.

[1] *Zur Summation der Fourierschen Reihe,* Acta Univ. Hungar., **3** (1927), pp. 17–24.

Taberski, R.

[1] *On the convergence of singular integrals,* Zesz. Nauk. Univ. Poznanin, Math. Fiz., **2** (1960), pp. 33–51.

[2] *Asymptotic formula for Cesàro singular integrals,* Bull. Acad. Polon. Sci., Ser. Sci. Math. Astr. Phys., **10** (1962), No. 12, pp. 637–640.

Taĭkov, L. V.

[1] *On divergence of Fourier series (Russian),* Dokl. Akad. Nauk SSSR, **137** (1961), No. 4, pp. 782–785.

Tandori, K.

[1] *Über ein Problem von S. B. Stetschkin,* Acta Sci. Math. (Szeged), **26** (1965), No. 1-2, pp. 27–32.

[2] *Ein Divergenzsatz für Fourierreihen,* Acta Sci. Math. (Szeged), **30** (1969), No. 1-2, pp. 43–48.

[3] *Über eine Zusammenhang zwischen der Grossenordnung der Partialsummen der Fourierreihe und der Integrabilitäteigenschaft der Funktionen,* Acta Sci. Math., **45** (1983), No. 1–4, pp. 409–413.

Telyakovskiĭ, S. A.

[1] *Integrability of trigonometric series. Estimate of integral modulus of continuity (Russian),* Mat. Sbornik, **91** (1973), No. 4, pp. 537–553.

[2] *Approximation of functions of class $\overline{W}'$ by Cesàro means,* Trudy Mat. Inst. Akad. Nauk SSSR, **157**(1981), pp. 191-197.

Tevzadze, N. R.

[1] *On convergence of double Fourier series of squared integrable functions,* Bull. Acad. Sci. Georgian SSR, **58** (1970), No. 2, pp. 277–279.

Tevzadze, T. Sh.

[1] *Some classes of functions and trigonometric Fourier series (Russian),* In the collection: Some problems of the functions theory and functional analysis, vol. II, Tbilisi State Univ., Tbilisi, 1981, pp. 31–93.

Tikhomirov, V. M.

[1] *The theory of approximation (Russian)*, In: Itogi Nauki i Tekhniki, Sovrem. Probl. Mat., Fund. Napravl., vol. **14**, Akad. Nauk SSSR, Inst. Nauchn. i Tekhn. Inform., Moscow, 1987, pp. 103–260.

Timan, A. F.

[1] *The theory of approximation to functions of real variables (Russian)*,GIFML, Moscow, (1960).

Titchmarsh, E. C.

[1] *Conjugate trigonometrical integrals,* Proc. Lond. Math. Soc., **24** (1924), No. 2, 109-130.

[2] *On conjugate functions,* Proc. Lond. Math. Soc., **29** (1929), No. 1, pp. 49–80.

[3] *Introduction to the theory of Fourier integrals,* Oxford Univ. Press, Oxford, (1937).

[4] *The theory of functions,* Oxford Univ. Press, Oxford, (1939).

Tkebuchava, G. E.

[1] *Integral inequalities in reflexive weighted Orlicz spaces (Russian),* Bull. Acad. Sci. Georgian SSR, **121** (1986), No. 3, pp. 477–479.

[2] *On the multiple systems of functions in Orlicz spaces (Russian),* In the collection: Some problems of the functions theory and functional analysis, vol. IV, Tbilisi State Univ., Tbilisi, 1988, pp. 194–206.

[3] *On multiple Fourier, Fourier–Walsh and Fourier–Haar series in nonreflexive separable Orlicz spaces (Russian),* Bull. Acad. Sci. Georgian SSR, **149** (1994), No. 2, pp. 1–4.

Tolstov, G. P.

[1] *Fourier series,* Prentice-Hall, Englewood Cliffs, NJ, (1962).

Topuria, S. B.

[1] *Fourier–Laplace series on the sphere (Russian),* Tbilisi State Univ., Tbilisi, (1987).

Torchinsky, A.

[1] *Real-variable methods in harmonic analysis,* Academic Press, London, (1986).

Totik, V.

[1] *On the strong approximation by the (C, α)-means of Fourier series,* I, Anal. Math., **6** (1980), No. 1, pp. 57–85.

[2] *On the strong aroximation by the (C, α)-means of Fourier series,* II, Anal. Math., **6** (1980), No. 2, pp. 165–184.

[3] *On the divergence of Fourier series,* Publ. Math. Debrecen, **29** (1982), No. 3–4, pp. 251–264.

[4] *Note on a result of K. Tandori,* Analysis, **3** (1983), No. 1–4, pp. 135–142.

Trigub, R. M.

[1] *Linear methods of summability and absolute convergence of Fourier series (Russian),* Izv. Akad. Nauk SSSR, Ser. Mat., **32** (1968), No. 1, pp. 24–29.

Tsereteli, O. D.

[1] *Metric properties of conjugate functions (Russian),* In: Itogi Nauki i Tekhniki, Sovrem. Probl. Mat., vol. **7**, Akad. Nauk SSSR, Inst. Nauchn. i Tekhn. Inform., Moscow, 1975, pp. 18–43.

Tsikh, A. K.

[1] *Grothendieck residue and its applications to algebraic geometry (Russian),* In: Itogi Nauki i Tekhniki, Sovrem. Probl. Mat., Fund. Napravl., vol. **8**, Akad. Nauk SSSR, Inst. Nauchn. i Tekhn. Inform., Moscow, 1985, pp. 45–64.

Tumanov, A. E.

[1] *Geometry of CR-manifolds (Russian),* In: Itogi Nauki i Tekhniki, Sovrem. Probl. Mat., Fund. Napravl., vol. **9**, Akad. Nauk SSSR, Inst. Nauchn. i Tekhn. Inform., Moscow, 1985, pp. 225–246.

Ul'yanov, P. L.

[1] *On some equivalent conditions for convergence of series and integrals (Russian),* Uspekhi. Mat. Nauk., **8** (1953), No. 6, pp. 133–141.

[2] *On the divergence of Fourier series (Russian),* Uspekhi. Mat. Nauk., **12** (1957), No. 3, pp. 75–132.

[3] *On the approximation of functions (Russian),* Sibirsk. Mat. Zh., **5** (1964), No. 2, pp. 418–437.

[4] *Solved and unsolved problems of theory of trigonometric and orthogonal series (Russian),* Uspekhi. Mat. Nauk., **19** (1964), No. 1, pp. 3–69.

[5] *On the absolute and uniform convergence of Fourier series (Russian),* Mat. Sbornik, **72** (1967), No. 2, pp. 193–225.

[6] *Imbedding of some classes of functions (Russian),* Izv. Akad. Nauk SSSR, Ser. Mat., **32** (1968), No. 3, pp. 649–686.

[7] *A. N. Kolmogorov and divergent Fourier series (Russian),* Uspekhi. Mat. Nauk., **38** (1983), No. 4, pp. 51–90.

[8] *Metric theory of functions (Russian),* In the collection: The outlines of development of mathematics in the USSR, Nauk. Dumka, Kiev, 1983, pp. 217–236.

[9] *N. N. Luzin's works in the metric theory of functions (Russian),* Uspekhi. Mat. Nauk., **40** (1985), No. 3, pp. 15–70.

de la Vallée–Poussin, J. Ch.

[1] *Leçons sur l'approximation des fonctions d'une variable réelle,* Paris, (1919).

Vinogradova, I. A., Skvortsov, V. A.

[1] *Generalized integrals and Fourier series (Russian),* In: Itogi Nauki i Tekhniki, Sovrem. Probl. Mat., Mat. Anal., Akad. Nauk SSSR, Inst. Nauchn. i Tekhn. Inform., Moscow, 1971, pp. 65–107.

Vitushkin, A. G.

[1] *Remarkable facts in complex analysis (Russian),* In: Itogi Nauki i Tekhniki, Sovrem. Probl. Mat., Fund. Napravl., vol. **7**, Akad. Nauk SSSR, Inst.

Nauchn. i Tekhn. Inform., Moscow, 1985, pp. 5–22.

[2] *Holomorphic mappings and geometry of surfaces (Russian)*, In: Itogi Nauki i Tekhniki, Sovrem. Probl. Mat., Fund. Napravl., vol. **7**, Akad. Nauk SSSR, Inst. Nauchn. i Tekhn. Inform., Moscow, 1985, pp. 167–226.

Vladimirov, V. S.

[1] *Methods of theory of functions of several complex variables (Russian)*, Nauka, Moscow, (1964).

Vladimirov, V. S., Sergeev, A. G.

[1] *Complex analysis in a tube of future (Russian)*, In: Itogi Nauki i Tekhniki, Sovrem. Probl. Mat., Fund. Napravl., vol. **8**, Akad. Nauk SSSR, Inst. Nauchn. i Tekhn. Inform., Moscow, 1985, pp. 191–266.

Wang Si-lei

[1] *On some theorems on conjugate functions*, Chinese Math., **5** (1964), No. 3, pp. 447–459.

[2] *On some theorems on conjugate functions*, Scientia Sinica, **14** (1965), No. 4, pp. 507–521.

Weierstrass, K.

[1] *Über die analytische Darstellbarkeit Sogenannter willkürlicher Funktionen reeller Argumente*, Sitzungsberichte der Akad. Berlin, (1885), pp. 638–805.

Widom, H.

[1] *Singular integral equations in L^p spaces*, Trans. Amer. Math. Soc., **97** (1960), No. 1, pp. 131–160.

Yano, K.

[1] *On Hardy and Littlwood's theorem*, Proc. Jap. Acad., **33** (1957), No. 2, pp. 73–74.

Yanushauskas, A. I.

[1] *Multiple trigonometric series (Russian)*, Nauka, Novosibirsk, (1989).

Young, F.

[1] *Transformations of Fourier coefficients*, Proc. Amer. Math. Soc., **3** (1952), No. 5, pp. 783–791.

Young, W. H.

[1] *On the integration of Fourier series*, Proc. Lond. Math. Soc., **9** (1911), pp. 449–462.

[2] *On the Fourier series of bounded functions*, Proc. Lond. Math. Soc., **12** (1913), pp. 41–70.

Yuzhakov, A. P.

[1] *The methods of calculation of multidimensional residues (Russian)*, In: Itogi Nauki i Tekhniki, Sovrem. Probl. Mat., Fund. Napravl., vol. **8**, Akad. Nauk SSSR, Inst. Nauchn. i Tekhn. Inform., Moscow, 1985, pp. 6–28.

Zadereĭ, P. V., Stepanets, A. I.

[1] *On the approximation of conjugate functions of Hölder classes by Fejér sums,* Anal. Math., **5** (1979), No. 2, pp. 167–178.

Zaĭdenberg, M. G., Lin, V. Ya.

[1] *Finiteness theorems for holomorphic maps (Russian),* In: Itogi Nauki i Tekhniki, Sovrem. Probl. Mat., Fund. Napravl., vol. **9**, Akad. Nauk SSSR, Inst. Nauchn. i Tekhn. Inform., Moscow, 1986, pp. 127–193.

Zamansky, M.

[1] *Classes de saturation de certain procédés de l'approximation des séries de Fourier des fonctions continués et application à quelques problèmes d'approximation,* Ann. Scuola Norm. Sup. Pisa, **66** (1949), No. 1, pp. 19–93.

Zhizhiashvili, L. V.

[1] *Conjugate functions of two variables and double Fourier series (Russian),* Dokl. Akad. Nauk SSSR, **149** (1963), No. 4, pp. 765–768.

[2] *Conjugate functions of two variables and double conjugate trigonometric series (Russian),* Dokl. Acad. Nauk SSSR, **155** (1964), No. 3, pp. 521–523.

[3] *Some properties of (C,α)-means of Fourier series and conjugate trigonometric series (Russian),* Mat. Sbornik, **63** (1964), No. 4, pp. 489–504.

[4] *On some properties of Cesàro means of Fourier series (Russian),* Uspekhi Mat. Nauk, **21** (1966), No. 1, pp. 171–174.

[5] *On some properties of Cesàro means of Fourier series (Russian),* Trudy Moskov. Mat. Obshch., **15** (1966), pp. 145–180.

[6] *On conjugate functions (Russian),* Dokl. Akad. Nauk SSSR, **167** (1966), No. 2, pp. 278–281; English translation: Soviet. Math. Dokl., **7** (1966), pp. 368–371.

[7] *Conjugate functions and Fourier series (Russian),* Bull. Acad. Sci. Georgian SSR, **43** (1966), No. 2, pp. 183–186.

[8] *Fourier series and maximal theorems (Russian),* Dokl. Akad. Nauk SSSR, **174** (1967), No. 2, pp. 274–276.

[9] *Summation of double Fourier series (Russian),* Sibirsk. Mat. Zh., **8** (1967), No. 3, pp. 548–565.

[10] *Conjugate functions and trigonometric series (Russian),* Mat. Zametki, **2** (1967), No. 6, pp. 695–702.

[11] *Generalization of a theorem of Marcinkiewicz (Russian),* Izv. Akad. Nauk SSSR, Ser. Mat., **32** (1968), No. 5, pp. 1112–1122.

[12] *Generalization of a theorem of M. Riesz to the case of several variables (Russian),* Mat. Zametki, **4** (1968), No. 3, pp. 269–280.

[13] *Conjugate functions and trigonometric series (Russian),* Tbilisi State Univ., Tbilisi, (1969).

[14] *On divergence of Fourier series (Russian),* Dokl. Akad. Nauk SSSR, **194** (1970), No. 4, pp. 758–759; English translation: Soviet Math. Dokl., **11** (1970), pp. 1281–1283.

[15] *Convergence and divergence of Fourier series (Russian),* Dokl. Akad. Nauk

SSSR, **199** (1971), No. 6, pp. 1234–1236; English translation: Soviet Math. Dokl., **12** (1971), pp. 1261–1263.

[16] *Some problems in the theory of simple and multiple trigonometric and orthogonal series (Russian)*, Uspekhi Mat. Nauk, **28** (1973), No. 2, pp. 65–119; English translation: Russian Math. Surveys, **28** (1973), No. 2, pp. 65–127.

[17] *Conjugate functions of n variables (Russian)*, Dokl. Akad. Nauk SSSR, **218** (1974), No. 3, pp. 517–518 English translation: Soviet Math. Dokl., **15** (1974), pp. 1372–1373.

[18] *The summability of multiple trigonometric Fourier series (Russian)*, Bull. Acad. Sci. Georgian SSR, **75** (1974), No. 1, pp. 49–51.

[19] *Convergence and summability of trigonometric Fourier series (Russian)*, Bull. Acad. Sci. Georgian SSR, **79** (1975), No. 2, pp. 273–275.

[20] *Simple and multiple Fourier series (Russian)*, Dokl. Akad. Nauk SSSR, **225** (1975), No. 2, pp. 256–258; English translation: Soviet Math. Dokl., **16** (1975), pp. 1477–1479.

[21] *Convergence and divergence of trigonometric Fourier series (Russian)*, Dokl. Akad. Nauk SSSR, **225** (1975), No. 3, pp. 495–496; English translation: Soviet Math. Dokl., **16** (1975), pp. 1491–1493.

[22] *Summability of multiple trigonometric series (Russian)*, Bull. Acad. Sci. Georgian SSR, **79** (1975), No. 3, pp. 529–531.

[23] *Convergence of multiple trigonometric Fourier series (Russian)*, Bull. Acad. Sci. Georgian SSR, **80** (1975), No. 1, pp. 17–19.

[24] *Convergence and summability of trigonometric Fourier series (Russian)*, Mat. Zametki, **19** (1976), No. 6, pp. 887–898; English translation: Math. Notes, **19** (1976), Nos. 5–6, pp. 518–524.

[25] *Trigonometric Fourier series (Russian)*, Mat. Sbornik, **100** (1976), No. 4, pp. 580–609.

[26] *Hilbert transforms of a function of several variables (Russian)*, Bull. Acad. Sci. Georgian SSR, **88** (1977), No. 1, pp. 41–43.

[27] *Simple and multiple trigonometric Fourier series (Russian)*, Bull. Acad. Sci. Georgian SSR, **88** (1977), No. 2, pp. 293–295.

[28] *Integrability and continuity of conjugate functions of several variables (Russian)*, Bull. Acad. Sci. Georgian SSR, **97** (1980), No. 1, pp. 17–20.

[29] *Convergence of multiple trigonometric Fourier series and their conjugate series in the spaces C and L (Russian)*, Bull. Acad. Sci. Georgian SSR, **97** (1980), No. 2, pp. 277–279.

[30] *Generalization of a result of A. Zygmund (Russian)*, Bull. Acad. Sci. Georgian SSR, **102** (1981), No. 3, pp. 553–555.

[31] *Summability of multiple trigonometric Fourier series (Russian)*, Bull. Acad. Sci. Georgian SSR, **103** (1981), No. 1, pp. 17–19.

[32] *Validity of A. N. Kolmogorov's theorem for conjugate functions of several variables (Russian)*, Mat. Zametki, **32** (1982), No. 1, pp. 13–21.

[33] *Divergence of multiple trigonometric Fourier series (Russian)*, Bull. Acad. Sci. Georgian SSR, **110** (1983), No. 3, pp. 465–467.

[34] *On some approximating properties of Cesàro means of Fourier series (Russian)*, Dokl. Akad. Nauk SSSR, **269** (1983), No. 6, pp. 1307–1309.
[35] *Some problems in multidimensional harmonic analysis (Russian)*, Tbilisi State Univ., Tbilisi, (1983).
[36] *Summation of multiple Fourier series (Russian)*, Bull. Acad. Sci. Georgian SSR, **115** (1984), No. 3, pp. 465–467.
[37] *Conjugate functions of several variables and multiple conjugate trigonometric series (Russian)*, Dokl. Akad. Nauk SSSR, **284** (1985), No. 1, pp. 34–35.
[38] *Multiple power series in complex domain (Russian)*, Bull. Acad. Sci. Georgian SSR, **123** (1986), No. 1, pp. 17–19.
[39] *Almost everywhere summability by Cesàro method of a negative order of multiple trigonometric Fourier series (Russian)*, Bull. Acad. Sci. Georgian SSR, **123** (1986), No. 3, pp. 465–467.
[40] *Convergence of multiple conjugate trigonometric series in the spaces C and L (Russian)*, Bull. Acad. Sci. Georgian SSR, **125** (1987), No. 2, pp. 253–255.
[41] *Integrability of majorants connected with multiple Fourier series (Russian)*, Bull. Acad. Sci. Georgian SSR, **125** (1987), No. 3, pp. 469–471.
[42] *On solution of a problem of P. L. Ul'yanov (Russian)*, Bull. Acad. Sci. Georgian SSR, **140** (1990), No. 2, pp. 141–143.
[43] *On the summation of multiple conjugate trigonometric series in the metric of the space L^p, $p \in]0,1[$ (Russian)*, Bull. Acad. Sci. Georgian SSR, **140** (1990), No. 3, pp. 465–467.
[44] *Solution of a problem of D. E. Men'shov (Russian)*, Dokl. Akad. Nauk SSSR, **320** (1991), No. 2, pp. 273–274; English translation: Soviet Math. Dokl., **44** (1992), No. 2, pp. 427–429.
[45] *On almost everywhere summation of multiple conjugate trigonometric series (Russian)*, Dokl. Akad. Nauk SSSR, **320** (1991), No. 3, pp. 529–530; English translation: Soviet Math. Dokl., **44** (1992), No. 2, pp. 471–473.
[46] *Some problems of multidimensional harmonic analysis, second edition (Russian)*, Tbilisi State University, Tbilisi, (1996).

Zygmund, A.

[1] *O module ciaglosci sumy szeregu soprezonego z szeregiem Fouriera,* Prace Mat. Fiz., **33** (1924), pp. 125–132.
[2] *Sur la sommation des séries conjuguées aux séries de Fourier,* Bull. de l'Acad. Polon., (1924), pp. 251–258.
[3] *Sur la sommabilité de séries de Fourier des fonctions vérifiant la condition de Lipschitz,* Bull. de l'Acad. Polon., (1925), pp. 1–9.
[4] *Sur les fonctions conjuguées,* Fund. Math., **13** (1929), pp. 284–303.
[5] *A remark on conjugate series,* Proc. Lond. Math. Soc., **34** (1932), No. 4, pp. 435–446.
[6] *Trigonometrical series,* Monografje Matematyczne, Warsaw, (1935).
[7] *The approximation of functions by typical means of their Fourier series,* Duke Math. J., **12** (1945), No. 4, pp. 685–754.
[8] *On the degree of approximation of function by Fejér means,* Bull. Amer.

Math. Soc., **51** (1945), pp. 274–276.

[9] *Smooth functions,* Duke Math. J., **12** (1945), No. 1, pp. 47–76.

[10] *On the summability of multiple Fourier series,* Amer. J. Math., **69** (1947), No. 4, pp. 836–850.

[11] *On the boundary values of functions of several complex variables,* Fund. Math., **36** (1949), No. 2, pp. 207–235.

[12] *A remark on functions of several complex variables,* Acta Sci. Math. (Szeged), **12** (1950), No. 1, pp. 66–68.

[13] *On a theorem of Marcinkiewicz concerning interpolation of operations,* J. de Math., **35** (1956), No. 1, pp. 223–248.

[14] *On singular integrals,* Rendic. Mat. e Appl., **16** (1957), No. 3–4, pp. 468–505.

[15] *Trigonometric series, vol.* **1,** Cambridge Univ. Press, New York, (1959).

[16] *Trigonometric series, vol.* **2,** Cambridge Univ. Press, New York, (1959).

Index

Other *Mathematics and Its Applications* titles of interest:

J. Chaillou: *Hyperbolic Differential Polynomials and their Singular Perturbations.* 1979, 184 pp. ISBN 90-277-1032-5

S. Fucik: *Solvability of Nonlinear Equations and Boundary Value Problems.* 1981, 404 pp. ISBN 90-277-1077-5

V.I. Istratescu: *Fixed Point Theory. An Introduction.* 1981, 488 pp. *out of print,* ISBN 90-277-1224-7

F. Langouche, D. Roekaerts and E. Tirapegui: *Functional Integration and Semiclassical Expansions.* 1982, 328 pp. ISBN 90-277-1472-X

N.E. Hurt: *Geometric Quantization in Action. Applications of Harmonic Analysis in Quantum Statistical Mechanics and Quantum Field Theory.* 1982, 352 pp. ISBN 90-277-1426-6

F.H. Vasilescu: *Analytic Functional Calculus and Spectral Decompositions.* 1983, 392 pp. ISBN 90-277-1376-6

W. Kecs: *The Convolution Product and Some Applications.* 1983, 352 pp. ISBN 90-277-1409-6

C.P. Bruter, A. Aragnol and A. Lichnerowicz (eds.): *Bifurcation Theory, Mechanics and Physics. Mathematical Developments and Applications.* 1983, 400 pp. *out of print,* ISBN 90-277-1631-5

J. Aczel (ed.) : *Functional Equations: History, Applications and Theory.* 1984, 256 pp. *out of print,* ISBN 90-277-1706-0

P.E.T. Jorgensen and R.T. Moore: *Operator Commutation Relations.* 1984, 512 pp. ISBN 90-277-1710-9

D.S. Mitrinovic and J.D. Keckic: *The Cauchy Method of Residues. Theory and Applications.* 1984, 376 pp. ISBN 90-277-1623-4

R.A. Askey, T.H. Koornwinder and W. Schempp (eds.): *Special Functions: Group Theoretical Aspects and Applications.* 1984, 352 pp. ISBN 90-277-1822-9

R. Bellman and G. Adomian: *Partial Differential Equations. New Methods for their Treatment and Solution.* 1984, 308 pp. ISBN 90-277-1681-1

S. Rolewicz: *Metric Linear Spaces.* 1985, 472 pp. ISBN 90-277-1480-0

Y. Cherruault: *Mathematical Modelling in Biomedicine.* 1986, 276 pp. ISBN 90-277-2149-1

R.E. Bellman and R.S. Roth: *Methods in Approximation. Techniques for Mathematical Modelling.* 1986, 240 pp. ISBN 90-277-2188-2

R. Bellman and R. Vasudevan: *Wave Propagation. An Invariant Imbedding Approach.* 1986, 384 pp. ISBN 90-277-1766-4

A.G. Ramm: *Scattering by Obstacles.* 1986, 440 pp. ISBN 90-277-2103-3

Other *Mathematics and Its Applications* titles of interest:

C.W. Kilmister (ed.): *Disequilibrium and Self-Organisation.* 1986, 320 pp.
ISBN 90-277-2300-1

A.M. Krall: *Applied Analysis.* 1986, 576 pp.
ISBN 90-277-2328-1 (hb), ISBN 90-277-2342-7 (pb)

J.A. Dubinskij: *Sobolev Spaces of Infinite Order and Differential Equations.* 1986, 164 pp. ISBN 90-277-2147-5

H. Triebel: *Analysis and Mathematical Physics.* 1987, 484 pp.
ISBN 90-277-2077-0

B.A. Kupershmidt: *Elements of Superintegrable Systems. Basic Techniques and Results.* 1987, 206 pp. ISBN 90-277-2434-2

M. Gregus: *Third Order Linear Differential Equations.* 1987, 288 pp.
ISBN 90-277-2193-9

M.S. Birman and M.Z. Solomjak: *Spectral Theory of Self-Adjoint Operators in Hilbert Space.* 1987, 320 pp. ISBN 90-277-2179-3

V.I. Istratescu: *Inner Product Structures. Theory and Applications.* 1987, 912 pp.
ISBN 90-277-2182-3

R. Vich: *Z Transform Theory and Applications.* 1987, 260 pp.
ISBN 90-277-1917-9

N.V. Krylov: *Nonlinear Elliptic and Parabolic Equations of the Second -Order.* 1987, 480 pp. ISBN 90-277-2289-7

W.I. Fushchich and A.G. Nikitin: *Symmetries of Maxwell's Equations.* 1987, 228 pp. ISBN 90-277-2320-6

P.S. Bullen, D.S. Mitrinovic and P.M. Vasic (eds.): *Means and their Inequalities.* 1987, 480 pp. ISBN 90-277-2629-9

V.A. Marchenko: *Nonlinear Equations and Operator Algebras.* 1987, 176 pp.
ISBN 90-277-2654-X

Yu.L. Rodin: *The Riemann Boundary Problem on Riemann Surfaces.* 1988, 216 pp.
ISBN 90-277-2653-1

A. Cuyt (ed.): *Nonlinear Numerical Methods and Rational Approximation.* 1988, 480 pp. ISBN 90-277-2669-8

D. Przeworska-Rolewicz: *Algebraic Analysis.* 1988, 640 pp. ISBN 90-277-2443-1

V.S. Vladimirov, YU.N. Drozzinov and B.I. Zavialov: *Tauberian Theorems for Generalized Functions.* 1988, 312 pp. ISBN 90-277-2383-4

G. Morosanu: *Nonlinear Evolution Equations and Applications.* 1988, 352 pp.
ISBN 90-277-2486-5

Other *Mathematics and Its Applications* titles of interest:

A.F. Filippov: *Differential Equations with Discontinuous Righthand Sides.* 1988, 320 pp. ISBN 90-277-2699-X

A.T. Fomenko: *Integrability and Nonintegrability in Geometry and Mechanics.* 1988, 360 pp. ISBN 90-277-2818-6

G. Adomian: *Nonlinear Stochastic Systems Theory and Applications to Physics.* 1988, 244 pp. ISBN 90-277-2525-X

A. Tesar and Ludovt Fillo: *Transfer Matrix Method.* 1988, 260 pp. ISBN 90-277-2590-X

A. Kaneko: *Introduction to the Theory of Hyperfunctions.* 1989, 472 pp. ISBN 90-277-2837-2

D.S. Mitrinovic, J.E. Pecaric and V. Volenec: *Recent Advances in Geometric Inequalities.* 1989, 734 pp. ISBN 90-277-2565-9

A.W. Leung: *Systems of Nonlinear PDEs: Applications to Biology and Engineering.* 1989, 424 pp. ISBN 0-7923-0138-2

N.E. Hurt: *Phase Retrieval and Zero Crossings: Mathematical Methods in Image Reconstruction.* 1989, 320 pp. ISBN 0-7923-0210-9

V.I. Fabrikant: *Applications of Potential Theory in Mechanics. A Selection of New Results.* 1989, 484 pp. ISBN 0-7923-0173-0

R. Feistel and W. Ebeling: *Evolution of Complex Systems. Selforganization, Entropy and Development.* 1989, 248 pp. ISBN 90-277-2666-3

S.M. Ermakov, V.V. Nekrutkin and A.S. Sipin: *Random Processes for Classical Equations of Mathematical Physics.* 1989, 304 pp. ISBN 0-7923-0036-X

B.A. Plamenevskii: *Algebras of Pseudodifferential Operators.* 1989, 304 pp. ISBN 0-7923-0231-1

N. Bakhvalov and G. Panasenko: *Homogenisation: Averaging Processes in Periodic Media. Mathematical Problems in the Mechanics of Composite Materials.* 1989, 404 pp. ISBN 0-7923-0049-1

A.Ya. Helemskii: *The Homology of Banach and Topological Algebras.* 1989, 356 pp. ISBN 0-7923-0217-6

M. Toda: *Nonlinear Waves and Solitons.* 1989, 386 pp. ISBN 0-7923-0442-X

M.I. Rabinovich and D.I. Trubetskov: *Oscillations and Waves in Linear and Nonlinear Systems.* 1989, 600 pp. ISBN 0-7923-0445-4

A. Crumeyrolle: *Orthogonal and Symplectic Clifford Algebras. Spinor Structures.* 1990, 364 pp. ISBN 0-7923-0541-8

V. Goldshtein and Yu. Reshetnyak: *Quasiconformal Mappings and Sobolev Spaces.* 1990, 392 pp. ISBN 0-7923-0543-4

Other *Mathematics and Its Applications* titles of interest:

I.H. Dimovski: *Convolutional Calculus.* 1990, 208 pp. ISBN 0-7923-0623-6

Y.M. Svirezhev and V.P. Pasekov: *Fundamentals of Mathematical Evolutionary Genetics.* 1990, 384 pp. ISBN 90-277-2772-4

S. Levendorskii: *Asymptotic Distribution of Eigenvalues of Differential Operators.* 1991, 297 pp. ISBN 0-7923-0539-6

V.G. Makhankov: *Soliton Phenomenology.* 1990, 461 pp. ISBN 90-277-2830-5

I. Cioranescu: *Geometry of Banach Spaces, Duality Mappings and Nonlinear Problems.* 1990, 274 pp. ISBN 0-7923-0910-3

B.I. Sendov: *Hausdorff Approximation.* 1990, 384 pp. ISBN 0-7923-0901-4

A.B. Venkov: *Spectral Theory of Automorphic Functions and Its Applications.* 1991, 280 pp. ISBN 0-7923-0487-X

V.I. Arnold: *Singularities of Caustics and Wave Fronts.* 1990, 274 pp. ISBN 0-7923-1038-1

A.A. Pankov: *Bounded and Almost Periodic Solutions of Nonlinear Operator Differential Equations.* 1990, 232 pp. ISBN 0-7923-0585-X

A.S. Davydov: *Solitons in Molecular Systems. Second Edition.* 1991, 428 pp. ISBN 0-7923-1029-2

B.M. Levitan and I.S. Sargsjan: *Sturm-Liouville and Dirac Operators.* 1991, 362 pp. ISBN 0-7923-0992-8

V.I. Gorbachuk and M.L. Gorbachuk: *Boundary Value Problems for Operator Differential Equations.* 1991, 376 pp. ISBN 0-7923-0381-4

Y.S. Samoilenko: *Spectral Theory of Families of Self-Adjoint Operators.* 1991, 309 pp. ISBN 0-7923-0703-8

B.I. Golubov A.V. Efimov and V.A. Scvortsov: *Walsh Series and Transforms.* 1991, 382 pp. ISBN 0-7923-1100-0

V. Laksmikantham, V.M. Matrosov and S. Sivasundaram: *Vector Lyapunov Functions and Stability Analysis of Nonlinear Systems.* 1991, 250 pp. ISBN 0-7923-1152-3

F.A. Berezin and M.A. Shubin: *The Schrödinger Equation.* 1991, 556 pp. ISBN 0-7923-1218-X

D.S. Mitrinovic, J.E. Pecaric and A.M. Fink: *Inequalities Involving Functions and their Integrals and Derivatives.* 1991, 588 pp. ISBN 0-7923-1330-5

Julii A. Dubinskii: *Analytic Pseudo-Differential Operators and their Applications.* 1991, 252 pp. ISBN 0-7923-1296-1

V.I. Fabrikant: *Mixed Boundary Value Problems in Potential Theory and their Applications.* 1991, 452 pp. ISBN 0-7923-1157-4

Other *Mathematics and Its Applications* titles of interest:

A.M. Samoilenko: *Elements of the Mathematical Theory of Multi-Frequency Oscillations.* 1991, 314 pp. ISBN 0-7923-1438-7

Yu.L. Dalecky and S.V. Fomin: *Measures and Differential Equations in Infinite-Dimensional Space.* 1991, 338 pp. ISBN 0-7923-1517-0

W. Mlak: *Hilbert Space and Operator Theory.* 1991, 296 pp. ISBN 0-7923-1042-X

N.Ja. Vilenkin and A.U. Klimyk: *Representation of Lie Groups and Special Functions. Volume 1: Simplest Lie Groups, Special Functions, and Integral Transforms.* 1991, 608 pp. ISBN 0-7923-1466-2

N.Ja. Vilenkin and A.U. Klimyk: *Representation of Lie Groups and Special Functions. Volume 2: Class I Representations, Special Functions, and Integral Transforms.* 1992, 630 pp. ISBN 0-7923-1492-1

N.Ja. Vilenkin and A.U. Klimyk: *Representation of Lie Groups and Special Functions. Volume 3: Classical and Quantum Groups and Special Functions.* 1992, 650 pp. ISBN 0-7923-1493-X
(Set ISBN for Vols. 1, 2 and 3: 0-7923-1494-8)

K. Gopalsamy: *Stability and Oscillations in Delay Differential Equations of Population Dynamics.* 1992, 502 pp. ISBN 0-7923-1594-4

N.M. Korobov: *Exponential Sums and their Applications.* 1992, 210 pp. ISBN 0-7923-1647-9

Chuang-Gan Hu and Chung-Chun Yang: *Vector-Valued Functions and their Applications.* 1991, 172 pp. ISBN 0-7923-1605-3

Z. Szmydt and B. Ziemian: *The Mellin Transformation and Fuchsian Type Partial Differential Equations.* 1992, 224 pp. ISBN 0-7923-1683-5

L.I. Ronkin: *Functions of Completely Regular Growth.* 1992, 394 pp. ISBN 0-7923-1677-0

R. Delanghe, F. Sommen and V. Soucek: *Clifford Algebra and Spinor-valued Functions. A Function Theory of the Dirac Operator.* 1992, 486 pp. ISBN 0-7923-0229-X

A. Tempelman: *Ergodic Theorems for Group Actions.* 1992, 400 pp. ISBN 0-7923-1717-3

D. Bainov and P. Simenov: *Integral Inequalities and Applications.* 1992, 426 pp. ISBN 0-7923-1714-9

I. Imai: *Applied Hyperfunction Theory.* 1992, 460 pp. ISBN 0-7923-1507-3

Yu.I. Neimark and P.S. Landa: *Stochastic and Chaotic Oscillations.* 1992, 502 pp. ISBN 0-7923-1530-8

H.M. Srivastava and R.G. Buschman: *Theory and Applications of Convolution Integral Equations.* 1992, 240 pp. ISBN 0-7923-1891-9

Other *Mathematics and Its Applications* titles of interest:

A. van der Burgh and J. Simonis (eds.): *Topics in Engineering Mathematics*. 1992, 266 pp. ISBN 0-7923-2005-3

F. Neuman: *Global Properties of Linear Ordinary Differential Equations*. 1992, 320 pp. ISBN 0-7923-1269-4

A. Dvurecenskij: *Gleason's Theorem and its Applications*. 1992, 334 pp. ISBN 0-7923-1990-7

D.S. Mitrinovic, J.E. Pecaric and A.M. Fink: *Classical and New Inequalities in Analysis*. 1992, 740 pp. ISBN 0-7923-2064-6

H.M. Hapaev: *Averaging in Stability Theory*. 1992, 280 pp. ISBN 0-7923-1581-2

S. Gindinkin and L.R. Volevich: *The Method of Newton's Polyhedron in the Theory of PDE's*. 1992, 276 pp. ISBN 0-7923-2037-9

Yu.A. Mitropolsky, A.M. Samoilenko and D.I. Martinyuk: *Systems of Evolution Equations with Periodic and Quasiperiodic Coefficients*. 1992, 280 pp. ISBN 0-7923-2054-9

I.T. Kiguradze and T.A. Chanturia: *Asymptotic Properties of Solutions of Nonautonomous Ordinary Differential Equations*. 1992, 332 pp. ISBN 0-7923-2059-X

V.L. Kocic and G. Ladas: *Global Behavior of Nonlinear Difference Equations of Higher Order with Applications*. 1993, 228 pp. ISBN 0-7923-2286-X

S. Levendorskii: *Degenerate Elliptic Equations*. 1993, 445 pp. ISBN 0-7923-2305-X

D. Mitrinovic and J.D. Kečkić: *The Cauchy Method of Residues, Volume 2*. Theory and Applications. 1993, 202 pp. ISBN 0-7923-2311-8

R.P. Agarwal and P.J.Y Wong: *Error Inequalities in Polynomial Interpolation and Their Applications*. 1993, 376 pp. ISBN 0-7923-2337-8

A.G. Butkovskiy and L.M. Pustyl'nikov (eds.): *Characteristics of Distributed-Parameter Systems*. 1993, 386 pp. ISBN 0-7923-2499-4

B. Sternin and V. Shatalov: *Differential Equations on Complex Manifolds*. 1994, 504 pp. ISBN 0-7923-2710-1

S.B. Yakubovich and Y.F. Luchko: *The Hypergeometric Approach to Integral Transforms and Convolutions*. 1994, 324 pp. ISBN 0-7923-2856-6

C. Gu, X. Ding and C.-C. Yang: *Partial Differential Equations in China*. 1994, 181 pp. ISBN 0-7923-2857-4

V.G. Kravchenko and G.S. Litvinchuk: *Introduction to the Theory of Singular Integral Operators with Shift*. 1994, 288 pp. ISBN 0-7923-2864-7

A. Cuyt (ed.): *Nonlinear Numerical Methods and Rational Approximation II*. 1994, 446 pp. ISBN 0-7923-2967-8

Other *Mathematics and Its Applications* titles of interest:

G. Gaeta: *Nonlinear Symmetries and Nonlinear Equations.* 1994, 258 pp. ISBN 0-7923-3048-X

V.A. Vassiliev: *Ramified Integrals, Singularities and Lacunas.* 1995, 289 pp. ISBN 0-7923-3193-1

N.Ja. Vilenkin and A.U. Klimyk: *Representation of Lie Groups and Special Functions.* Recent Advances. 1995, 497 pp. ISBN 0-7923-3210-5

Yu. A. Mitropolsky and A.K. Lopatin: *Nonlinear Mechanics, Groups and Symmetry.* 1995, 388 pp. ISBN 0-7923-3339-X

R.P. Agarwal and P.Y.H. Pang: *Opial Inequalities with Applications in Differential and Difference Equations.* 1995, 393 pp. ISBN 0-7923-3365-9

A.G. Kusraev and S.S. Kutateladze: *Subdifferentials: Theory and Applications.* 1995, 408 pp. ISBN 0-7923-3389-6

M. Cheng, D.-G. Deng, S. Gong and C.-C. Yang (eds.): *Harmonic Analysis in China.* 1995, 318 pp. ISBN 0-7923-3566-X

M.S. Livšic, N. Kravitsky, A.S. Markus and V. Vinnikov: *Theory of Commuting Nonselfadjoint Operators.* 1995, 314 pp. ISBN 0-7923-3588-0

A.I. Stepanets: *Classification and Approximation of Periodic Functions.* 1995, 360 pp. ISBN 0-7923-3603-8

C.-G. Ambrozie and F.-H. Vasilescu: *Banach Space Complexes.* 1995, 205 pp. ISBN 0-7923-3630-5

E. Pap: *Null-Additive Set Functions.* 1995, 312 pp. ISBN 0-7923-3658-5

C.J. Colbourn and E.S. Mahmoodian (eds.): *Combinatorics Advances.* 1995, 338 pp. ISBN 0-7923-3574-0

V.G. Danilov, V.P. Maslov and K.A. Volosov: *Mathematical Modelling of Heat and Mass Transfer Processes.* 1995, 330 pp. ISBN 0-7923-3789-1

A. Laurinčikas: *Limit Theorems for the Riemann Zeta-Function.* 1996, 312 pp. ISBN 0-7923-3824-3

A. Kuzhel: *Characteristic Functions and Models of Nonself-Adjoint Operators.* 1996, 283 pp. ISBN 0-7923-3879-0

G.A. Leonov, I.M. Burkin and A.I. Shepeljavyi: *Frequency Methods in Oscillation Theory.* 1996, 415 pp. ISBN 0-7923-3896-0

B. Li, S. Wang, S. Yan and C.-C. Yang (eds.): *Functional Analysis in China.* 1996, 390 pp. ISBN 0-7923-3880-4

P.S. Landa: *Nonlinear Oscillations and Waves in Dynamical Systems.* 1996, 554 pp. ISBN 0-7923-3931-2

Other *Mathematics and Its Applications* titles of interest:

A.J. Jerri: *Linear Difference Equations with Discrete Transform Methods.* 1996, 462 pp. ISBN 0-7923-3940-1

I. Novikov and E. Semenov: *Haar Series and Linear Operators.* 1996, 240 pp. ISBN 0-7923-4006-X

L. Zhizhiashvili: *Trigonometric Fourier Series and Their Conjugates.* 1996, 312 pp. ISBN 0-7923-4088-4

MIX
Papier aus verantwortungsvollen Quellen
Paper from responsible sources
FSC® C105338

If you have any concerns about our products,
you can contact us on
ProductSafety@springernature.com

In case Publisher is established outside the EU,
the EU authorized representative is:
Springer Nature Customer Service Center GmbH
Europaplatz 3, 69115 Heidelberg, Germany

Printed by Libri Plureos GmbH
in Hamburg, Germany